Modern Recording Techniques

Fifth Edition

Modern Recording Techniques

Fifth Edition

David Miles Huber

Robert E. Runstein

Focal Press

Boston   Oxford   Auckland   Johannesburg   Melbourne   New Delhi

Focal Press is an imprint of Butterworth–Heinemann.

∞ Recognizing the importance of preserving what has been written, Butterworth–Heinemann prints its books on acid-free paper whenever possible.

Library of Congress Cataloging-in-Publication Data
Huber, David Miles.
   Modern recording techniques / David Miles Huber, Robert E. Runstein.—5th ed.
      p. cm.
   Includes index.
   ISBN 0-240-80456-2 (pbk. : alk. paper)
      1. Magnetic recorders and recording. 2. Sound—Recording and reproducing. I. Runstein, Robert E. II. Title.

TK7881.6 .H85 2001
621.389'3—dc21   2001023198

British Library Cataloguing-in-Publication Data
A catalogue record for this book is available from the British Library.

The publisher offers special discounts on bulk orders of this book.
For information, please contact:
Manager of Special Sales
Butterworth–Heinemann
225 Wildwood Avenue
Woburn, MA 01801-2041
Tel: 781-904-2500
Fax: 781-904-2620

For information on all Focal Press publications available, contact our World Wide Web home page at: http://www.focalpress.com

10 9 8 7 6 5 4 3 2 1

Printed in the United States of America

Table of Contents

◆

Acknowledgments

First off all, I'd like you to know that the subtle, yet seriously deranged guy on the cover is me . . . Now down to biz . . .

I'd like to thank my partner, Daniel Eric Butler, for putting up with the general rantin', ravin', and all 'round craziness that goes into writing a never-ending epic. Same goes for my best buddies: Hector La Torre (modernrecording.com); Scott Colburn, Gravelvoice Productions (www.gravelvoice.com); Mische Eddins, Microsoft Studios; Steve "Stevo" L. Royea as well as Phil & Vivian Williams (www.voyagerrecords.com). Special thanks also go to Bob Ellison and all the folks at Syntrillium software (www.syntrillium.com) . . . makers of Cool Edit Pro and Cool Edit 2000.

A few special mentions go to Marie Lee and the great folks at Focal Press for getting this puppy off the ground and into your hands. Last but not least, I'd like to thank my production editor Maura Kelly, as well as to Kristin Landon for her great copyediting, Marie Lee for spearheadin' the deal and being a very understanding friend . . . and a very special thanks to my main contact and good buddy, Terri Jadick, for being the best there is and for being an understanding friend that shows no remorse in supporting a lifelong vice—Chocoholism! Major time thanx to y'all!

I'd also like to thank the following individuals and companies who've assisted in the preparation of this book by providing invaluable photographs and technical information:

Larry Villella, ADK, www.a-dk.com; Cindy Parker, Acoustical Solutions, Inc., www.acousticalsolutions.com; Acoustical Physics Laboratories; Marcus Thompson, Acoustic Sciences Corporation, www.tubetrap.com; Erikk Lee and David Harbison, Auralex Acoustics, www.auralex.com; Wes Dooley, Audio Engineering Associates, www.wesdooley.com; Kevin Madden and Sarita M. Stewart, AKG Acoustics, Inc., www.akg-acoustics.com; Cliff Castle, Audix Corporation, www.audixusa.com; Bruce Borgerson, ATR Service Company, www.atrservice.com; Peter Chaikin, Alesis Studio Electronics, Inc., www.alesis.com; Dominic Cramp, Arboretum Systems, Inc., www.arboretum.com; Richard Giannini, Asimware Innovations, www.asimware.com; Randy Hargis, Akai Professional, www.akaipro.com; Dan Sheingold, Analog Devices, www.analog.com; Jennifer Schanhals and Chris Ellerby, Amek, www.amek.com; Marvin Ceasar, Aphex Systems, Inc., www.aphex.com; Altec Lansing Technologies, Inc., www.alteclansing.com; Russ Berger, Russ Berger Design Group, Inc., www.rbdg.com; Peter Grueneisen, studio bau:ton, www.bauton.com; Alexis D. Kurtz, Beyerdynamic, www.beyerdynamic.com; Fujiko Kameda and Kurt Bujack, Bujack Audio; Dan Gallagher, Behringer International GMBH, www.behringer.de; Paul Bass, BSS Audio USA, www.bss.co.uk; Zac Wheatcroft, Bias Software, www.bias-inc.com; Daniel Buckley and James Tanner, Bryston Ltd, www.bryston.ca; Mary C. Bell, Jeff Palmer and Brian and Caron Smith, ClearSonic Mfg. Inc., www.clearsonic.com; Mick Whelan, Crown International, Inc., www.crownaudio.com; Carl Malone, CM Automation, www.cmautomation.com; Tom Johnson and Bonnie Anderson, Coda Music Technology, www.codamusic.com; Dan Stout, Colossal Mastering, Chicago, IL, www.colossalmastering.com; Paula Salvatore, Capitol Studios, Hollywood, CA, www.capitolstudios.com; Bob Katz, Digital Domain, www.digido.com; Tom Dambly, Digidesign, www.digidesign.com; Jim Fiore, dissidents Software, www.dissidents.com; Luke Giles, Drawmer, www.transaudiogroup.com; Kenton Smith, Digitech, www.digitech.com; Connie Nomann and Kenton Smith, dbx Professional Products, www.dbxpro.com; Dolby Laboratories, Inc, www.Dolby.com; Wendy Clayton, Theresa Grant and Wendy Blakeway, Euphonix Inc., www. euphonix.com; Bill Dooley, Extasy Recording Studios, www.extasyrecordingstudios.com; Derk Hagedorn, E-MU/Ensoniq, www.emu.com; Richard Factor, Eventide Inc., www.eventide.com; Juels Thomas, Event Electronics, www.event1.com; Tom Dambly, Focusrite, www.focusrite.com; Molly Carter, Furman Sound, Inc., www.furmansound.com; David Miles Huber, 51bpm, www.51bpm.com; Focal Press, www.focalpress.com; Fostex Corp. of America, www.fostex.com; Scott Colburn, Gravelvoice Studios, www.gravelvoice.com; Will Eggleston, Genelec OY, www.genelec.com; Jack W. Gilfoy and Charlie Leib, hafler, www.hafler.com; Tomiko Jones Photography, Seattle, WA, www.tomikojonesphoto.com; Chuck Thompson, JLCooper Electronics, www.jlcooper.com; Julian Colbeck, Keyfax Software, www.keyfax.com; KRK Systems, Inc., www.krksys.com; John Rotondo, Lexicon, Inc., www.lexicon.com; Keith Medley, Christopher Buttner, Steve Eborall and Andrew Gruner, Mackie Designs Inc., www.mackie.com; David Miles Huber, modrec.com, www.modrec.com; Jose Guillen, Marshall Electronics, Inc., www.mars-cam.com; Jim Cooper, Mark of the Unicorn, Inc., www.motu.com; Jeff Wilson, Minnetonka Audio Software, www.minnetonkaaudio.com; Michael "Hucky" Huckler, Midiman, Inc., www.midiman.net; Ron Stein, midibrainz software, www.midibrains.com; Jim McGraw, McGraw Publishing Peripherals, www.sittingmachine.com; John La Grou, Millennia Music & Media Systems, www.mil-media.com; Aaron H. Pratt, MicroBoards Technology, Inc., www.microboards.com; John T. Mullin and Jim Van Buskirk, Nemesys Music Technology, Inc., www.nemesysmusic.com; Neumann USA, www.neumannusa.com; David Kennaugh, Neato LLC, www.neato.com; Chris Steinwand, Otari Corporation, www.otari.com; Orban Associates, Inc.; Kelly Erwin, Ocean Way Recording, www.oceanwayrecording.com; Adrienne Thompson, Primera Technology, Inc., www.primeratechnology.com; Kirt Kim, QSC Audio Products, Inc., www.qscaudio.com; Roger D'Arcy, Recording Architecture, www.aaa-design.com; John Jennings, Royer Labs, www.royerlabs.com; Sara Griggs, Roland Corp. US, www.rolandus.com; Ellen Allhands, Rane Corporation, www.rane.com; Kathryn Kelly and Elena Bernardo, Roxio, www.roxio.com; Record Plant Recording Studios; Sal Schandon, Sheffield Audio-Video Productions, www.sheffieldav.com; Debra Pagen and Cathi Simpson, Solid State Logic, www.solid-state-logic.com; Walters-Storyk, Walters-Storyk Design Group, www.wsdg.com; Sandy Schroeder, Shure Incorporated, www.shure.com; Studer North America, www.studer.ch; Courtney Spencer and Bob Tamburri, Sony Professional Audio, www.sony.com/proaudio; Bob Ellison, Syntrillium Software Corp., www.syntrillium.com; Rebecca Grow, Sonic Foundry, www.sonicfoundry.com; Marsha Vdovin, Steinberg, www.steinberg.de; Ken Dewar, Soundscape Digital Technology LTD., www.soundscape-digital.com; Dave, Soundtrek, www.soundtrek.com; Michael Lambie, Sound Quest Inc., www.squest.com; Kimberly J. Cahail, Symetrix Inc., www.symetrixaudio.com; Phil Sanchez, Tascam, www.tascam.com; Larry Anschell, Turtle Recording, www.turtlerecording.com; Karyn Ormiston, tc electronic, www.tcelectronic.com; TC Works, www.tcworks.de; Chris Rice, Twelve Tone Systems, www.cakewalk.com; Clivia Schiebel, Tannoy/TGI North America, Inc., www.Tannoy.com; Terri Aberg, Telex Communications, Inc., www.telex.com; Bryan Ekus and Tony Denning, Tapematic USA, Inc., www.tapematic.com; Suz Howells, Universal Audio, www.uaudio.com; Bob Kratt and Ken Camozzi, Versadyne International, www.versadyne.com; Marsha Vdovin and Sheri Ragan, Waves, www.waves.com; Ken Centofante, Westlake Audio, www.westlakeaudio.com; Bonnie Reed and Reed Ruddy, Studio X, www.studioxinc.com; Susan Hart, Yamaha Corporation of America, www.yamaha.com/proaudio.

About the Authors

David Miles Huber is widely acclaimed in the recording industry as a digital audio consultant, author and guest lecturer on the subject of digital audio and recording technology. As well as being a regular contributing writer for numerous magazines and websites, Dave has written such books as *The MIDI Manual* (Focal Press) and *Professional Microphone Techniques* (Artist-Pro.com). He also manages the Educational Outreach Program for Syntrillium software (www.syntrillium.com), makers of Cool Edit 2000 and Cool Edit Pro.

In addition to all this, he's a professional musician in the ambient dance/relaxational field, having written, produced and engineered CDs that have sold over the million mark. His latest stuff can be checked out at 51bpm.com.

Robert E. Runstein has been associated with all aspects of the recording industry, working as a performer, sound mixer, electronics technician, A&R specialist and record producer. He has served as chief engineer and technical director of a recording studio and has taught several courses in modern recording techniques. He is a member of the Audio Engineering Society.

Trademarks

All terms mentioned in this book that are known to be trademarks or service marks have been appropriately capitalized. Focal Press cannot attest to the accuracy of this information. Use of a term in this book should not be regarded as affecting the validity of any trademark or service mark.

ADAT and QuadraVerb are registered trademarks of Alesis Studio Electronics.

Aphex Compellor is a trademark of Aphex Systems Ltd.

Apple; Macintosh Plus, SE and II; Hypercard; Videoworks; and MacRecorder are registered trademarks of Apple Computer, Inc.

Cakewalk and Cakewalk Pro Audio are trademarks of Twelve Tone Systems.

dbx is a registered trademark of dbx, Newton, MA, USA, Division of Harmon International.

Dolby, Dolby SR, Dolby A, Dolby B, Dolby C, Dolby Surround Sound and Dolby Tone are registered trademarks of Dolby Laboratories Licensing Corporation.

Event 20/20bas is a trademark of Event Electronics.

Harmonizer and Ultra-Harmonizer are registered trademarks of Eventide, Inc.

PZM is a registered trademark of Crown International, Inc.

Sony is a registered trademark of Sony Corporation of America.

Sound Designer II, Sound Tools, Pro Tools, NuBus and DINR are registered trademarks of Digidesign.

Tannoy is a registered trademark of Tannoy LTD. (North America Inc.).

Tube Trap is a trademark of Acoustic Sciences Corp.

XGEN is a trademark of Telex Communications, Inc.

Yamaha is a registered trademark of Yamaha Corporation of America.

CHAPTER 1

Introduction

The world of modern music and sound production is multifaceted. It's a world of creative individuals: musicians, engineers, producers, manufacturers and businesspeople who are experts in such fields as music, acoustics, electronics, production, broadcast media, multimedia, marketing, graphics, law and the day-to-day workings of the business of music. The combined efforts of these talented people work together to create a single end product: marketable music. The process of turning a creative spark into a final product takes commitment, talent, a creative production team, a marketing strategy and often . . . money. Over the history of recorded sound, the process of capturing music and transforming it into a marketable product has radically changed. In the past, the process of turning one's own music into a final product required the use of a commercial recording studio, which was (and still is) equipped with professional staff and a specialized facility that was often hired for big bucks. With the introduction of the LSI (large-scale integrated circuit), mass production and mass-marketing (three of the most powerful forces in the "information age), another option has come onto the scene . . . the radical idea that musicians, engineers and/or producers could have their own project or desktop (computer-based) studios. Along with this concept comes the realization that almost anyone can afford, construct and learn to master a personal production facility. . . . In short, we're living in the midst of a techno-artistic revolution that puts more power, artistic control and knowledge in the hands of creative individuals from all walks of life in the music biz.

On the techno side, for those who are new to the world of modern multitrack recording, MIDI (musical instrument digital interface), digital audio and their production environments, years of dedicated practice are often required to develop the skills that are needed to successfully master the art and the application of these technologies. A person new to the recording or project studio environment (Figure 1.1) might easily be awestruck by the amount and variety of equipment that's involved in the process. However, when we become familiar with this environment, a definite order to the studio's makeup soon begins to appear, with each piece of equipment being designed to have a role in the overall scheme of music and audio production.

Figure 1.1. *Extasy Recording South, Los Angeles. (Courtesy of Extasy Recording Studios, www.extasyrecordingstudios.com)*

The goal of this book is to serve as a guide and reference tool to help you become familiar with the recording and production process. When used in conjunction with mentors, lots of hands-on experience, further reading and simple common sense . . . this book will, I hope, help you understand the toys, tools and day-to-day practices of music recording and production.

Although it's taken the modern music studio about 80 years to evolve to its present level of technological sophistication, we've just undergone an important, evolutionary stage in the business of music and music production: the dawning of the digital age. In a time when digital audio and its related technologies are often taken for granted, it's easy to forget that we're in the process of being given an amazing array and choice of cost-effective and powerful tools for fully realizing our creative and human potential. As always, patience and a nose-to-the-grindstone attitude are needed in order to learn how to use them effectively. However, this knowledge can help free you for the really important stuff—making music and sound. In my opinion, these are definitely the good ol' days.

The Recording Studio

The commercial music studio (Figures 1.2 through 1.4) is made up of one or more acoustic environments that are specially designed and tuned for the purpose of capturing the best possible sound onto tape or other recording medium. Commercial studios are structurally isolated in order to keep outside sounds from entering the room and getting onto tape (as well as keeping inside sounds from leaking out and disturbing the surrounding neighbors).

Figure 1.2. Four Seasons Media Productions, St. Louis. (Courtesy of Russ Berger Design Group, Inc., www.rbdg.com)

Figure 1.3. CTS Studio 2, London. (Courtesy of Recording Architecture, www.aaa-design.com)

Figure 1.4. Synchrosound Studio A, Kuala Lumpur, Malaysia. (Courtesy of Walters-Storyk Design Group—designed by Beth Walters and John Storyk, www.wsdg.com, photo credit—Robert Wolsch)

Studios vary in size, shape and acoustic design (Figures 1.5 through 1.7) and usually reflect the owner's personal taste. In addition to this, they're often designed to accommodate certain music styles and/or production needs, as shown by the following examples:

- A studio that records a wide variety of music (ranging from classical to rock) might have a large main room with smaller, isolated rooms off to the side for loud or softer instruments, vocals, etc.

- A studio designed for orchestral film scoring might be larger than other studio types. Such a studio will often have high ceilings to accommodate the large sound buildups that are often generated by a large number of studio musicians.

- A studio used for audio-for-video, film dialog, vocals and mixdown might only incorporate a single small recording space off the control room for overdubs.

There's no secret formula for determining what the perfect studio design is. Each studio has its own sonic character, layout, "feel" and decor that are based on the personal tastes of its owners, the designer (if any) and studio rates (based on the studio's investment return and the supporting market conditions).

During the 1970s, studios were generally small. Because of the advent of (and overreliance on) artificial effects devices, these rooms tended to be acoustically absorptive. The basic concept was to eliminate as much of the original acoustic environment as possible and replace it with artificial ambience.

Fortunately, since the mid-1980s, many commercial studios have begun to move back to the original design concepts of the 1930s and 1940s when studios were much larger. This increase in size (along with the addition of one or more smaller, iso-rooms) has repopularized the art of

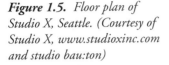

Figure 1.5. *Floor plan of Studio X, Seattle. (Courtesy of Studio X, www.studioxinc.com and studio bau:ton)*

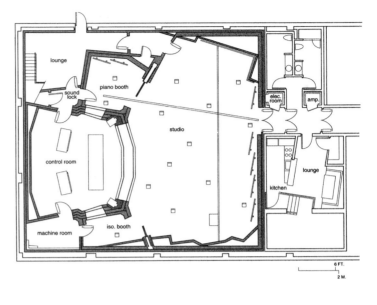

Figure 1.6 *Floor plan of Sony/ Tree's Music Studio, Nashville. (Courtesy of Russ Berger Design Group, Inc., www.rbdg.com)*

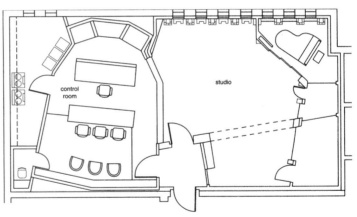

capturing the room's original acoustic ambience along with the actual sound pickup. In fact, through improved studio design techniques, we've learned how to have the best of both eras by building a room that disperses sound in a controlled manner (thereby reducing unwanted sound leakage from an instrument to other mics in the room), but at the same time has a well-developed sonic and reverberant "personality" of its own. The effect of combining direct and natural room acoustics is often used as a tool for "livening up" an instrument . . . a popular technique with live rock drums, string sections, electric guitars, choirs and so on.

Figure 1.7 Floor plan of Paisley Park's Studio A. (Courtesy of Paisley Park Studios)

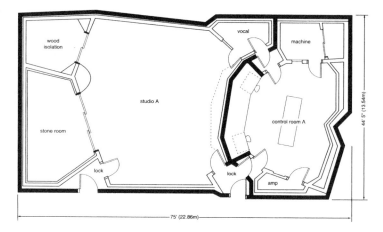

The Control Room

A recording studio's control room (Figures 1.8 through 1.10) serves several purposes. Ideally, the control room is acoustically isolated from the sounds that are produced in the studio and the surrounding areas. These rooms are optimized to act as a critical listening environment using carefully balanced and placed monitor speakers. It also houses the majority of the studio's recording, control and effects-related equipment. At the heart of the control room is the recording console.

Figure 1.8. Four Seasons Media Productions, St. Louis. (Courtesy of Russ Berger Design Group, Inc., www.rbdg.com)

Figure 1.9. *George Martin at Air Lyndhurst, London. (Courtesy of Solid State Logic, www.solid-state-logic.com)*

Figure 1.10. *Lower East Side audio-for-video posting suite. (Courtesy of Walters-Storyk Design Group—designed by Beth Walters and John Storyk, www.wsdg.com, photo credit— Robert Wolsch)*

The recording console (also referred to as the board or desk) can be thought of as an artist's palette for the recording engineer, producer and artist. The console allows the engineer to combine, control and distribute the input and output signals of most (if not all) of the devices found in the control room. The console's basic function is to allow for any variable combination of mixing (control over relative amplitude and signal blending between channels), spatial positioning (left/right, as well as possible surround sound control over front, back and sub), routing (the ability to send any input from a source to a signal destination) and switching for the multitude of audio input/output signals that are commonly encountered in an audio production facility.

Tape machines generally are located at the rear or to the side of a control room. Because of the added noise and heat generated by recorders, computers, power supplies, amplifiers and other devices, it's become increasingly common to house this equipment in an isolated machine room that has an adjoining window and door for easy access and visibility. In either case, remote control and autolocator devices (which are used for locating tape and media position cue points) are often situated in the control room, near the engineer, so that he or she has easy access to all tape and electronics functions. Effects devices (used to electronically alter and/or augment the character of a sound) and other signal processors are also often placed nearby for easy accessibility.

As with recording studio designs, every control room will have its own unique sound, "feel," comfort factor and studio booking rate. Commercial control rooms often vary in design and amenities—from a room that's basic in form and function to one that's lavishly outfitted with the best toys, design layouts and fully-stocked kitchens in the business. Again, the style and layout is up to personal choice. However, as you'll see throughout this book, there are numerous guidelines that can help you make the most of a recording space. In addition to the layout, feel and equipment, it's important to remember that the people (the staff, musicians and you) will often play a huge role in capturing the feel of a performance . . . not just the equipment.

Recording Studio Marketing Techniques

In recent years, the role of the recording studio has begun to change as a result of upsurges in project studios, audio-for-video/film, multimedia and Internet audio. These market forces have made it necessary for certain facilities to rethink their operational business strategies. Often these changes have been met with a degree of success, as is the case in the following examples:

- Personal production and home project studios have deeply reduced the need for an artist or producer to have constant and costly access to a professional facility. Many pro studios now cater to artists and project studio owners who might have an occasional project that calls for a larger space or better-equipped recording facility. In addition, after an important project has been completed, a professional facility might be needed to mix the tapes down into their final form. Most business-savvy studios are only too happy to capitalize on these new and constantly changing market demands.

- Upsurges in audio-for-video and audio-for-film postproduction have created new markets that allow the professional recording studio to provide services to the local, national and international broadcast and visual production communities. Studios with both the equipment and a creative staff to enter into lasting relations with the audio-for-visual and broadcast production are often able to thrive in the tough business of music, when music production alone might not provide enough income to keep a studio afloat.

- Studios are also taking advantage of Internet audio distribution techniques by offering Web development, distribution and other services as an incentive to their clients.

These and other market niches (many of which may be unique to your particular area) have been adopted by commercial music and recording facilities (as well as personal project studios) to meet the changing market demands of the new millennium (Figure 1.11). No longer is there only one game in town. Tapping into changes in market forces and meeting them with new solutions is an important factor for making it in the business of music production and distribution. I'd like to say that all-important word again . . . "*business*." Make no mistake about it . . . starting, staffing and maintaining a production facility and/or getting your or your clients music heard is serious work that requires dedication, stamina, innovation and guts.

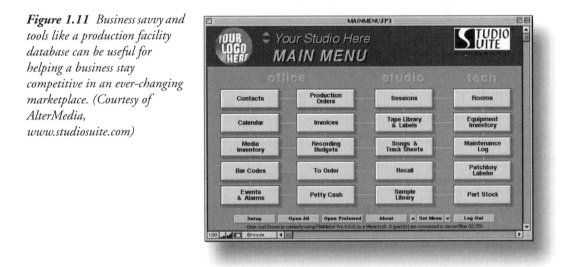

Figure 1.11 *Business savvy and tools like a production facility database can be useful for helping a business stay competitive in an ever-changing marketplace. (Courtesy of AlterMedia, www.studiosuite.com)*

The Recording Process

Many people can approach the recording process in many different ways. With the advent of cost-effective equipment and choices in equipment and media, it's become an accepted fact that the process of recording has become much more personal, often involving different facets of style, production and equipment choice. Setting aside the process of creating music using sound design, digital groove editors and MIDI composition/production techniques (for the moment), the process of capturing sound onto a recorded medium will often take one of two fundamental forms:

- Multitrack recording
- Live recording

Multitrack Recording

The role of multitrack recording technology is to add a degree of flexibility to the recording process by allowing multiple sound sources to be captured onto and played back from isolated tracks in a non-real-time production environment. Because the recorded tracks are isolated from each other (with track capabilities usually being offered in groups of eight . . . i.e., 8, 16,

24, 32 and 48), any number of instruments can be recorded and rerecorded without affecting other instruments. In addition, recorded tracks can be altered, added or erased at any time in order to augment or "sweeten" the original soundtrack.

The common phases of the multitrack recording include:

- Recording
- Overdubbing
- Mixdown

Recording

The first phase in multitrack production is the *recording* process. In this phase, one or more sound sources are picked up by a microphone or are recorded directly (as often occurs when recording electric or electronic instruments) to one or more of the isolated tracks of a multi-track recorder.

Isolation is a key concept in multitrack production. By recording a single instrument to a dedicated track (or group of tracks), it's possible to vary the level, spatial positioning (such as L/R panning) and routing without affecting adjacent tracks (Figure 1.12). This isolation also makes it possible for tracks to be replaced or independently processed at a later time without affecting adjacent tracks, thus giving the multitrack process an amazing degree of flexibility.

***Figure 1.12** Basic representation of how isolated sound sources can be recorded to a multitrack recorder.*

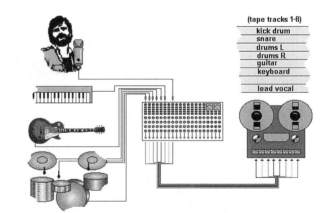

Overdubbing

One of the most important aspects of multitrack production is the *overdubbing* process. Because of the isolated nature of the recorded tracks, it's possible for one or more of the previously recorded tracks to be monitored, while simultaneously recording new instrument and/or vocal parts onto other available tracks (Figure 1.13). This overdub process lets you add individual parts to an existing project until the song or soundtrack has been built up. If a mistake is made, it's generally a simple matter to recue the recording device and repeat the process until you've captured the best take onto tape, disk or disc.

Figure 1.13 *Basic representation of the overdubbing process.*

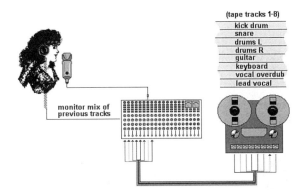

Mixdown

After the phases of recording and overdubbing have been completed, the *mixdown* process can begin (Figure 1.14). During mixdown, the separate audio tracks of a multitrack recorder can be combined and routed through the recording console. At this point, volume, tone, spatial positioning (panning) and special effects can be artistically balanced by the engineer to create a stereo or surround-sound mix that can be recorded as a final master. Once all the songs in a music project have been "mixed down," they can be edited into a final, sequenced order. The resulting program can then be mastered and manufactured into a final, commercially saleable product.

Figure 1.14 *Basic representation of the mixdown process.*

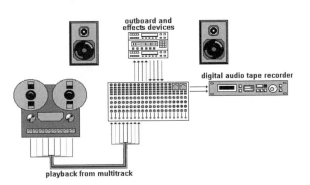

Live/On-Location Recording: A Different Animal

Unlike the traditional multitrack recording environment where large amounts of overdubbing are often used to build up a song over time, *live recording* is created on the spot, in real time . . . often during a single in-studio or on-stage performance, with little or no studio postproduction other than mixing.

A live recording might be very simple . . . possibly recorded using only a few mics that are mixed directly to two tracks, or a "gig" might call for a more elaborate multitrack setup, requiring the use of a temporary control room or fully equipped mobile recording van or truck (Figure 1.15). The latter type will obviously require a great deal of preparation and expertise; however, isolating instruments onto their own tracks will often give you more control over the individual instruments during the mixdown phase.

Figure 1.15 Sheffield mobile recording truck. (Courtesy of Sheffield Audio-Video Productions, www.sheffieldav.com) *a.* Control room. *b.* Truck.

a

b

Although the equipment and setup will be familiar to any studio engineer, live recording differs from its more controlled studio counterpart in that it exists in a world where the motto is "you only get one chance." When you're recording an event where the artist is spilling his or her guts to hundreds or even thousands of fans, it's critical that everything run smoothly. Live recording usually requires a unique system setup, degree of preparedness and . . . above all, experience.

The Project Studio

In recent decades, the business of manufacturing cost-effective professional and semiprofessional recording equipment has virtually exploded. As a result, it's now possible (if not downright common) for musicians, engineers or producers to have a high-quality recording and/or MIDI facility in their homes, apartments or personal places of business. This proliferation has grown to such a point that these facilities (known as *project studios*) have become a strong driving force in the music, multimedia and communication industries (Figures 1.16 and 1.17).

Figure 1.16 *Alex Lifeson's project studio (guitarist with RUSH). (Photo credit: Andrew MacNaughton, Courtesy of Mackie Designs Inc., www.mackie.com)*

Figure 1.17 *Bruno Ravel in the B.O.M.B. Faktory, www.bombfaktory.com. (Photo credit: Christopher Buttner, Courtesy of Mackie Designs Inc., www.mackie.com)*

Project studios have become increasingly important for the following reasons:

- Cost-effective power—The obvious reason for the proliferation of these facilities is that with the advent of the VLSI (very large scale integrated-circuit), the price of mass-duplicating highly sophisticated electronic systems has dropped significantly. Studio equipment that only a few decades ago would've easily cost hundreds of thousands of dollars can now be purchased at one-tenth and even one-hundredth the price. The cost of booking a pro studio can add up fast, so having your own facility and learning its technology and techniques can quickly pay off in financial savings. Knowing when to make full use of your own facility and when to blend the services of an outside professional studio can result in a high-quality product that will cost significantly less to produce.

- Setting your own schedule and saving money while you're at it!—An equally obvious advantage is the ability to create your own music on your own schedule. The expense incurred in using a professional studio requires that you be practiced and ready to roll on a specific date or range of days. A project studio can free you up to record when the mood hits, without having to worry about punching the studio's time clock.

- Creative and functional advantages—With the advent of MIDI, hard-disk recorders, modular digital multitrack recorders and so on . . . the project studio offers creative and functional advantages over the commercial studio for creating and producing your own personal style of music. It must be stressed, however, that certain projects will require the guidance and experience of a professional. Being aware of the production needs of the overall project is an important aspect in ensuring its overall success.

Audio-for-Video and Film

In recent decades, audio has become a much more important part of video, film and broadcast production. Previous to this, broadcast audio was almost an afterthought. With the advent of stereo television, the music video and the MTV/VH1 generation, audio has grown from its relatively obscure position to being a highly respected part of audio-for-video production (Figure 1.18).

Figure 1.18 *Betelgeuse, NYC. (Courtesy of Walters-Storyk Design Group—designed by Beth Walters and John Storyk, www.wsdg.com. Photo credit—Robert Wolsch)*

With the common use of surround-sound in the creation of movie sound tracks (Figure 1.19), along with the growing popularity of surround in home entertainment systems (and an ever-growing number of playback system for sound, visual media and computer media), the public has come to expect higher levels in audio production quality. In this day and age, MIDI, hard-disk recording, time code, automated mixdown and advanced processing have become everyday parts of the production environment . . . requiring professionals to be highly specialized and skilled in order to meet the demanding schedules and production complexities.

Figure 1.19 *Film mixing room, Lansdowne Studios, London. (Courtesy of Recording Architecture, www.aaa-design.com)*

Multimedia

With the integration of text, graphics, MIDI, digital audio and digitized video into almost every facet of the personal computer environment, the field of *multimedia audio* has become a fast-growing, established industry that represents an important and lucrative source of income for both creative individuals and production facilities alike. In addition to the use of audio in multimedia products for education, business and entertainment, most of the robot-zappin', dare-devil flyin' addicts are probably aware that one of the largest and most lucrative areas of multimedia audio production is the field of scoring, designing and producing audio for computer games (Figure 1.20). Zaaaaaapppppppppp!

Figure 1.20 *Electronic Arts, Vancouver, Canada. (Courtesy of Walters-Storyk Design Group—designed by Beth Walters and John Storyk, www.wsdg.com. Photo credit—Robert Wolsch)*

The Web

It almost goes without saying that the World Wide Web has changed almost every facet of distributing and acquiring information on the Internet. Many of the top magazines have some sort of Web presence for reading back and even current issue articles, and tons of sites can be found that are dedicated to even the most offbeat ideas and equipment. It's a rare person who isn't aware of the fact that the list of resources, opinions, insights, tips and articles relating to every imaginable facet of almost anything continues to grow at a staggering pace!

I'm hesitant to name specific recording resource sites here . . . for the obvious fact that the Web is ever-changing. Those of you who'll be reading this book 5 years, a year or even a few months from now could easily find that cyberspace has totally changed and that some or all the named sites have disappeared, have been swallowed up or have been made obsolete by the biggest, latest and greatest site.

This leaves me with the task of pointing out the resources that can best lead you to the site or informational resource that you might need for equipment, selling/buying gear, mags, software downloads, leads on independent music . . . you name it. (Figure 1.21 shows one such site.) These include:

- companyname.com—Adding ".com" onto the company's name will (more often than not) lead you to their site.
- Search engines—Any of the popular search engines can help you to track down the info or site that you need.
- Industry magazines—Looking through trade advertisements can be one of the best ways to find a resource. Often the classifieds ads at the back of a magazine can be a resource that's both enlightening and full of fun things.

Figure 1.21. *One of the many sites that can help you with your search and reference needs in the field of recording is modrec.com. (Courtesy of modrec.com)*

The People Who Make It All Happen

When you get right down to it, the recording field is built around pools of talented individuals and service industries, who work together for a common goal—producing and selling music. As such, it's the *"people"* in the recording industry who make the business of music happen. Recording studios, as well as the other businesses in the industry, aren't known only for the kind of equipment they have, but often are judged by the quality, knowledge, vision and personalities of their staffs. The following areas represent but a few of the ways in which a person can be involved in this multifaceted industry:

The Artist

The strength of a recorded performance begins and ends with the artist. All the technology in the world is of little use without the existence of the central ingredients of human creativity, emotion and technique.

Just as the overall sonic quality of a recording is no better than its weakest link, it's the performer's job to see that the foundation of all music—its inner soul—is laid out for all to experience and hear. After this has been done, a well-produced, carefully planned recording can act as a framework for the music's original drive, intention and emotion.

Studio Musicians

A project often needs additional musicians to add extra spice and depth to the artist's recorded performance. This can take a number of forms. For example, a project might require musical ensembles (such as a choir, string section or background vocals) to add a necessary part or to give a piece a "fuller" sound. If a large ensemble is required, it might be necessary to call in a professional music contractor to coordinate all the musicians and make all the financial arrangements, as well as a music arranger to notate and possibly conduct the various musical parts.

It's also possible that a specific member of a group might not be available or might not be up to the overall musical standards that are required by the project. In such situations, it's not uncommon for a professional *studio musician* to be called in to fill the part. An entire group of studio musicians might even be called on to provide the best possible musical support for a high-profile artist or vocalist.

The Producer

Beyond the scheduling and budgetary factors that go into coordinating a recording project, it's the job of a *producer* to help the artist and/or record company create the best possible recorded performance and final product (according to his/her vision).

In reality, a producer can be hired onto a project to fulfill a number of possible duties. He/she might be contracted to have complete control over a project's artistic, financial and programmatic content. More likely, however, the producer will act collaboratively with an artist or group to guide them through the recording process. The producer assists in the selection and focus of musical arrangements to best fit the targeted audience, brings out the best performance possible and then translates that performance (through the recording medium) into a final, saleable product.

A producer can also be chosen for his/her ability to understand the process of creating a final recorded project from several perspectives: business, musical performance, creative insight and mastery of the recording process. Since engineers spend much of their working time with musicians and industry professionals with the intention of making their clients sound good, it's not uncommon for an engineer to take on the role of producer or coproducer (by default or by mutual agreement). Conversely, as producers become increasingly more knowledgeable about recording technology, it's also become more common to find them sitting behind the controls of a console.

The Engineer

The engineer's job can best be described as "an interpreter in a techno-artistic field." It's the engineer's job to express the artist's music and the producer's concepts through the medium of recording. This job actually is an art form because both music and recording are subjective in nature and rely on the tastes and experience of everyone involved.

During a recording session, the *engineer* generally places the musicians in the desired studio positions, chooses and places the microphones, sets levels and balances on the recording console and records the performance onto tape. In an electronic music setting, the engineer might also set up the MIDI sequencing equipment, musical instruments, hard-disk recorders and so on. During an overdubbing or mixdown session, the engineer uses his or her talent and knowledge of the art and technology of the recording media to get the best possible sound.

Assistant Engineer

Larger studios often train future staff engineers by allowing them to work as assistants to full-fledged engineers. The *assistant engineer* often does microphone and headphone setups, runs tape machines, does session breakdowns and (in certain cases) performs rough mixes and balance settings for the engineer on the console.

With the proliferation of freelance engineers (those not employed by the studio, but who are retained by the artist or record company to work on a particular project), the role of the assistant engineer has become even more important. It's often the assistant engineer's role to help freelance engineers from getting lost amidst the technical aspects and quirks that are peculiar to that studio.

Maintenance Engineer

The maintenance engineer's job is to see that the equipment in the studio is maintained in top condition, regularly aligned and repaired when necessary. Larger organizations (those with more than one studio) might employ a full-time staff *maintenance engineer*; smaller studios are more commonly serviced by freelance maintenance engineers on an on-call basis.

Studio Management

Running a music or production studio is a serious business that requires the special talents of businesspeople who are knowledgeable about the inner workings of the music studio, the music business and (above all) . . . people. It requires the constant attention of a studio manager (who might or might not be the owner), booking department (who keeps track of most of the details relating to studio usage, billing and possibly marketing) and, last but not least, a competent secretarial staff. Although any or all of these functions can easily vary from studio to studio, these and other equally important roles are required in order to successfully operate a commercial production facility on a day-to-day basis.

Women in the Industry

Ever since its inception, males have dominated the recording industry. I remember many a session in which the only women on the scene were female artists or studio groupies. Fortunately, over the years, women have begun to play a more prominent role "behind the glass" and in every facet of studio production and the business of music as a whole. The fact is, no matter who you are, where you're from, or what your race, gender or planetary orientation is, remember this universal truth: If your heart is in it and you're willing to work hard enough, you'll make it.

Behind the Scenes

In addition to studio and production staff, there are scores of professionals who serve as a backbone for keeping the business of music alive and functioning. Without the many different facets of music business, technology, production, distribution and law . . . the biz of music would be very, very different. A small sampling of professionals that help make it happen include:

- Studio management
- Music law
- Graphic arts and layout
- Artist management
- A&R (artist and repertoire)
- Manufacturing
- Music and print publishing
- Distribution

This incomplete listing serves as a reminder that the business of making music is full of diverse possibilities, and extends far beyond the notion that in order to "make it" in the biz you'll have to sell your soul or be who or what you're not. In short, there are many paths that can be taken in this techno-artistic business. Once you've found the one that best suits your own personal style, you can then begin the lifelong task of gaining knowledge and experience and pulling together a network of those who are currently working in the field. As the advert says . . . Just do it!

The Transducer

Before we jump into the heart of this book, I'd like to take a moment to look at a concept that's central to all music, sound, electronic and the "art" of sound recording—the transducer. If any conceptual tool can help you to understand the technological underpinnings of the art and process of recording, this is it!

Quite simply, a *transducer* is any device that changes one form of energy into another, corresponding form of energy. For example, a guitar is a transducer in that it takes the vibrations of picked or strummed strings (the medium), amplifies them through a body of wood and converts these vibrations into corresponding sound-pressure waves . . . which are then perceived as sound (Figure 1.22).

Figure 1.22. The guitar and microphone as transducers.

A microphone is another example of a transducer. Sound-pressure waves (the medium) act on the mic's diaphragm and are converted into corresponding electrical voltages. The electrical signal from the microphone can then be amplified (not a process of transduction because the medium stays in its electrical form) and is then fed to a recording device. The recorder will then change these voltages into analogous magnetic flux signals on magnetic tape or into representative digital data that can be encoded onto tape, hard disk or disc.

On playback, the stored magnetic signal or digital data can be reconverted back to its original electrical form, is amplified and then is fed to a speaker system. The speakers convert the electrical signal into a mechanical motion (by way of magnetic induction), which, in turn, re-creates the original air-pressure variations that were sensed by the microphone.

As can be seen from Table 1.1, transducers can be found practically everywhere in the audio environment. In general, transducers and the media they use are often the weakest link in the

Table 1.1. Media used by transducers in the studio to transfer energy.

Transducer	From	To
Ear	Sound waves in air	Nerve impulses in the brain
Microphone	Sound waves in air	Electrical signals in wires
Record head	Electrical signals in wires	Magnetic flux on tape
Playback head	Magnetic flux on tape	Electrical signals in wires
Phono cartridge	Grooves cut in disc surface	Electrical signals in wires
Speaker	Electrical signals in wires	Sound waves in air

audio system chain. Given our present technology, the process of changing the energy in one medium into a corresponding form of energy in another medium can't be accomplished perfectly. Noise, distortion and (often) coloration of the sound are introduced to some degree. These effects can only be minimized, but not eliminated. Differences in design are another major factor that can affect sound quality. Even a slight design change between two mics, speaker systems, digital audio converters, guitar pickups, speaker or other transducers can cause them to sound quite different. This factor, combined with the complexity of music and acoustics, helps makes the field of recording the subjective and personal art form that it is.

Digital recordings are often lower in noise and distortion because fewer transducers are introduced into the playback chain. In a totally digital system (Figure 1.23), the acoustic waveforms are picked up by a microphone and are converted into electrical signals. These signals are then converted into digital form by an analog-to-digital (A/D) converter. The A/D converter changes these continuous electrical waveforms into corresponding discrete numeric values that represent the waveform's instantaneous, analogous voltage levels.

Figure 1.23. *The all-digital recording chain.*

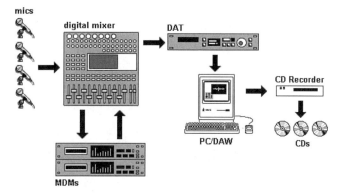

Arguably, digital information has the distinct advantage over analog in that data can be transferred between electrical, magnetic and optical media with virtually no degradation in quality because the information isn't transduced; it continues to be stored in its original, discrete binary form. In other words, only the medium changes—the data containing the actual information doesn't. Therefore, if you're listening to a digital recording on a home compact disc (CD) player over decent speakers, the actual playback will yield a sonic clarity of a quality near or equal to that of the original master recording.

Does this mean that digital's better? Nope! It's just another way of expressing sound through a medium, which in the end, is just one of the many possible artistic and technological choices in making and recording music.

CHAPTER 2

Sound and Hearing

When making a recording, what we're actually interested in doing is capturing and storing sound so that an original event can be re-created at a later date. If we begin with the concept that "sound" is a word that describes the brain's perception and interpretation of a physical stimulus that arrives at the ears . . . the examination of sound can be divided into three areas:

- The nature of the stimulus
- The characteristics of the ear as a transducer
- The psychoacoustics of hearing

The area of *psychoacoustics* deals with how and why the brain interprets a particular sound stimulus in a certain way. By understanding the physical nature of sound and how the ears change it from a physical phenomenon to a sensory one, we can discover what's needed to make a recording convey a particular effect.

Sound-Pressure Waves

Sound arrives at the ear in the form of periodic variations in atmospheric pressure (called *sound-pressure waves*). This is the same atmospheric pressure that's measured by the weather service with a barometer; however, the changes are too small in magnitude and fluctuate too rapidly to be observed on a barometer.

An analogy of how sound waves travel in air can be demonstrated by dropping a stone into a relatively calm pool of water. In this case, the motion of the waves on the water's surface (Figure 2.1) can be used to represent the motion of sound-pressure waves as they move away from a sound source in air. The only difference between these analogies is that sound-pressure waves radiate outwardly in a three-dimensional, spherical pattern, not only on a two-dimensional surface.

Figure 2.1. *A representation of wave movement on the surface of water as it moves away from its point of origin.*
a. Top view.
b. Side view of the water's surface.

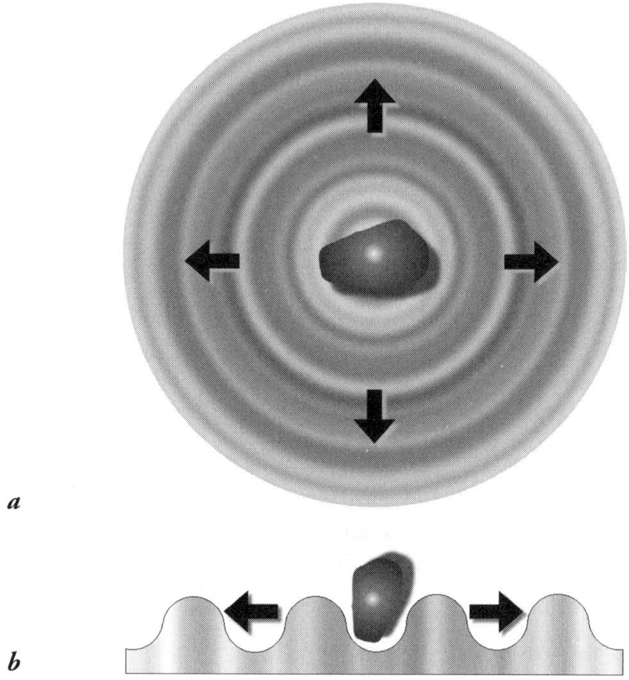

a

b

Likewise, sound-pressure waves are generated by a vibrating body that's in contact with air (such as a loudspeaker, a person's vocal cords or a guitar string that vibrates an instrument's body). A vibrating mass serves to squeeze additional air molecules into a given area as it moves outward, away from the sound source. This causes the area being acted upon to have a greater than normal atmospheric pressure, called a *compression* (Figure 2.2a). As the compressed molecules continue to move away from the sound source, an area with a lower-than-normal atmospheric pressure will be created, called a *rarefaction* (Figure 2.2b). It's interesting to note that the molecules themselves don't move through air at the velocity of sound—only the sound wave moves through the atmosphere in the form of high-pressure compression waves that push against areas of lower pressure (in an outward direction), in a process that's known as *wave propagation*.

Figure 2.2. *Effects of a vibrating mass on air molecules and their propagation.*
a. Compression: Air molecules are forced together to form a compression.
b. Rarefaction: As the compression moves away from the originating source, a rarefacted area of lower atmospheric pressure is created.

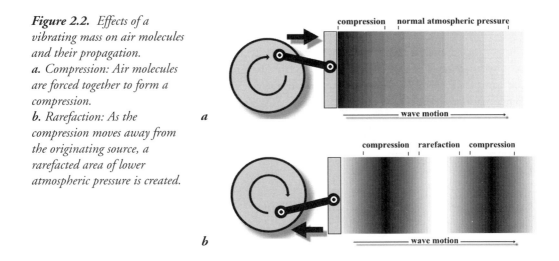

Waveform Characteristics

A *waveform* is essentially the graphic representation of a signal's sound-pressure level or voltage level as it moves through a medium over time. In short, a waveform lets us see and explain the actual phenomena of wave propagation in our physical environment, and will generally have the following fundamental characteristics:

- Amplitude
- Frequency
- Velocity
- Wavelength
- Phase
- Harmonic content
- Envelope

These characteristics allow one waveform to be distinguished from another. The most fundamental of these are *amplitude* and *frequency.* The following sections describe each of these characteristics . . . Although several math formulas have been included, it's by no means important that you memorize or worry about them. I feel that it's far more important that you grasp the basic principles of acoustics, rather than fret over the underlying math. More often than not, acoustics is an artistic science that melds science with the art of intuition and experience. Don't worry, be happy. . . .

Amplitude

The distance above or below the centerline of a waveform (such as a pure sine wave) represents the *amplitude level* of that signal (Figure 2.3). The greater the distance or displacement from that centerline, the more intense the pressure variation, electrical signal or physical displacement within a medium will be.

Figure 2.3. *Graph of a waveform showing various ways to measure amplitude.*

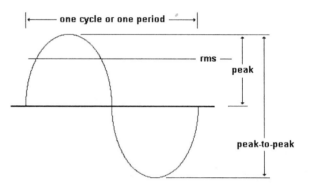

Waveform amplitudes can be measured in several ways. For example, the measurement between the maximum positive and negative signal levels of a wave is called its *peak amplitude value* (or *peak value*). The difference between the positive and negative peak signal levels is called the *peak-to-peak value*. The *root-mean-square (rms)* value was developed to determine the meaningful average level of a waveform over time (which more closely approximates the level that's perceived by our ears). For a sine wave, the rms value is arrived at by squaring the amplitude of the wave at each point along the waveform and then taking the mathematical average of the combined results (which is equal to 0.707 times the instantaneous peak amplitude level). Because the square of a positive or negative value is always positive, the rms value will always be positive. The following simple equations show the relationship between a waveform's peak and rms values:

$$\text{rms voltage} = 0.707 \times \text{peak voltage}$$

$$\text{peak voltage} = 1.414 \times \text{rms voltage}$$

Frequency

The rate at which an acoustic generator, electrical signal or vibrating mass repeats a cycle of positive and negative amplitude is known as the *frequency* of that signal. One completed excursion of a wave (which is plotted over the 360° axis of a circle) is known as a *cycle* (Figure 2.4). The number of cycles that occur within a second (frequency) is measured in *hertz (Hz)*.

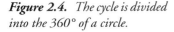

Figure 2.4. *The cycle is divided into the 360° of a circle.*

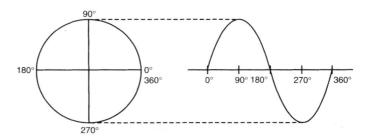

The diagram in Figure 2.5 shows the value of a waveform as starting at zero. At time t = 0, this value increases to a positive maximum value and then decreases back through zero, where the process begins all over again in a repetitive fashion. A cycle can begin at any angular degree point on the waveform . . . however, to be complete, it must pass through the zero line and end at a point that has the same value (either positive or negative) as its starting level. For example, the waveform that starts at t = 0 and ends at t = 2 constitutes a cycle, as does the waveform from t = 1 to t = 3.

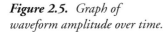

Figure 2.5. *Graph of waveform amplitude over time.*

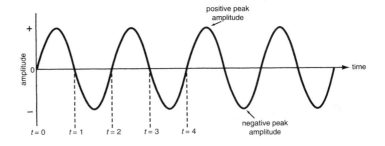

Velocity

The velocity of a sound wave as it travels through air at 68°F (20°C) is approximately 1130 feet per second (ft/sec) or 344 meters per second (m/sec). This speed is temperature-dependent and increases at a rate of 1.1 ft/sec for each Fahrenheit degree increase in temperature (2 ft/sec per Centigrade degree).

Wavelength

The wavelength of a wave (frequently represented by the Greek letter lambda, λ) is the actual distance in a medium between the beginning and the end of a cycle (or between corresponding points on adjacent cycles). The physical length of a wave can be calculated using:

$$\lambda = V/f$$

where:

    λ is the wavelength in the medium

    V is the velocity in the medium

    f is the frequency in hertz

The time it takes to complete 1 cycle is called the period of the wave. To illustrate, a 30-Hz sound wave completes 30 cycles each second or 1 cycle every one-30th of a second. The period of the wave is expressed using the symbol T:

$$T = 1/f$$

where T is the number of seconds per cycle.

Assuming that sound propagates at the rate of 1130 ft/sec, all that's needed is to divide this figure by the desired frequency. For example: the simple math for calculating the wavelength of a 30-Hz waveform would be 1130/30 = 37.6 feet long; a 1000-Hz wave form would work out as 1130/1000 = 1.13 feet long; and a 10,000-Hz waveform would be: 1130/10,000 = 0.113 feet long. From this you can see that whenever the frequency is increased, the wavelength will decrease (Figure 2.6).

Figure 2.6. *Two cyclic wavelengths.*

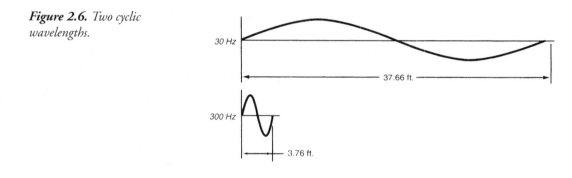

Reflection of Sound

Much like light waves, sound reflects off a surface boundary at an angle that's equal to (and in an opposite direction of) its initial angle of incidence. This basic property is one of the cornerstones of the complex study of acoustics. For example, Figure 2.7a shows how a sound wave reflects off a solid smooth surface in a simple and straightforward manner. Figure 2.7b shows how a convex surface will splay the sound outward from its surface, radiating the sound in a wide dispersion pattern. In Figure 2.7c, a concave surface is used to focus a sound at a single point, while a 90° corner (as shown in Figure 2.7d) reflects patterns back in their original incident direction. This holds true both for the 90° corners of a wall and for the intersections where the wall and floor meet. These corner reflections can also provide insight into the volume build-ups that often occur in corners (particularly at wall-to-floor corner intersections). Of course, corners or wall intersections with angles less than or greater than 90° will reflect signals according to their designed angle.

Figure 2.7. *Incident sound waves striking surfaces with varying shapes.*
a. A single-planed, solid, smooth surface.
b. A convex surface.
c. A concave surface.
d. A 90° corner reflector.

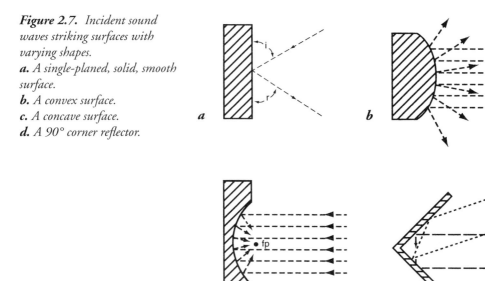

Diffraction of Sound

Sound has the inherent ability to diffract around or through a physical acoustic barrier. In other words, sound can bend around an object in a manner that reconstructs the original waveform in both frequency and amplitude. For example, in Figure 2.8a, we can see how a small obstacle will scarcely impede a larger acoustic waveform. Figure 2.8b shows how a larger obstacle can obstruct a larger portion of the waveform. However, past the obstruction, the signal bends around the area in the barrier's wake and begins to reconstruct the waveform. In Figure 2.8c, we can see how a signal is able to radiate through an opening in a large barrier. Although the signal is greatly impeded, it nevertheless begins to reconstruct itself in wavelength and amplitude (relative to the size of the opening) and begins to radiate outward from the opening as though it were a new point of origin. Finally, Figure 2.8d shows how a large opening in a barrier will let much of the waveshape pass through unimpeded. On passing, however, the waveform will begin to bend outwards in order to reconstruct itself back into its original three-dimensional shape.

Figure 2.8. *The effects of obstacles on sound radiation and diffraction.*
a. A sound having a large wavelength compared to a small obstacle size will scarcely be impeded.
b. An obstacle that's large compared to the signal's wavelength will impede the signal.

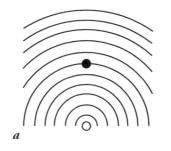

Figure 2.8. *(continued)*
c. *A small opening will impede the signal and will thereafter act as a new source point.*
d. *A larger opening will allow sound to pass through and readily diffract back into its original shape.*

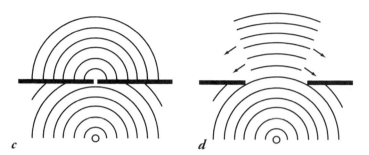

c   *d*

Frequency Response

The charted output of a sound-producing device over a range of frequencies (usually the 20–20,000 Hz range of human hearing) is known as the measured device's *frequency response.*

As an example, let's take a look at the *frequency-response curve* of several mics (Figure 2.9). In these and all cases, the y-axis represents the average signal amplitude of the signal at the output of the device that's being measured. The x-axis represents the frequency (or pitch) of the signal. When the input to the device is fed a constant-amplitude signal that sweeps over the entire frequency spectrum, we can graph the results of the device's changes in output level over frequency on a chart that can be easily read. If the output amplitude is the same at all frequencies, the drawn curve will be a flat, straight line from left to right (this is where the term *flat frequency response* comes from). This indicates that the device passes all frequencies equally (no frequency is emphasized or de-emphasized). If the curve were to dip at certain frequencies, you would know that those frequencies have lower amplitudes than the other frequencies (and vice versa). A frequency-response curve graphically indicates the effect that a device has on the tonal quality that's passed or generated by a device or instrument.

Figure 2.9. *Frequency-response curves.*
a. *Curve showing a bass boost.*
b. *Curve showing a boost at the upper-end response.*

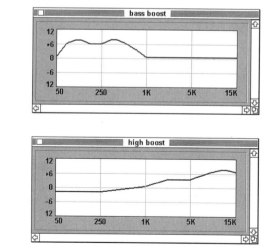

a

b

Phase

Since we know that a cycle can begin at any point on a waveform, it follows that whenever two or more waveforms are involved in producing a sound, their relative amplitudes can (and often will) be different at any one point in time. For simplicity's sake, let's limit our example to two pure tone waveforms (sine waves), which have equal amplitudes and frequency . . . but start their cyclic period at different times. Waveforms such as this are said to be *out-of-phase* with respect to each other.

Phase is measured in degrees (°) and can be described as being the relative phase degree angle with another wave(s) over 360°. The sine wave (so named because its amplitude follows a trigonometric sine function) is usually considered to begin at 0° with an amplitude of zero; the waveform then increases to a positive maximum at 90°, decreases back to having a zero amplitude at 180°, and then increases to a maximum (but in a negative direction) at 270° . . . finally returning back to its original level at 360° . . . simply to begin all over again.

Whenever two or more waveforms arrive at a location out-of-phase, their relative signal levels can be added together to create a combined amplitude level at that one point in time. Whenever two waveforms (having the same frequency, shape and peak amplitude) are completely in-phase (meaning that they have no relative time difference) are added, the resulting waveform will have the same frequency, phase and shape . . . but will double in amplitude (Figure 2.10a). If the same two waves are combined completely out-of-phase (having a phase difference of 180°), they will cancel each other when added. This results in a straight line of zero amplitude (Figure 2.10b). If the second wave is only partially out-of-phase (not exactly 180°), the levels will constructively interfere at points where the amplitudes of the two waves have the same sign (that is, both are positive or both are negative) and destructively interfere at points where the signs of the two wave amplitudes are opposing (Figure 2.10c).

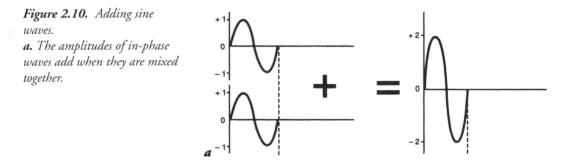

Figure 2.10. *Adding sine waves.*
a. *The amplitudes of in-phase waves add when they are mixed together.*

Figure 2.10. *(continued)*
b. Waves of equal amplitude cancel completely when mixed 180° out-of-phase.
c. When partial phase angles are mixed together, the combined signals will add in certain places and subtract in others.

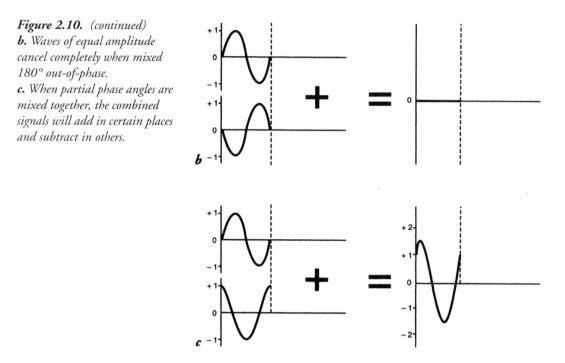

Phase shift is a term that describes one waveform's lead or lag in time with respect to another. Basically, it results from a time delay between the two (or more) waveforms (with differences in distance being the most common source of this type of delay). For example, a 500-Hz wave completes one cycle every 0.002 sec. If you start with two in-phase 500-Hz waves and delay one of them by 0.001 sec (half the wave's period), the delayed wave will lag the other by one-half a cycle . . . or 180°. Another example might include a single source that's being picked up by two microphones, which have been placed at different distances (Figure 2.11), thereby creating a corresponding time delay. Such a delay can also occur when a reflected sound is picked up by the same mic that is used to pick up the direct sound. These signals may be in-phase at frequencies where the path-length difference is equal to the signal's wavelength, and out-of-phase at those frequencies where the multiples fall at or near the half-wavelength distance. These boosts and cancellations will then combine to alter the signal's overall frequency response at the pickup.

Figure 2.11. *Cancellations can occur when a single source is picked up by two mics.*

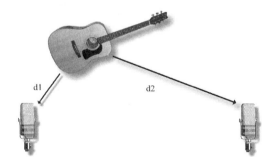

In order to keep these interferences above 20 kHz (and thus out of the audio range), the path-length difference should be less than 0.34 inch (which corresponds to a time delay of 0.03 ms). Because this is such a small distance, you can see that virtually any reflection or time-delayed signal of sufficient level could potentially interfere with a pickup's response. To avoid this form of distortion, you must either eliminate the reflections or reduce their level to points that these cancellations will be negligible. This is one of the reasons why leakage between instruments should be kept to a minimum whenever possible.

Harmonic Content

Up to this point, the discussion has centered on the sine wave, which is composed of a single frequency that produces a pure sound at a specific pitch. Musical instruments, however, rarely produce pure sine waves, and it's good thing they don't. If they did, all instruments would basically sound the same, and music would be pretty boring. The factor that allows us to differentiate between instruments is the presence of various frequencies (called *partials*) that exists in addition to the pitch that's being played (which is called the *fundamental*). Partials that are higher than the fundamental frequency are called upper partials or *overtones*. Overtone frequencies that are whole-number multiples of the fundamental frequency are called *harmonics*. For example, the frequency that corresponds to concert A is 440 Hz (Figure 2.12a). An 880-Hz wave is a harmonic of the 440-Hz fundamental because it is twice the frequency (Figure 2.12b). In this case, the 440-Hz fundamental is technically the first harmonic because it is 1 times the fundamental frequency, and the 880-Hz wave is called the second harmonic because it is 2 times the fundamental. The third harmonic would be 3 × 440 Hz or 1320 Hz (Figure 2.12c). Some instruments, such as bells, xylophones, and other percussion instruments, will have partials that aren't harmonically related to the fundamental at all.

Figure 2.12. *An illustration of harmonics.*
a. *Fundamental waveform (first harmonic).*

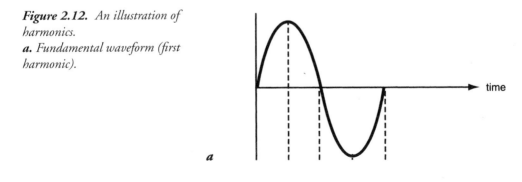

a

Figure 2.12. *(continued)*
b. *Second harmonic.*
c. *Third harmonic.*

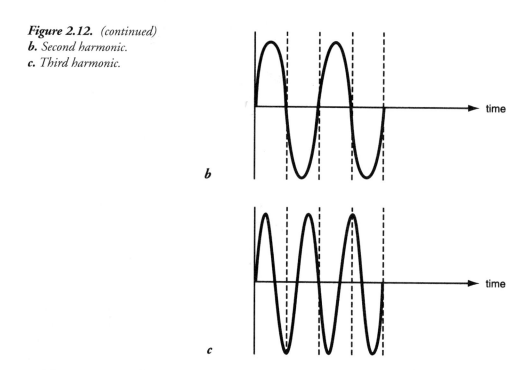

b

c

The ear perceives frequencies that have ratios which are whole multiples of the fundamentals as being specially related (a condition known as the musical *octave*). For example, since concert A is 440 Hz, the ear hears 880 Hz as being the first tone higher than concert A that sounds most like concert A. The next note above 880 Hz that bears this relation to concert A is 1760 Hz. Therefore, 880 Hz is said to be one octave above 440 Hz, and 1760 Hz is said to be two octaves above 440 Hz, etc. Two notes that have the same fundamental frequency and are played at the same time are said to be in unison, even if they have different harmonic structures.

Because sound waves that are produced by musical instruments contain harmonics in various amplitude and phase relationships, the resulting waveforms bear little resemblance to the shape of the single-frequency sine wave. Musical waveforms can be divided into two categories: *simple* and *complex*. Square waves, triangle waves, and sawtooth waves are examples of simple waves that contain harmonics (Figures 2.13). They're said to be simple because they are continuous and repetitive in nature. One cycle of a square wave looks exactly like the next, and are all symmetrical about the zero line.

Figure 2.13. *Simple waveforms.*
a. *Square waves.*

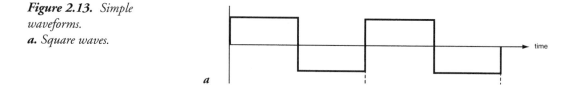

a

b. Triangle waves.
c. Sawtooth waves.

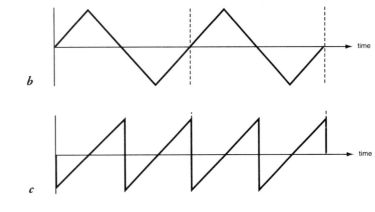

Complex waves, on the other hand, don't necessarily repeat and often aren't symmetrical about the zero line. An example of a complex waveform (Figure 2.14) is one that's created by any naturally created sound (such as music and speech). Complex waves don't always repeat, so it might be difficult to divide them into cycles or categorize them by frequency simply by looking at their waveshape.

Figure 2.14. *Example of a complex waveform.*

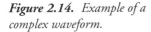

Regardless of the shape or complexity of a waveform that reaches the eardrum, the inner ear separates the sound into its component sine waves before transmitting the stimulus to the brain. For this reason, it isn't the shape of a waveform that should interest us as much as the component breakdown of the frequencies that are perceived by the brain. This can be illustrated by passing a square wave through a bandpass filter that's set to pass only a narrow band of frequencies at any one time. Doing this would show that the square wave is composed of a fundamental frequency plus a number of harmonics that are made up of odd-number multiple frequencies (whose amplitude decreases as their frequency increases). In Figure 2.15, we see how individual sine-wave harmonics combine to form a square wave.

Figure 2.15. *Breaking a square wave down into its odd-harmonic components.*
a. *A square wave with frequency f.*

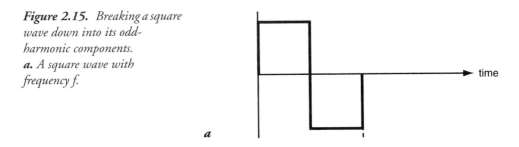

Figure 2.15. *(continued)*
b. *A sine wave with frequency f.*
c. *Sum of a sine wave with frequency f and a lower amplitude sine wave of frequency 3f.*
d. *Sum of a sine wave of frequency f and lower amplitude sine waves of 3f and 5f, beginning to resemble a square wave.*

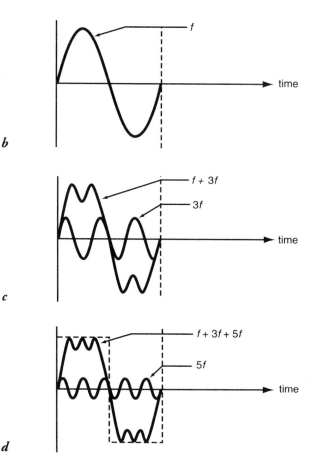

If we were to analyze the harmonic content of sound waves that are produced by a violin and compare them to the content of the waves that are produced by a viola (with both playing concert A, 440 Hz), we would come up with results like those shown in Figure 2.16. Notice that the violin's harmonics differ in both degree and intensity from those of the viola. The harmonics and their relative intensities (which determine an instrument's characteristic sound) are called the *timbre* of an instrument. If we changed an instrument's harmonic balance, the sonic character of the instrument would also be changed. For example, if the violin's upper harmonics were reduced, the violin would sound a lot like the viola.

Figure 2.16. *Harmonic structure of concert A-440.*
a. *Played on a violin.*

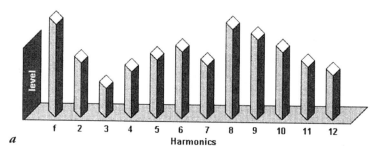
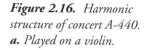

b. *Played on a viola.*

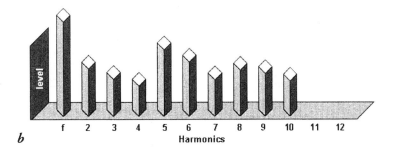

Because the relative harmonic balance is so important to an instrument's sound, the frequency response of a mic, amp, speaker and all other elements in the signal path can have an effect on the "timbre" (harmonic balance) of a sound. If the frequency response isn't flat, the timbre of the sound will be changed. For example, if the high frequencies are amplified less than the low and middle frequencies, the sound will be duller than it should be. Equalizers can be used to vary the timbre of an instrument, thereby changing its subjective sound.

In addition to variations in harmonic balance that can exist between instruments and their instrument families, it's common for both the fundamental and harmonic balance to vary with respect to the direction as they radiate from an instrument. Figure 2.17 shows the principal radiation patterns as they emanate from a cello (as seen from both the side and top views).

Figure 2.17. The radiation patterns of a cello as viewed from the side (left) and top (right).

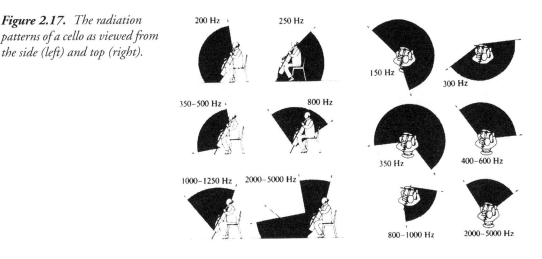

Acoustic Envelope

Timbre is not the only characteristic that lets us differentiate between instruments. Every instrument produces a unique *envelope* that works in combination with its timbre to determine its subjective sound. The envelope of a waveform can be described as characteristic variations in level that occurs over the duration of a played note.

The envelope of an acoustic signal is composed of three sections:

- Attack—the level or build-up that occurs when a note is initially sounded.
- Sustain—describes volume increases, decreases and sustain that occurs after the initial attack.
- Decay—the fade or reduction in level over time once the note has stopped playing.

Each of these sections has three variables: time duration, amplitude and amplitude variation with time. Figure 2.18a illustrates the envelope of a trombone note. The attack, decay times and internal dynamics consist of a sustain that produces a smooth, flowing sound. Figure 2.18b illustrates the envelope of a snare drum beat. Notice that the initial attack is much louder than the internal dynamics . . . while the final decay trails off very quickly, resulting in a sharp, percussive sound. A cymbal crash (Figure 2.18c) combines a high-level, fast attack with a longer sustain and decay that creates a smooth, lingering shimmer.

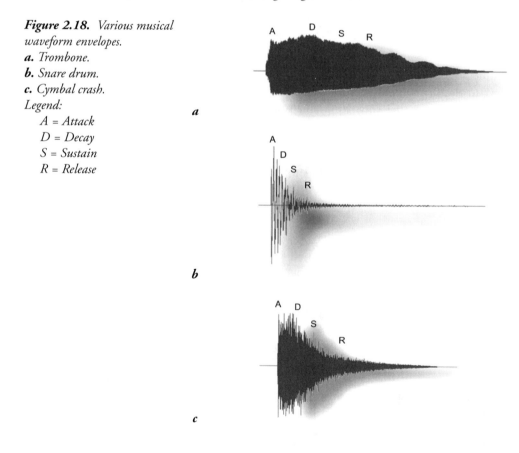

Figure 2.18. *Various musical waveform envelopes.*
a. *Trombone.*
b. *Snare drum.*
c. *Cymbal crash.*
Legend:
   A = Attack
   D = Decay
   S = Sustain
   R = Release

It's important to note that the concept of envelopes use peak waveform values, while human perception of loudness is proportional to the average wave intensity over a period of time (rms value). Therefore, high-amplitude portions of the envelope won't make an instrument sound loud unless the amplitude is maintained for a long enough period. Short high-amplitude sections tend to contribute to sound character rather than to loudness. Through the use of dynamic range controllers (such as compressors, limiters and expanders), the sound character of an instrument can be modified by changing its envelope without changing its timbre.

The waveform envelope of an electronic music instrument is similar in most respects to its acoustic counterpart and is modified with respect to its initial attack time, decay time (from the initial attack), sustain time and final release time. It's most commonly referred to in its abbreviated form, ADSR (attack, decay, sustain, and release).

Loudness Levels: The Decibel (dB)

The ear operates over an energy range of approximately 10^{13}:1 (10,000,000,000,000:1)—that's an extremely wide range. Because such a range is difficult for humans to deal with easily, a logarithmic scale has been adopted to compress the measurements into figures that are more workable. The system used for measuring sound-pressure level (SPL), signal level and relative changes in signal level is the *decibel (dB)*, a term that literally means 1/10th of a bell . . . a measurement unit that was named after Alexander Graham Bell, the inventor of the telephone.

In order to develop an understanding of the decibel, we first need to examine logarithms and the logarithmic scale (Figure 2.19). The *logarithm (log)* is a mathematical function that reduces large numeric values into smaller, more manageable numbers. Because log numbers increase exponentially in a way that's similar to how we perceive loudness (i.e., 1, 2, 4, 16, 128, 256, 65,536 . . .), it expresses our perceived sense of volume more precisely than a linear curve can.

Figure 2.19. *Linear and logarithmic scales.*
a. *Linear.*
b. *Logarithmic.*

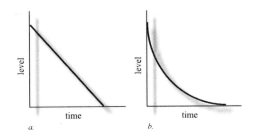

Before we delve into a deeper study of this important concept for understanding many of our perceptual senses, I'd like to take a moment out to stress the fact that much confusion has surrounded way in which the "log" scale relates to sound and hearing . . . especially when readers are confronted with tons of formulas that frankly don't seem to relate to their working lives.

In light of this, I'd like to do my best to present you with an understanding of the basic concepts and building block ideas behind the log scale and help you to understand just what examples like "+3 dB at 10,000 Hz" really mean. Be patient with yourself if you still don't get it . . .

over time the idea of the decibel will become as much a part of your working vocabulary as ounces, gallons and miles-per-hour.

Logarithmic Basics

In audio, we use logarithmic values to express the differences in intensities between two levels (often, but not always, comparing a measured level to a standard reference level). Because the differences between these two levels can be really, really big, we have to break these huge numbers down into values that are mathematical *exponents* of 10.

To begin with, finding the log of a number like 17,386 without a calculator isn't only difficult . . . in audio, it's unnecessary! All that's really important is that you memorize three simple guidelines:

- The log of the number 2 is 0.3.
- When a number is an integral power of 10 (i.e., 100, 1,000, 10,000 . . . etc.) the log can be found, simply by adding up the number of zeros.
- Numbers that are greater than 1 will have a positive log value, while those less than 1 will have a negative log value.

The first one is an easy fact: the log of 2 is 0.3 . . . this will make more sense shortly. The second one is even easier: the logs of numbers like 100, 1,000 or 10,000,000,000,000 can be arrived at by counting up the zeros. The last guideline relates to the fact that if the measured value is less than the reference value, the log will be negative.

For example:

$$\log 2 = 0.3$$
$$\log \tfrac{1}{2} = \log 0.5 = -0.3$$

and . . .

$$\log 10{,}000{,}000{,}000{,}000 = 13$$
$$\log 1000 = 3$$
$$\log 100 = 2$$
$$\log 10 = 1$$
$$\log 1 = 0$$
$$\log 0.1 = -1$$
$$\log 0.01 = -2$$
$$\log 0.001 = -3$$

All other numbers can be arrived at by using a calculator . . . however, it's unlikely that you'll ever need to know any log values, beyond understanding the "concepts" behind those that are listed above.

The dB

Now that we've gotten past this, I'd again like to break with tradition and attempt an explanation of the dB in a way that's less complex and relates more to our day-to-day needs in the sound biz.

First off, the dB is a logarithmic value that "expresses differences in intensities between two levels." From this, we can infer that these "levels" can be expressed in several units of measure . . . the most common level units are sound pressure level, voltage, and power (wattage). Now, let's look at the basic math for these three categories:

Sound-Pressure Level

Sound-pressure level (SPL) is the acoustic pressure that's built up within a defined atmospheric area (usually a square centimeter—cm^2). Quite simply, the higher the sound-pressure level, the louder the sound (Figure 2.20). In this instance, our measured reference (SPL_{ref}) is the threshold of hearing, which is defined as being the softest sound that an average person can hear. An average conversation usually has an SPL of about 70 dB, while average home stereo levels have a loudness of between 80 and 90 dB SPL. Sounds that are so loud as to be painful have SPLs of about 130 dB (10,000,000,000,000 times louder than the 0-dB reference).

We can arrive at an SPL rating by using the formula:

$$dB\ SPL = 20\ log\ SPL/SPL_{ref}$$

where SPL is the measured sound pressure in $dyne/cm^2$.

SPL_{ref} is a reference sound pressure ($0.0002\ dyne/cm^2$—the threshold of hearing).

From this, I feel that the only concept that needs to be understood is the idea that SPL levels change with the square of the distance (hence the 20 log part of the equation). This means that whenever a source/pickup distance is doubled, the SPL level will reduce by 6 dB (20 log 2/1 = 20 × .3 = 6 dB SPL); as the distance is halved, it will increase by 6 dB (20 log 1/2 = 20 × 0.3 = 6dB SPL), as shown in Figure 2.21.

Figure 2.20. *Chart of sound-pressure levels. (Courtesy of General Radio Company)*

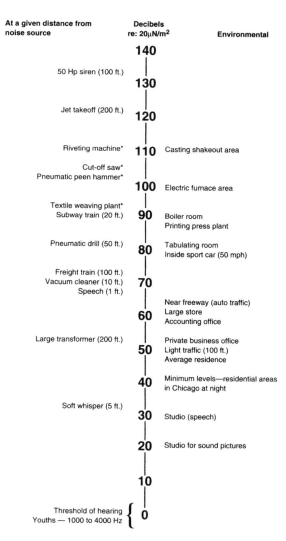

Figure 2.20. *Chart of sound-pressure levels. (Courtesy of General Radio Company)*

Typical A-Weighted Sound Levels

At a given distance from noise source	Decibels re: 20μN/m²	Environmental
	140	
50 Hp siren (100 ft.)	**130**	
Jet takeoff (200 ft.)	**120**	
Riveting machine*	**110**	Casting shakeout area
Cut-off saw* Pneumatic peen hammer*	**100**	Electric furnace area
Textile weaving plant* Subway train (20 ft.)	**90**	Boiler room Printing press plant
Pneumatic drill (50 ft.)	**80**	Tabulating room Inside sport car (50 mph)
Freight train (100 ft.) Vacuum cleaner (10 ft.) Speech (1 ft.)	**70**	
	60	Near freeway (auto traffic) Large store Accounting office
Large transformer (200 ft.)	**50**	Private business office Light traffic (100 ft.) Average residence
	40	Minimum levels—residential areas in Chicago at night
Soft whisper (5 ft.)	**30**	Studio (speech)
	20	Studio for sound pictures
	10	
Threshold of hearing Youths — 1000 to 4000 Hz	**0**	

*Operator's position

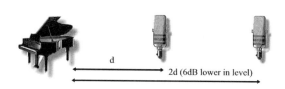

Figure 2.21. *Doubling the distance of a pickup will lower the direct signal level by 6 dB SPL.*

d

2d (6dB lower in level)

Voltage

As with acoustic energy, comparing one voltage level to another level (or reference level) can be expressed as dBv using the equation:

$$dBv = 20 \log V/V_{ref}$$

where:

  V is the measured voltage

  V_{ref} is a reference voltage (0.775 volt)

Power

Power is usually a measure of wattage or current . . . both of which are generally associated with signals that are carried throughout audio signal paths. Unlike SPL and voltage, the equation for signal level (which is often expressed in dBm) is:

$$dBm = 10 \log P/P_{ref}$$

where:

  P is the measured voltage

  P_{ref} is referenced to 1 milliwatt (0.001 watt = 1mW) across a 600? line

The Simple Heart of the Matter

I'm going to stick my neck out and state that, when dealing with dBs, it's far more common for working professionals to deal with the concept of power when it comes to understanding the dB. The dBm equation expresses the "spirit" of the term dB when dealing with the markings on a device or numeric values in a computer dialog box. This is due to the fact that power is the unit of measure that's most often expressed when dealing with audio equipment controls. Therefore, it's my personal opinion that the average working stiff only needs to grasp the following basic concepts, when relating the dB to audio equipment and their markings:

- A 1-dB change will be barely noticeable to most ears.
- Turning something up by 3 dB will double the signal's level (believe it or not, doubling the signal level won't increase the perceived loudness as much as you might think).
- Turning something down by 3 dB will halve the signal's level (likewise, halving the signal level won't decrease the perceived loudness as much as you might think).
- The log of an exponent of ten can be easily figured by simply counting the zeros (i.e., the log of 1000 is 3). Given that this figure is multiplied by 10 (10 log P/P_{ref}) . . . turning something up by 10 dB will increase the signal's level tenfold, 20 will yield a hundredfold increase, 30 dB will yield a thousandfold increase, etc.

Most pros know that turning a level fader up by 3 dB will effectively double its energy output (and vice versa). Beyond this, it's unlikely that anyone will ever ask "Would you please turn that up a thousand times?" . . . It just won't happen. However, when a pro asks his/her assistant to turn up the gain by 20 dB, that assistant will often instinctively know what 20 dB is . . . and what it sounds like. I guess I'm saying that the math really isn't *nearly* as important as getting an instinctive "feel" for the dB and how it relates to relative levels within audio production.

The Ear

A sound source produces acoustic waves by alternately compressing and rarefying the air between it and the listener, causing fluctuations that fall above and below normal atmospheric pressure. The human ear is a sensitive transducer that responds to these pressure variations by way of a series of related processes that occur within the auditory organs . . . our ears.

When these variations arrive at the listener, sound-pressure waves are collected into the aural canal by way of the outer ear's pinna and are then directed to the eardrum, a stretched drum-like membrane (Figure 2.22). The sound waves are then changed into mechanical vibrations, which are transferred to the inner ear by way of three bones called the hammer, anvil and stirrup. These bones act both as an amplifier (by significantly increasing the vibrations that are transmitted by the eardrum) and as a limiting protection device (by reducing the level of loud, transient sounds, such as thunder or fireworks explosions). The vibrations are then applied to the inner ear (cochlea)—a tubular, snail-like organ that contains two fluid-filled chambers. Within these chambers are tiny hair receptors that are lined in a row along the length of the cochlea. These hairs respond to certain frequencies depending on their placement along the organ, which results in neural stimulation that gives us the sensation of hearing. Permanent hearing loss generally occurs when they're damaged or as they deteriorate with age.

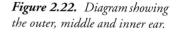

Figure 2.22. Diagram showing the outer, middle and inner ear.

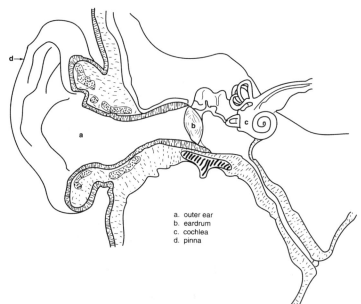

a. outer ear
b. eardrum
c. cochlea
d. pinna

Threshold of Hearing

In the case of SPL, a convenient pressure-level reference is the *threshold of hearing*, which is the minimum sound pressure that produces the phenomenon of hearing in most people and is equal to 0.0002 microbar. One microbar is equal to one-millionth normal atmospheric pressure, so it's apparent that the ear is an extremely sensitive instrument. In fact, if the ear were any more sensitive, the thermal motion of molecules in the air would be audible. When referencing sound-pressure levels to 0.0002 microbar, this threshold level usually is denoted as 0 dB SPL, which is defined as the level that an average person can hear a specific frequency only 50 percent of the time.

Threshold of Feeling

An SPL that causes discomfort in a listener 50 percent of the time is called the *threshold of feeling*. It occurs at a level of about 118 dB SPL between the frequencies of 200 Hz and 10 kHz.

Threshold of Pain

The SPL that causes pain in a listener 50 percent of the time is called the *threshold of pain* and corresponds to an SPL of 140 dB in the frequency range between 200 Hz and 10 kHz.

Auditory Perception

The ear is a nonlinear device; as a result, it produces *harmonic distortion* whenever it's subjected to sound waves that are above a certain loudness. Harmonic distortion is the production of waveform harmonics that don't exist in the original signal. For example, the ear can cause a loud 1 kHz sine wave to be heard as a combination of waves of 1 kHz, 2 kHz, 3 kHz, and so on. Although the ear might receive the overtone structure of a violin, if the listening level is loud enough, it might also produce additional harmonics (thus changing the perceived timbre of the instrument). This implies that sound monitored at very loud levels might sound quite different when played back at low levels.

The terms *linear* and *nonlinear* are used to describe the input versus output level characteristics of a transducer or device. A linear device or medium is one whose input and output amplitudes have the same input/output ratio at all signal levels. For example, if the input of a linear amplifier is doubled, the output signal amplitude will also double. If at certain signal levels, however, doubling the input signal amplitude will increase the output signal amplitude by more or less than by a factor of two, the amplifier would be said to be nonlinear at those amplitudes.

In addition to being nonlinear with respect to perceived tones, the ear's frequency response (that is, its perception of timbre) changes with the loudness of the perceived signal. The loudness compensation switch found on many hi-fi preamplifiers is an attempt to compensate for the decrease in the ear's sensitivity to low- and high-frequency sounds at low levels.

The *Fletcher–Munson equal-loudness contour curves* (Figure 2.23) indicate the ear's average sensitivity to different frequencies at various levels. The horizontal curves indicate the sound-pressure levels that are required for our ears to hear them as being equal in level to a 1000-Hz reference level. Thus, to equal the loudness of a 1-kHz tone at 110 dB SPL (a level typically created by a trumpet-type car horn at a distance of 3 feet), a 40-Hz tone has to be about 6 dB louder, while a 10-kHz tone must be 4 dB greater in order to be perceived as being equally loud. At 50 dB SPL (the noise level present in the average private business office), the level of a 40-Hz tone must be 30 dB greater and a 10-kHz tone must be 3 dB greater than a 1-kHz tone to be perceived as having the same volume. Thus, if a piece of music is monitored and sounds great at a level of 110 dB SPL, it will sound both bass and treble shy when played at 50 dB SPL. It has generally been found that changes in apparent frequency balance are less apparent at levels of 85 dB SPL, which appears to be the best average sound monitoring level.

Figure 2.23. *The Fletcher–Munson equal loudness contour for pure tones as perceived by humans having average hearing acuity. These perceived loudness levels are charted relative to sound pressure levels at 1000 Hz.*

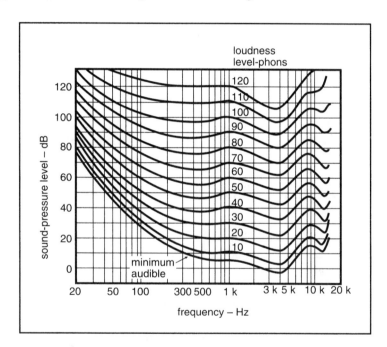

The loudness of a tone can also affect our ear's perception of pitch. For example, if the intensity of a 100-Hz tone is increased from 40 to 100 dB SPL, the ear will hear a pitch decrease of about 10 percent. At 500 Hz, the pitch will change about 2 percent for the same increase in sound-pressure level. This is one reason why musicians find it difficult to tune their instruments when listening through loud headphones.

As a result of the nonlinearities in the ear's response, tones will often interact with each other rather than being perceived separately. Three types of interaction effects can occur:

- Beats
- Combination tones
- Masking

Beats

Two tones that differ only slightly in frequency and have approximately the same amplitude will produce an effect known as *beats* (repetitive volume surges), which are equal in frequency to the difference between the two tones. This phenomenon is often used as an aid for tuning instruments, because the beats slow down and stop as the two notes approach and reach the same pitch. Beats are actually a result of the ear's inability to separate closely pitched notes. This results in the creation of a third wave that is the sum of the two notes when they are in-phase and the difference when they're out-of-phase.

Combination Tones

Combination tones result when two loud tones differ by more than 50 Hz. The ear produces an additional set of tones that's equal to both the sum and the difference between the two original tones and is also equal to the sum and difference of their harmonics. The simple formulas for computing these tones are:

$$\text{sum tone} = f_1 + f_2$$

$$\text{difference tone} = f_1 - f_2$$

Difference tones can be easily heard when they're below the frequency of both the tone's fundamentals. For example, 2000 and 2500 Hz produce a difference tone of 500 Hz.

Masking

Masking is the phenomenon by which loud signals prevent the ear from hearing softer sounds. The greatest masking effect occurs when the frequency of the sound and the frequency of the masking noise are close to each other. For example, a 4-kHz tone will mask a softer 3.5-kHz tone, but has little effect on the audibility of a quiet 1000-Hz tone. Masking can also be caused by harmonics of the masking tone (i.e., a 1-kHz tone with a strong 2-kHz harmonic might mask a 1900-Hz tone). This phenomenon is one of the main reasons that stereo placement and equalization are so important in the mixdown process. An instrument that sounds fine by itself can be completely hidden or changed in character by louder instruments with a similar timbre. Equalization might be required to make the instruments sound different enough to overcome any masking effects.

Perception of Direction

Although one ear can't discern the direction of a sound's origin, two ears can. This capability of two ears to localize a sound source within an acoustic space is called *spatial* or *binaural localization*. This effect results from using the following three cues received by the ears:

• Interaural intensity differences

• Interaural arrival-time differences

• The effects of the pinnae (outer ears)

Middle- to higher-frequency sounds originating from the right side will reach the right ear at a higher intensity level than the left ear, causing an *interaural intensity difference*. This difference occurs because the head casts an acoustic block or shadow, which allows only reflected sounds from surrounding surfaces to reach the opposite ear (Figure 2.24). Because the reflected sound travels farther and loses energy at each reflection, in our example, the intensity of sound perceived by the left ear will be reduced, resulting in a signal that's perceived as originating from the right.

Figure 2.24. The head casts an acoustic shadow that helps with localization at middle to upper frequencies.

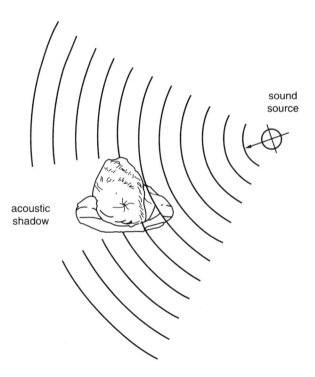

sound source

acoustic shadow

This effect is relatively insignificant at lower frequencies, where wavelengths are large compared to the head's diameter and the wave easily bends around its acoustic shadow. A different method of localization (known as *interaural arrive-time differences*) is employed at lower frequencies. In both Figures 2.24 and 2.25, small time differences occur because the acoustic path length to the left ear is slightly longer than the path to the right ear. The sound pressure there-fore arrives at the left ear at a later time than the right ear. This method of localization (in com-bination with interaural intensity differences) helps to give us lateral localization cues over the entire frequency spectrum.

Figure 2.25. Interaural arrival-time differences occurring at lower frequencies.

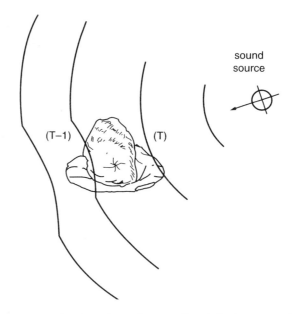

sound source

(T−1)   (T)

Intensity and delay cues allow us to perceive the direction of a sound's origin, but not whether the sound originates from front, behind or below. The pinna (Figure 2.26), however, makes use of two ridges that reflects sound into the ear. These ridges introduce minute time delays between the direct sound (which reaches the entrance of the ear canal) and the sound that's reflected from the ridges (which varies according to source location).

It's interesting to note that beyond 130° from the front of our face, there can be no ridge reflec-tions because they're blocked by the pinna. Unreflected sounds that are delayed between 0 and 80 microseconds (μsec) will still be perceived as originating from the rear. Ridge number two produces delays of between 100 and 330 μsec, which help us to locate sources in the vertical plane. The delayed reflections from both ridges combine with the direct sound to produce fre-quency response colorations that are compared at the brain to determine source location. Small movements of the head can also provide additional position information.

If there are no differences between what the left and right ears hear, the brain assumes that the source is the same distance from each ear. This phenomenon allows us to position sound not only in the left and right loudspeakers, but also monophonically between them. If the same sig-nal is fed to both loudspeakers, the brain perceives the sound identically in both ears and deduces that the source must originate from directly in the center. By changing the propor-tional level that's sent to each speaker, the engineer changes the interaural intensity differences and thus creates the illusion of physical positioning between the speakers. This placement tech-nique is known as *panning* (Figure 2.27).

Figure 2.26. *The pinna and its reflective ridges for determining vertical location information.*

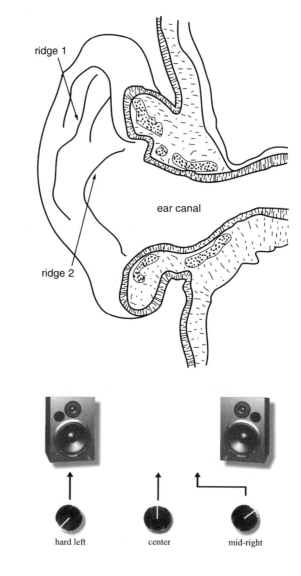

ridge 1

ear canal

ridge 2

Figure 2.27. *Pan pot settings and their relative spatial positions.*

hard left   center   mid-right

Perception of Space

In addition to perceiving the direction of sound, the ear and brain combine to help us perceive distance and a physical sense of the acoustic space in which a sound occurs. When a sound is generated, it simultaneously propagates away from the source in directions that are determined by the nature of the source. A percentage of the sound reaches the listener directly, without encountering any obstacles. A larger portion, however, is propagated to the many surfaces of an acoustic enclosure. If these surfaces are reflective, the sound is bounced back into the room, where some of them will reach the listener. If the surfaces are absorptive, less energy will be reflected back to the listener.

In air, sound travels at a constant speed of about 1130 feet per second, so a wave that travels from the source to the listener will follow the shortest path and arrive at the listener's ear first. This is called the *direct sound*. Those waves that bounce off surrounding surfaces must travel further to reach the listener and therefore arrive after the direct sound from a multitude of directions. These form what are called the *early reflections*. As a result of these additional longer path lengths, these reflections can often be heard after the source stops. A highly reflective surface absorbs less of the wave energy at each reflection and allows the sound to persist longer after the source stops in the form of *reverberation* (Figure 2.28).

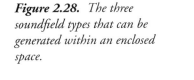

Figure 2.28. *The three soundfield types that can be generated within an enclosed space.*

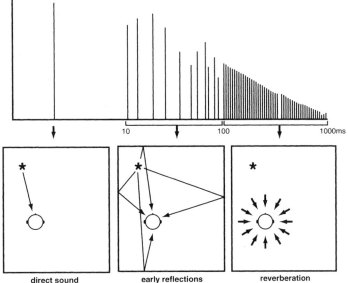

direct sound   early reflections   reverberation

Direct sound determines our perception of a sound source's location and size and conveys the true timbre of the source.

Early reflections give us clues as to the reflectivity, size and general nature of an acoustic space. These sounds generally arrive at the ears less than 50 msec after the brain perceives the direct sound and are the result of reflections off of the largest, most prominent boundaries within a room. The time elapsed between hearing the direct sound and the beginning of the early reflections help to provide information about the size of the performance room. Basically, the farther the boundaries are from the source and listener, the longer the delay will be before the sound can reach it and be reflected back.

Another aspect that occurs with early reflections is called *temporal fusion*. Early reflections arriving at the listener within 30 msec of the direct sound not only are audibly suppressed, but are fused with the direct sound. In effect, the ear can't distinguish the closely occurring reflections and considers them to be part of the direct sound. The 30-msec time limit for temporal fusion isn't absolute; rather, it depends on the sound's envelope. Fusion breaks down at 4 msec for transient clicks, whereas it can extend beyond 80 msec for slowly evolving sounds (such as a sustained organ note or legato violin passage). Despite the fact that the early reflections are suppressed and fused with the direct sound, they still modify our perception of the sound, making it both louder and fuller.

Sounds reaching the listener more than about 50 msec after the direct sound have been reflected off of so many different surfaces that they begin to reach the listener as a virtually continuous stream and from all directions. These densely spaced reflections are called *reverberation* or *reverb*. Reverb is characterized by its gradual decrease in amplitude and a sense of added warmth and body that it gives the sound. Because it has undergone multiple reflections, the timbre of the reverberation is quite different from that of the direct sound (often, with the most notable difference being a rolloff of high frequencies and a slight bass emphasis). The time it takes for the persisting sound to decrease to 60 dB below its original level is called its *decay time* or *reverb time*, which is generally determined by the room's absorption characteristics. The brain perceives the reverb time and the timbre of the reverberation and then uses this information to form an opinion on the hardness or softness of the surrounding surfaces. The loudness of the perceived direct sound increases rapidly as the listener moves closer to the source. Meanwhile, the loudness of the reverberation remains the same because of its diffusion throughout the room is roughly constant. This ratio of the direct sound's loudness to the reflected sound's level helps the listener judge his/her distance from the sound source.

Through the use of artificial reverb and delay units, the engineer can generate the necessary cues to convince the brain that a sound was recorded in a huge, stone-walled cathedral . . . when it was, in fact, recorded in a small, absorptive room. To do this, the engineer mixes the original unreverberated signal with a signal processor to provide the necessary delays and artificial reflections. Adjusting the number and amount of delays on the effects processor gives the engineer control over all of the necessary parameters to determine the perceived room size, while adjusting the decay time and frequency balance can help determine the room surfaces. By changing the proportional mix of direct to processed sound, the listener can be fooled into believing the sound source is placed either at the front or the rear of the artificially created space. Time delays can also be used to simulate single or multiple echo effects or to create the illusion of multiple instruments (or groups of performing instruments).

By repeating a signal using a short delay of 4–20 msec (or so), the brain can be fooled into thinking that the apparent number of instruments being played is doubled. This process is called *doubling* or *automatic double tracking (ADT)*. Often, acoustic doubling and tripling can be acoustically recreated during the overdub phase by recording a track and then going back and laying down one or more "passes" while listening to the original track. When this isn't possible, delay devices can be cost-effectively and easily used to simulate this effect. If a longer delay is chosen (more than about 35 msec), the repeat will be heard as discrete echoes . . . causing the delay (or series of repeated delays) to create a *slap echo* or *slap back*. This and other effects can be used to double or "thicken" up a sound . . . anybody wanna sound like Elvis or Annette Funicello?

CHAPTER 3

Studio Acoustics
and Design

The *Audio Cyclopedia* defines the term acoustics as ". . . a science dealing with the production, effects, and transmission of sound waves; the transmission of sound waves through various mediums, including reflection, refraction, diffraction, absorption, and interference; the characteristics of auditoriums, theaters, and studios, as well as their design."

We can see from this description that the proper acoustic design of a music recording, project, audio-for-visual or broadcast studio is no simple matter. A wide range of complex variables and interrelationships often come into play in the making of a successful studio design. The following basic requirements should be considered:

- Acoustic isolation—Prevents external noises from entering the studio environment, as well as preventing feuds that can occur when excessive volume levels leak out into the surrounding neighborhood.

- Frequency balance—The frequency components of a room shouldn't adversely affect the acoustic balance of instruments and/or speakers. Simply stated, the acoustic environment shouldn't alter the sound quality of the originally recorded or reproduced performance.

- Acoustic separation—The acoustical environment shouldn't interfere with intelligibility and should offer a high degree of acoustic separation (often a requirement for ensuring that sounds emanating from an instrument aren't unduly picked up by another instrument's mic).

- Reverberation—The control of sonic reflections within a space is an important factor for maximizing the intelligibility of music and speech. In addition, reverb adds an important psychoacoustic sense of "space," in the sense that it can give our brain subconscious cues as to a room's size, number of reflective boundaries, distance between the source and listener . . . and so on.

- Cost factors—Not the least of all design and construction factors is cost. Multimillion-dollar facilities often employ studio designers and construction teams to create a plush decor that's been acoustically "tuned" to fit the needs of both the owners and their clients. Owners of project studios and budget-minded production facilities, however, can also take full advantage of basic acoustic principles and construction techniques and apply them in more cost-effective ways.

This chapter discusses many of the basic acoustic principles and construction techniques that should be considered in the design of a music or sound production facility. I'd like to emphasize that any or all of these acoustical topics can be applied to any type of audio production facility; and aren't only limited to professional studio designs. For example, owners of modest project studios should know the importance of designing a control room that's symmetrical. It doesn't cost anything to know that if one speaker is in a corner and the other is on a wall, the perceived center image will be off balance.

As with many techno-artistic endeavors, studio acoustics and design are a mixture of fundamental physics (in this case, mostly dimensional mathematics) and an equally large dose of common sense and luck.

The Project Studio

The design and construction considerations for creating a privately owned recording or MIDI project studio (Figures 3.1 and 3.2) often differ from the design considerations for a professional music facility in two fundamental ways:

- Building constraints
- Cost

Generally, a project studio's room or series of rooms are built into an artist's home or a rented space where the construction and dimensional details are already defined. This fact (combined with inherent cost considerations) often leads the owner/artist to employ cost-effective techniques for sonically treating a room. Even if the room has little or no treatment (or if it isn't deemed necessary), keep in mind that a basic knowledge of acoustical physics and room design can be a handy and cost-effective tool as your experience, production needs and (you hope) your business grow.

Figure 3.1. *Gravelvoice Studio. (Courtesy of Gravelvoice Studios, www.gravelvoice.com; photo: Copyright, Tomiko Jones 2000, www.tomikojonesphoto.com)*

Figure 3.2. *Shaquille O'Neal's project studio. (Courtesy of Walters-Storyk Design Group— designed by Beth Walters and John Storyk, www.wsdg.com. Photo credit—Robert Wolsch)*

The Music Studio

The professional recording studio (Figures 3.3 through 3.5) is first and foremost a commercial business, so its design, decor and acoustical construction requirements are often much more demanding than those for a privately owned project studio. In some cases, a professional acoustical designer and construction team are placed in charge of the overall building phase of a professional facility. In many cases, however, the studio's budget precludes the hiring of such professionals, which places the studio owners and staff squarely in charged of designing and constructing the facility.

Figure 3.3. *Studio X, Seattle.*
(Courtesy of Studio X,
www.studioxinc.com and studio
bau:ton, www.bauton.com)
a. *Control room.*
b. *Recording studio.*

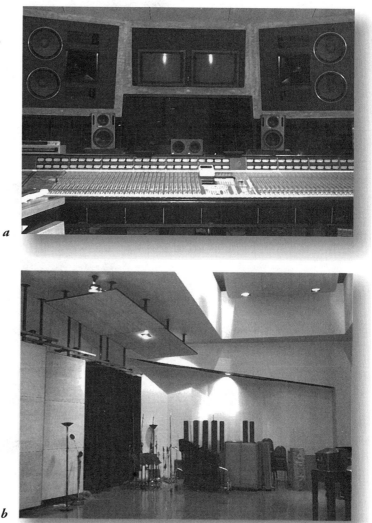

a

b

Figure 3.4. *Whitney Houston's home studio. (Courtesy of Russ Berger Design Group, Inc., www.rbdg.com)*
a. *Control room.*
b. *Recording studio.*

If you happen to have the luxury of building a new facility from the ground up or within an existing shell, you'd probably benefit from a professional studio designer's experience and skills—even if you're building or renovating the facility yourself. Such expert advice often proves to be cost-effective in the long run, as errors in design judgment can lead to cost over-runs, lost business due to unexpected delays or the unfortunate state of living with mistakes that could've been avoided.

Figure 3.5. *Turtle Recording Studio. (Courtesy of Turtle Recording, www.turtlerecording.com)*
a. *Control room.*
b. *Stage.*

The Audio-for-Visual Production Environment

An audio-for-visual production facility (Figure 3.6) that's used for video, film and game post-production (often simply called "post") includes such facets as *Foley* (the replacement and creation of on- and off-screen sound effects), *automatic dialog replacement (ADR)*, *music recording for film or other media (scoring)* and mixdown. As with music studios, audio-for-visual production facilities can range from being a simple budget-minded room that's equipped with project-studio-class equipment to a high-end, specially designed facility that can accommodate the posting needs of network video or feature film productions. As with the music studio, audio-for-visual construction and design techniques often span a wide range of styles and scope in order to fit the budget requirements at hand.

Figure 3.6. One Union Recording Studios. (Courtesy of Euphonix Inc., www. euphonix.com, Photo by Edward Colver)

Primary Factors Governing Studio and Control Room Acoustics

Regardless of which type of studio facility is being designed, built and used . . . there are a number of primary concerns that should be addressed in order to achieve the best possible acoustic results. In this section, we'll take a close look at the most important and relevant aspects of acoustics as they pertain to the sound studio design (acoustic spaces used for recording music or speech), as well as that of control room design (acoustic spaces used for the production and mixdown of music or speech). Important factors in this discussion include:

- Acoustic isolation
- Frequency balance
- Absorption
- Reflection
- Reverberation

As was said near the beginning of Chapter 2 . . . Although several mathematical formulas have been included in these sections, it's by no means important that you memorize or worry about them. By far, I feel that it's more important that you grasp the basic principles of acoustics, rather than fret over the underlying math. Remember—more often than not, acoustics is an artistic science that blends math with the art of intuition and experience.

Acoustic Isolation

Under certain conditions, it might be necessary for modern day recording studios and audio-for-visual sound stages to make use of effective isolation techniques to reduce external noise—whether that noise is transmitted through the medium of air (as might occur with nearby auto, train, or jet traffic) or through solids (such as noise transmission from air conditioner rumble or underground subways through building foundations). In order to reduce such extraneous noises, special construction techniques will often be required (Figure 3.7).

Figure 3.7. Various isolation and acoustical treatments for a recording/monitoring environment. (Courtesy of Auralex Acoustics, www.auralex.com)

If you have the luxury of building a studio facility from the ground up, a great deal of thought should be put into determining the studio's location. If a location has considerable neighborhood noise, you might have to resort to extensive (and expensive) construction techniques in order to isolate the sound. If there's absolutely no choice in a studio's location and the studio happens to be located next to a factory, just under the airport's main landing path, or over the subway's uptown line . . . you simply have to give in to destiny and build acoustical barriers to these outside interferences.

The reduction in the sound pressure level (SPL) of a sound source as it passes through an acoustic barrier of a certain physical mass (Figure 3.8) is termed the *transmission loss (TL)* of a signal. This attenuation can be expressed (in dB) as:

$$TL = 14.5 \log M + 23$$

where

- TL is the transmission loss in decibels
- M is the surface density (or combined surface densities) of a barrier in pounds per square foot (lb/ft^2).

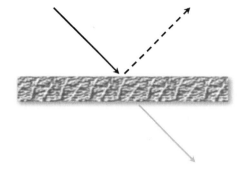

Figure 3.8. *Transmission loss refers to the reduction of a sound signal (in dB) as it passes through an acoustic barrier.*

Because transmission loss is frequency dependent, this second equation can be used to calculate transmission loss at various frequencies, with a relative degree of accuracy:

$$TL = 14.5 \log Mf - 16$$

where f is the frequency in hertz.

Both common sense and the preceding two equations tell us that heavier acoustic barriers will yield a higher transmission loss. For example, Table 3.1 tells us that a 12-inch-thick wall of dense concrete (yielding a surface density of 150 lb/ft^2) offers a much greater resistance to the transmission of sound than a 4-inch cavity filled with sand (which yields a surface density of 32.3 lb/ft^2). From the second equation (TL = 14.5 log Mf – 16), we can also draw the conclusion that for a given acoustic barrier, higher transmission losses will be encountered as the frequency rises. This can be easily proven by closing the door of a car that has its sound system turned up or by shutting a single door to a music studio's control room. In both instances, the high frequencies will be reduced in level, while the bass frequencies will be impeded to a much lesser extent.

Table 3.1. The surface densities (lbs/ft^2) of common building materials.

Material	Thickness (inches)	Surface Density (lbs/ft^2)
Brick	4	40.0
	8	80.0
Concrete (light weight)	4	33.0
	12	100.0
Concrete (dense)	4	50.0
	12	150.0
Glass	¼	3.8
	½	7.5
	¾	11.3
Gypsum wallboard	½	2.1
	⅝	2.6

(continues)

Table 3.1. (continued)

Material	Thickness (inches)	Surface Density (lbs/ft^2)
Lead	$\frac{1}{16}$	3.6
Particle board	$\frac{3}{4}$	1.7
Plywood	$\frac{3}{4}$	2.3
Sand	1	8.1
	4	32.3
Steel	$\frac{1}{4}$	10.0
Wood	1	2.4

From this, the goal would seem to be to build a studio wall, floor, ceiling, window or door out of the thickest and densest material that we can get. However, expense and physical space often play roles in determining just how much of a barrier is needed to achieve the desired isolation. In most cases, a balance can be struck to get the job done right, while using both space- and cost-effective building materials.

Walls

When building a studio wall or reinforcing an existing structure, your primary goal should be to reduce leakage (increase the transmission loss) through a wall at most or all frequencies. Generally, this is done by building a wall structure that's as massive as possible (in terms of both cubic and square foot density), as well as one that's highly damped (well supported by its reinforcement structures and relatively free of resonances), as shown in Figure 3.9.

Figure 3.9. *Typical gypsum wallboard constructions.* **a.** *Single stud design.*

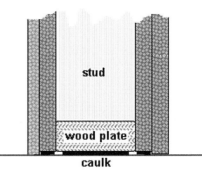

a

b. *Double, staggered stud construction having a higher transmission loss value.*

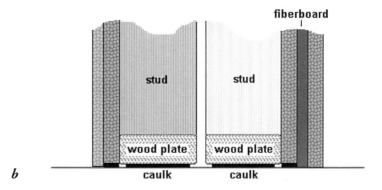

b

The following guidelines can be beneficial in the construction of framed walls that have high transmission losses:

- If at all possible, the inner and outer wallboards shouldn't be directly attached to the same stud. The best way to avoid this is to attach each wall facing to its respective staggered studs (Figure 3.9b).

- Each wall facing should have a different density to reduce the likelihood of increased transmission due to resonant frequencies that might be sympathetic to both sides. For example, one wall might be constructed of two, 5/8" thick gypsum wallboards, while the other wall might be composed of soft fiberboard that's surfaced with two, 1/2" thick gypsum wallboards.

- If you're going to attach gypsum wallboards to a single wall face, you can increase transmission loss by mounting the additional layers (not the first layer) with adhesive caulking rather than screws or nails.

- Spacing the studs 24" on center instead of using the traditional 16" spacing yields a slight increase in TL.

- To reduce leakage that might make it through the cracks, apply a bead of nonhardening caulk sealant to the inner gypsum wallboard layer at the wall-to-floor, wall-to-ceiling and corner junctions.

Generally, the same amount of isolation is required between the studio and the control room as is required between the studio's interior and exterior environments. The proper building of this wall and possible speaker soffit (a structure that's used to mount the main studio monitors directly into the control room or studio wall) is important, so that an accurate tonal balance can be heard over the control-room monitors without allowing leakage from the studio or resonances within the wall to color the audible signal.

When constructing a control room/studio wall/soffit combination (Figure 3.10), it's important that the wall be constructed using as large a mass as is physically possible in order to keep

Figure 3.10. *A control room/ studio wall/soffit design.*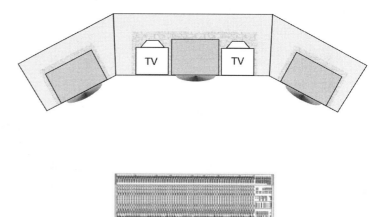

low-frequency reflections (and therefore cancellations) within the two rooms to a minimum. Typical construction materials include:

- Concrete—This is the best and most solid material. Besides the fact that it's often expensive, it's not always possible to pour cement into an existing design.
- Bricks (hollow form or solid facing bricks)—An excellent material that's often easier to place into an existing room.
- Gypsum "plaster" board—Multiple layers (2–3) are often needed to keep resonances and transmission losses to a minimum. It's often a good idea to reduce resonances by filling the wall cavities with rockwool or fiberglass, while bracing the internal structure to add an extra degree of stiffness. Once built, the surfaces can be decorated with any number of materials (such as wood or paint).
- Wood—This is a difficult material to work with (on an exclusive basis) as large quantities are required to attain the degree of stiffness that's needed.

Once the wall and soffit system has been constructed, the monitor speakers should be placed on rubber pads in order to keep wall resonances to a minimum.

Floors

For many recording facilities, the isolation of floor-borne noises from other rooms and building exteriors is an important consideration. For example, a building that's on a busy street and whose concrete floor rests directly on ground foundation might experience severe low-frequency rumble. Alternatively, a second-floor facility might have undue leakage from a noisy downstairs neighbor or, even more likely, might interfere with a quieter neighbor's business. In each of these situations, some form of isolation from floor-borne sound is essential. One of the most common ways to isolate floor-related noise is to construct a "*floating*" floor that's structurally decoupled from its subfloor foundation.

Two of the most common construction methods for floating a floor use either neoprene "hockey puck" isolation mounts (Figure 3.11a) or a continuous underlayment, such as rubberized floor mat coverings (Figure 3.11b). In the first approach, a flexible underlayment is laid over the existing floor foundation. The underlayment is then covered with 1/2" plywood, followed by plastic sheeting (for insulation and to prevent seepage), a layer of reinforcing wire mesh and finally a 4" top layer of concrete. This floating substructure is then ready for carpeting, wood finishing, painting or any other desired surface.

Figure 3.11. *Basic guidelines for building a floating floor.* *a.* *Using neoprene mounts.* *b.* *Roll-out matting can be substituted as an underlayment. (Courtesy of Acoustical Solutions, Inc., www.acousticalsolutions.com)*

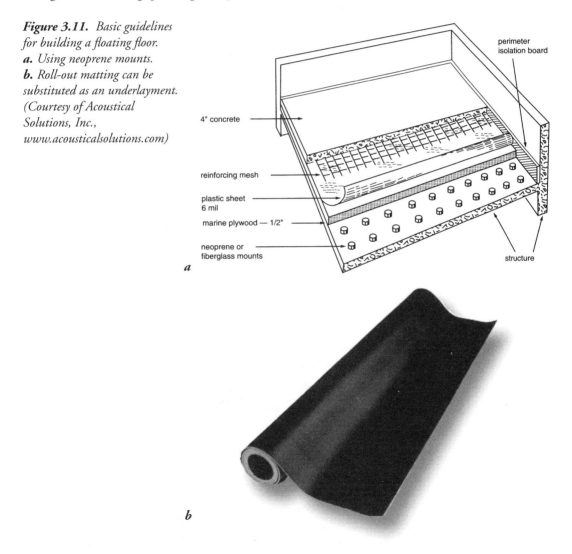

perimeter isolation board

4" concrete

reinforcing mesh

plastic sheet 6 mil

marine plywood — 1/2"

neoprene or fiberglass mounts

structure

a

b

It's important that the floating superstructure be isolated from both the underflooring and the outer wall. Failing to isolate these can transmit sound both through the walls to the floor and from the subfloor to the upper layer (thereby defeating the whole purpose of floating the floor). This wall perimeter isolation can be made from such pliable decoupling materials as widths of soft mineral fiberboard or neoprene.

A less expensive method for decoupling a floor is to layer the original floor with a carpet foam pad. A 1/2" or 5/8" layer of tongue-and-groove plywood or OSB boards can be laid on top of the pad (these shouldn't be nailed to the sub-floor; instead, they can be stabilized by locking the pieces together with thin, metal braces). Another foam pad can be laid over this structure and topped with carpeting or any other desired finishing (Figure 3.12).

Figure 3.12. An alternative, cost-effective way to float an existing floor.

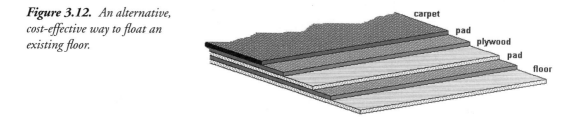

Risers

As we saw from the earlier equation (TL = 14.5 log Mf – 16), low-frequency sound travels through barriers much more easily than does high-frequency sound. It stands to reason that strong, low-frequency energy is transmitted more easily than highs between studios, from studio to the control room and/or to outside locations. In general, the drum set will most likely be the biggest leakage offender. By decoupling much of a drum set's low-frequency energy from a studio floor, we can greatly reduce low-frequency leakage problems. In most cases, the problem can be fixed by constructing a drum riser (Figure 3.13).

Figure 3.13. Construction detail for a typical drum riser.

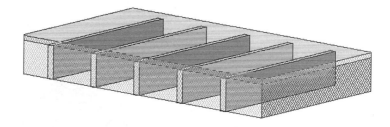

In order to reduce unwanted resonances, drum risers should be constructed using 2" × 6" or 2" × 8" beams for both the frame and the supporting joists (spaced at 16" or 12" on-center intervals). Sturdy 1/2" or 5/8" tongue-and-groove plywood panels should be glued to the supporting frames with carpenter's glue (or a similar wood glue) and then nailed down (using heavy-duty, galvanized nails). Once you've attached neoprene strips, neoprene coasters or (at the least) strips of carpeting to the bottom of the frame . . . the riser will be ready for action.

Ceilings

Foot traffic or other noises from above a sound studio or production room is another common source of external leakage. Ceiling noise can be isolated in a number of ways. If foot traffic is your problem and you own the floors above you, footstep noise can be reduced by simply carpeting the overhead hallway or floating the upper floor. If you don't have this luxury, one approach to deadening ceiling-borne sounds is to float a false structure from the existing ceiling or from the overhead ceiling joists (as is often done when a new room is being constructed). This technique can be fairly cost-effective when "Z" suspension channels are used (Figure 3.14). "Z" channels can be screwed to the ceiling joists to provide a flexible, yet strong support to which a floated wallboard ceiling can be attached. If necessary, you can even place fiberglass or other sound-deadening materials in the cavities between the overhead structures. Other more expensive methods are also available, which use spring support systems to hang false ceilings from an existing structure.

Figure 3.14. *"Z" channels can be used to hang a floating ceiling from an existing overhead structure.*

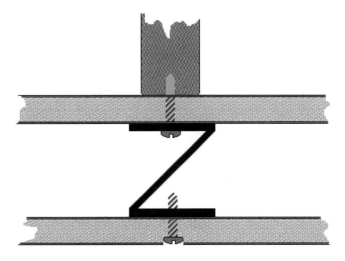

Windows and Doors

Access to and from a studio or production room area (in the form of windows and doors) can also be potential sources of sound leakage. For this reason, strict attention needs to be given to their design and construction.

Visibility within a studio is extremely important in a music production environment. It serves to promote good communication between the producer or engineer and the studio musician, when multiple rooms are involved (as well as between the musicians themselves). For this reason, windows have been an important factor in studio design for decades. The design and construction details for a window often varies with studio needs and budget requirements. These can range from being deep double-plate cavities that are built into double-wall constructions

(Figure 3.15a) to more modest constructions that are built into a single wall (Figure 3.15b). Other designs range from floor-to-ceiling windows that create a virtual "glass wall" to designs that have been built into studio/control room walls that are constructed from 3 feet of solid, poured concrete.

The glass panels used in window construction typically are 3/8" to 3/4" thick and are seated into the window frame with a rubber or similar type of elastic damping seal to prevent structure-borne oscillations. It's important that at least one of these panels be tilted at a 5° angle (minimum) with respect to the other, in order to eliminate standing waves within the sandwiched airspace. The existence of standing waves (which are discussed later in the "Frequency Balance" section) often serves to break down the transmission loss at specific frequencies. Similar to wall construction techniques, using glass panels of varying thickness will reduce the possibility of sympathetic vibration. Other windows within the facility (such as observation and tape machine room windows) should be designed with similar isolation considerations in mind.

Access doors to and from the studio, control room and exterior areas should be constructed using a double-door design to form a *sound lock* (Figure 3.16). This construction technique is used because of the high TL values that air has when it's sandwiched between two isolated barriers. Solid doors generally offer higher TL values than their cheaper, hollow counterparts. No matter which door type is used, the appropriate seals, weather stripping and door jambs should be used throughout so as to reduce leakage through the cracks.

Figure 3.15. *Details for practical window construction between the control room and studio.*
a. A window that's suitable for a high transmission loss wall.

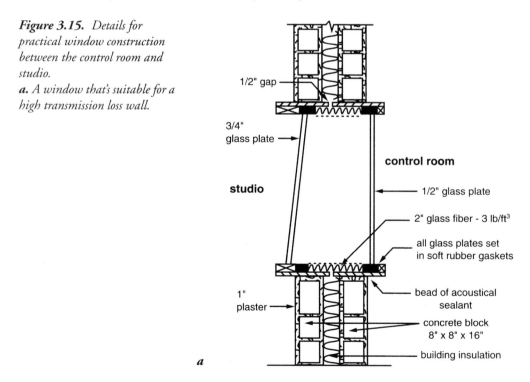

1/2" gap

3/4" glass plate

control room

studio

1/2" glass plate

2" glass fiber - 3 lb/ft³

all glass plates set in soft rubber gaskets

1" plaster

bead of acoustical sealant

concrete block 8" x 8" x 16"

building insulation

a

b. A more modestly framed window.

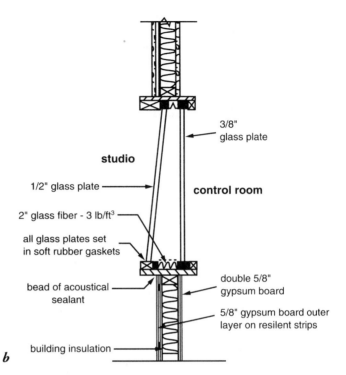

3/8"
glass plate

studio

1/2" glass plate

control room

2" glass fiber - 3 lb/ft³

all glass plates set
in soft rubber gaskets

bead of acoustical
sealant

double 5/8"
gypsum board

5/8" gypsum board outer
layer on resilent strips

building insulation

b

Figure 3.16. Sound lock design examples. (Courtesy of Russ Berger Design Group, Inc., www.rbdg.com)

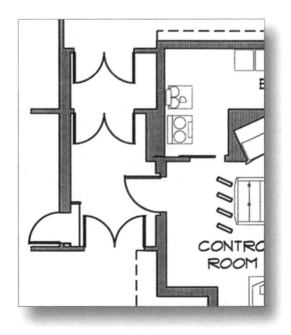

Iso-Rooms and Iso-Booths

Isolation rooms (iso-rooms) and the smaller iso-booths are acoustically sealed areas that are built into and easily accessible from the studio. These areas improve the separation that's often needed between loud and soft instruments, vocals and so on. The construction of these areas depends on the studio's size, design and sonic requirements (Figure 3.17). They can vary in acoustics from the main studio area, thereby offering an environment that's more "live" or more "dead" and can be used to better fit the instrument's acoustical needs. Iso-booths and iso-rooms can be designed as totally separate areas that can be accessed from the main studio or control room, or they might be tied directly to the main studio by way of sliding walls or glass sliding doors.

Figure 3.17. *"Iso-room" design located to the control room's side, Four Seasons Media Productions, St. Louis. (Courtesy of Russ Berger Design Group, Inc., www.rbdg.com)*

Movable partitions (known as *flats* or gobos) are commonly used in studios to provide on-the-spot sound barriers (Figure 3.18). These partitions generally provide the least amount of isolation, but are the most flexible and permit the isolation changes that many situations require. Often, an adequate degree of isolation can be attained by partitioning off a musician and/or instrument on one or more sides and by placing the mic inside. It's important, however, to be aware of the musician's need to see other musicians, the conductor and/or the producer. Musicality almost always takes precedence over technical protocol.

Frequency Balance

Another important factor in room design is the need for maintaining the original frequency balance of an acoustic signal. In other words, the room should exhibit a relatively flat frequency response over the entire audio range without adding its own particular sound coloration. The most common way to control the acoustic tonal character is to use materials and design techniques that govern the reflection and absorption factors within a room.

Figure 3.18. *Acoustic partition flats.*
a. *Guitar amp setup. (Courtesy of ClearSonic Mfg. Inc., www.clearsonic.com)*
b. *Drum panel setup. (Courtesy of Auralex Acoustics, www.auralex.com)*
c. *Piano panel setup. (Courtesy of Auralex Acoustics, www.auralex.com)*

a

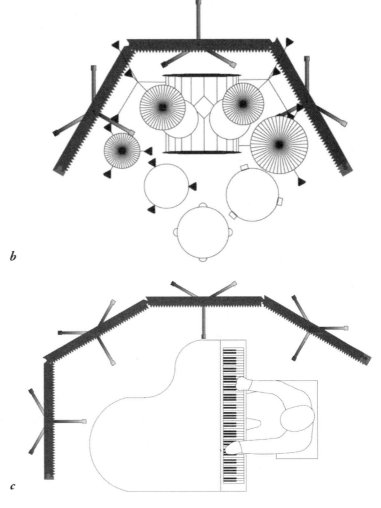

b

c

Figure 3.18. (continued)
d. *Setup within the control*
room. (Courtesy of Auralex
Acoustics, www.auralex.com)

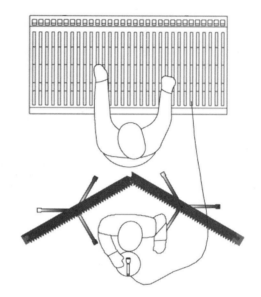

d

Reflections

One of the most important phenomena of sound as it travels through air is its ability to reflect off a boundary's surface at an angle that's equal to (and opposite of) its original angle of incidence (Figure 3.19). Just as light bounces off a mirrored surface or multiple reflections appear within a mirrored room, sound reflects throughout the various room surfaces in ways that are often equally complex. Nonetheless, sound can be controlled in ways so as to add to (or detract from) the room's sonic character.

Figure 3.19. *Sound reflects off*
a surface at an angle equal (and
opposite) to its original angle of
incidence.

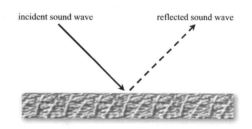

incident sound wave   reflected sound wave

In Chapter 2, "Sound and Hearing," we learned that sonic reflections could be controlled in ways that disperse the sound outward in a wide-angled pattern (through the use of a convex surface) or can focus sound to a specific point (through the use of a concave surface). Other surfaces shapes can reflect sound back at various other angles. For example, a 90° corner will reflect sound back in the same direction (a fact that accounts for the additive acoustic buildups at various frequencies at or near a wall-to-corner or corner-to-floor intersection). The all-time winner of the "avoid this at all possible costs award" goes to construction that includes opposing parallel walls into its design. Such conditions give rise to a phenomenon known as *standing waves*. Standing waves (also known as room modes) occur when sound is reflected off of parallel surfaces and travels back on its own path, thereby interfering with the amplitude response characteristics of the room. Walking around such a room produces the sensation of increases or decreases in the sound's perceived level at various frequencies. This perceived increase or decrease is due to cancellations and reinforcements of the combined reflected waveforms at the listener's position. The distance between parallel surfaces and the signal's wavelength determines the frequencies that produce standing waves. This effect can potentially cause sharp peaks or dips in the frequency curve (up to or beyond 19 dB) at the affected fundamental frequency (or frequencies) and harmonic intervals (Figure 3.20). This condition exists not only for opposing parallel walls, but also for all parallel surfaces (such as between floor and ceiling or between two reflective flats).

Figure 3.20. *The existence of reflective, parallel walls create an undue number of standing waves, which occur at various frequency intervals (f_1, f_2, f_3, f_4 and so on).*

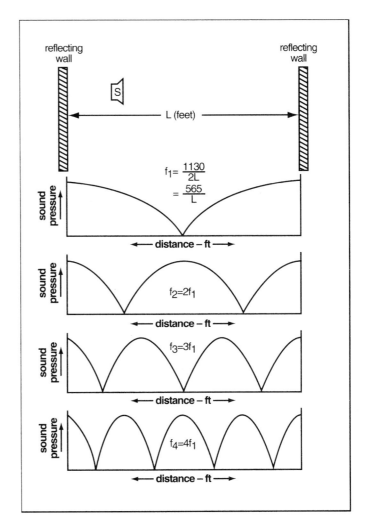

Nonparallel walls, however, generally won't eliminate low-frequency standing waves. Instead, a better solution is to use bass traps that are tuned to room resonance frequencies or proper room dimensions. Nonparallel walls will, however, prevent the buildup of flutter echoes.

The most effective way to ensure against standing waves is to construct walls, boundaries and ceilings that are nonparallel. Figure 3.21, for example, shows various standing wave modes as they occur within two rooms that have equal square footage areas. The rectangular room shows the unwanted standing waves as being uniform throughout the entire length or width of the room (and probably the ceiling for that matter), while the modes within the nonrectangular room have been broken up and appear at irregular spaces and intervals within the room.

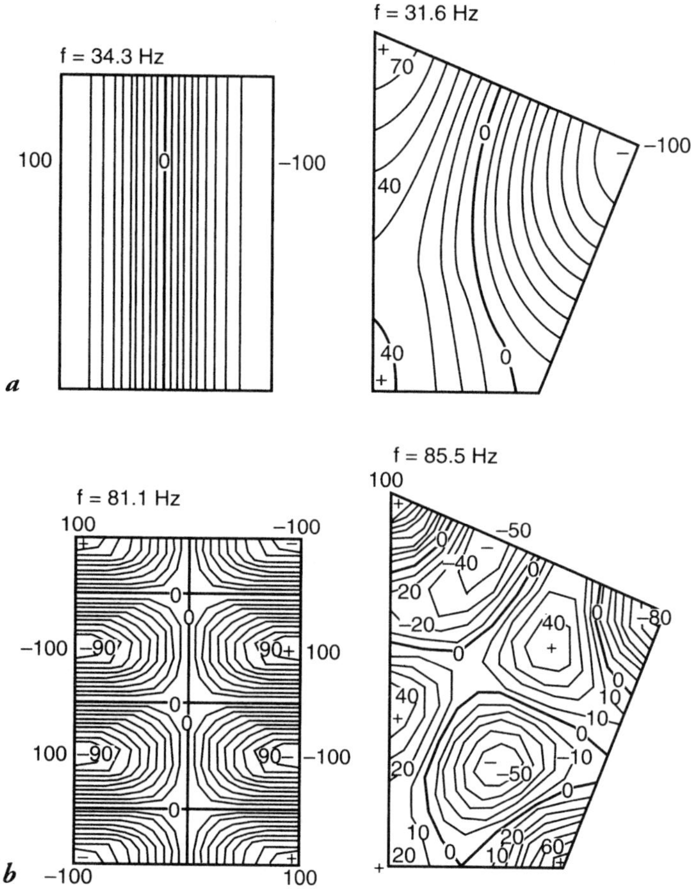

Figure 3.21. A comparison of two-dimensional soundfields in both rectangular and nonrectangular rooms that have the same surface area.
a. The 1,0,0 mode of the rectangular room (34.3 Hz) compared to the nonrectangular room (31.6Hz).
b. The 3,1,0 mode of the rectangular room (81.1 Hz) compared to the nonrectangular room (85.5 Hz).

c. The 4,0,0 mode in the rectangular room (98 Hz) compared to the nonrectangular room (95.3 Hz).

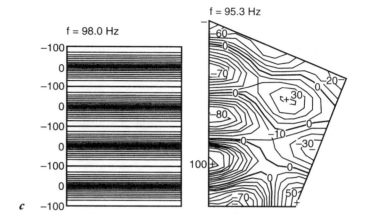

If the room is rectangular or if further sound wave dispersion within the space is desired, diffusers can be attached to the wall and/or ceiling boundaries. *Diffusers* (Figure 3.22) are acoustical boundaries that reflect the sound wave back at various angles, which are wider than the original incident angle (thereby breaking up the energy-destructive standing waves). In addition, the use of both nonparallel and diffusion wall construction can reduce a condition known as *flutter echo* and can smooth out the reverberation characteristics of a room by building further and more complex acoustical pathways.

Figure 3.22. *Diffuser examples.*
a. Various diffusion and absorption wall treatments. (Courtesy of Auralex Acoustics, www.auralex.com)

Figure 3.22. (continued)
b. *RPG Skyline Diffusor.*
(Courtesy of Acoustical
Solutions, Inc.,
www.acousticalsolutions.com)

b

Flutter echo (also called *slap echo*) is a condition that occurs when parallel boundaries are spaced far enough apart for the listener to discern a number of discrete echoes. Flutter echo often gives a smaller room a tube-like, hollow sound that affects the sound character as well as its frequency response. A larger room (which might contain delayed echo paths of 50 msec or more) has its echoes spaced far enough apart in time that the discrete reflections actually interfere with the intelligibility of the direct sound, often resulting in a jumble of noise.

Symmetry in Studio and Control Room Design

Although some prestigious designers of control rooms and studios have worked hard to create a successful, standardized acoustic environment, most control room and studio designs have as many forms and express as many design philosophies as there are studio owners. Certain ground rules of acoustical physics, however, must be followed in order to create a proper working environment. Of these ground rules, there's one guideline that's even more important than control over amplitude response . . . this is the need for *symmetrical reflections* in all axes within the design of a control room (and, one would hope, the studio's design as well). Should any primary boundaries of a control room (especially those wall or ceiling boundaries near the mixing position) be asymmetrical, sounds heard by one ear of the mixing engineer will receive one combination of direct and reflected sounds, while the other ear will hear a different balance (Figure 3.23). This condition alters the sound's center image characteristics, so when a sound is actually panned between the two monitor speakers, the sound will appear to be off-center. In order to avoid this problem, care should be take to ensure that both the side and ceiling boundaries are symmetrical with respect to each other. The splayed side wall (Figure 3.24) is one example of a symmetrical construction pattern that would reduce acoustic reflections, which could interfere with the direct acoustic energy at the listener's position.

Figure 3.23. *Asymmetrical side reflections cause an acoustic imbalance at the listener's position. (Courtesy of Acoustical Physics Laboratories)*

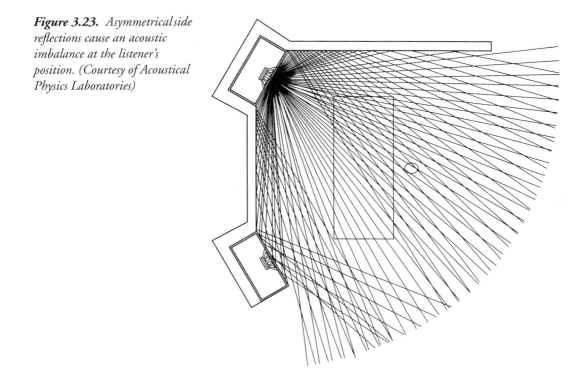

Large ceiling reflections also can interfere with the direct acoustic path. Figure 3.25 shows reflections radiating back to the listener from an average level ceiling, while the splayed ceiling design in Figure 3.26 shows the reduction of these unwanted reflections.

Absorption

Another method that can be used to change the soundfield within an acoustic space involves the use of surface materials and designs that can absorb unwanted sounds (either across the entire audible band or at specific frequencies).

The absorption of acoustic energy is, effectively, the inverse of reflection (Figure 3.27). Whenever sound strikes a material, the amount of acoustic energy that's absorbed relative to the amount that's reflected (often in the form of physical heat dissipation) can be expressed as a simple ratio known as the material's *absorption coefficient*. For a given material, this can be represented as:

$$A = I_a / I_r$$

Figure 3.24. *Symmetrical splayed side walls give a proper acoustic balance. (Courtesy of Acoustical Physics Laboratories)*

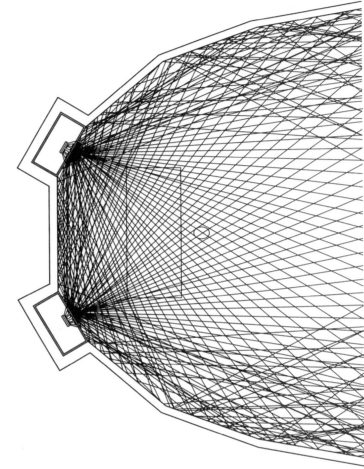

Figure 3.25. *Ceiling reflections cause acoustic interference at the listener's position. (Courtesy of Acoustical Physics Laboratories)*

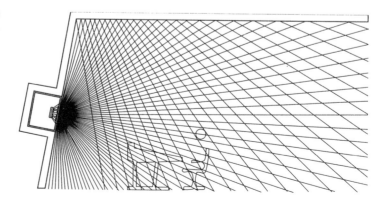

Figure 3.26. *Splayed ceiling reduces unwanted reflections. (Courtesy of Acoustical Physics Laboratories)*

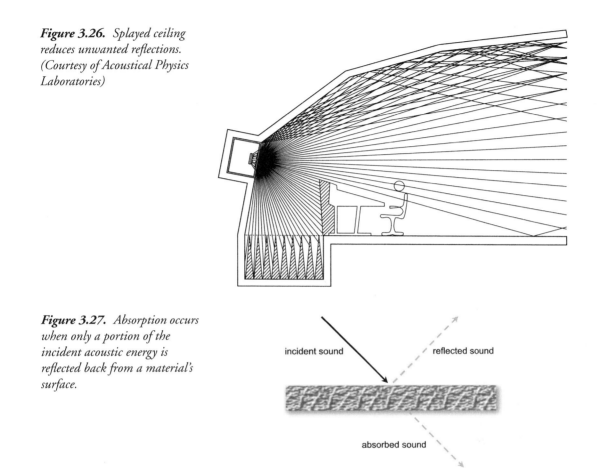

Figure 3.27. *Absorption occurs when only a portion of the incident acoustic energy is reflected back from a material's surface.*

where:

- I_a is the sound level (in dB) that is absorbed by the surface.
- I_r is the sound level (in dB) that is reflected back from the surface.

The factor $(1 - a)$ is the reflected sound. This makes the coefficient a decimal percentage value between 0 and 1. A sample listing of these coefficients is provided in Table 3.2.

Table 3.2. Absorption coefficients for various materials.

Material	Coefficients					
	125 Hz	*250 Hz*	*500 Hz*	*1000 Hz*	*2000 Hz*	*4000Hz*
Brick, unglazed	0.03	0.03	0.03	0.04	0.05	0.07
Carpet (heavy, on concrete)	0.02	0.06	0.14	0.37	0.60	0.65

(continues)

Table 3.2. (continued)

Material	125 Hz	250 Hz	500 Hz	1000 Hz	2000 Hz	4000Hz
			Coefficients			
Carpet (with latex backing, on 40 oz hair-felt of foam rubber)	0.08	0.27	0.39	0.34	0.48	0.63
Concrete block, coarse	0.36	0.44	0.31	0.29	0.39	0.25
Light velour (10 oz./sq. yd in contact with wall)	0.03	0.04	0.11	0.17	0.24	0.35
Concrete or terrazzo	0.01	0.01	0.015	0.02	0.02	0.02
Wood	0.15	0.11	0.10	0.07	0.06	0.07
Glass, large heavy plate	0.18	0.06	0.04	0.03	0.02	0.02
Glass, ordinary window	0.35	0.25	0.18	0.12	0.07	0.04
Gypsum board, nailed to 2 × 4 studs on 16-inch centers	0.29	0.10	0.05	0.04	0.07	0.09
Plaster, gypsum, or lime (smooth finish on tile or brick)	0.013	0.015	0.02	0.03	0.04	0.05
Plywood, 3/8 inch	0.28	0.22	0.17	0.09	0.10	0.11
Air, Sabins/1000 cu ft.					2.3	7.2
Audience seated in upholstered seats	0.44	0.54	0.60	0.62	0.58	0.50
Wooden pews	0.57	0.61	0.75	.86	0.91	0.86
Chairs, metal or wooden, seats unoccupied	0.15	0.19	0.22	0.39	0.38	0.30

If we say that a surface material has an absorption coefficient of 0.25, we're actually saying that the material absorbs 25 percent of the original acoustic energy, while reflecting 75 percent of the total sound energy at that frequency.

In order to determine the amount of absorption that's obtained by the total number of absorbers within a total volume area, it's necessary to calculate the average absorption coefficient for all the surfaces together. The average absorption coefficient (A_{ave}) of a room or total area can be expressed as:

$$A_{ave} = \frac{s_1 a_1 + s_2 a_2 + \ldots s_n a_n}{S}$$

where:

- $s_1, s_2 \ldots _n$ are the individual surface areas
- $a_1, a_2 \ldots _n$ are the individual absorption coefficients of the individual surface areas
- S is the total square surface area.

(Note: These coefficients were obtained by measurements in the laboratories of the Acoustical Materials Association. Coefficients for other materials may be obtained from Bulletin XXII of the Association.)

High-Frequency Absorption

The absorption of high frequencies is accomplished through the use of dense porous materials, such as cloth, fiberglass and carpeting. These materials are capable of having high absorption values at higher frequencies, thus allowing reflections in the room to be controlled in a frequency-dependent manner. Specially designed foam and acoustical treatments are also available that can be attached easily to recording studio, production studio or control room walls as a means for correcting multiple room reflections or to dampen high-frequency reflections (Figure 3.28).

Figure 3.28. AlphaSorb wall panels. (Courtesy of Acoustical Solutions, Inc., www.acousticalsolutions.com)

Low-Frequency Absorption

As was shown in Table 3.2, materials having a degree of absorption in the high frequencies often provide little resistance to the low-frequency end of the spectrum (and vice versa). This is due to the fact that low frequencies can be damped by pliable materials, meaning that the low-frequency energy is absorbed by the material's ability to bend and flex with the incident waveform (Figure 3.29).

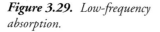

Figure 3.29. *Low-frequency absorption.*

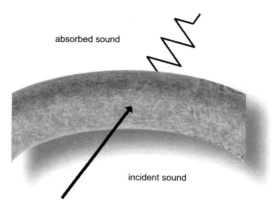

absorbed sound

incident sound

Another design type can be used to reduce low-frequency buildup at specific frequencies (and their multiples) within a room. These low-frequency attenuation devices (known as *bass traps*) are available in a number of design types. This section discusses three bass trap types:

- The quarter-wavelength trap
- The pressure zone trap
- The functional trap

The Quarter-Wavelength Trap

The quarter-wavelength bass trap (Figure 3.30) is an enclosure with a depth that's one-fourth the wavelength of the offending frequency's fundamental frequency and is often built into a wall (generally the rear wall), ceiling or floor structure (covered by a metal grating to allow foot traffic). The physics behind the absorption of a calculated frequency (and many of the harmonics that fall above it) lie in the fact that at the rear boundary of the trap, the pressure component of a sound wave will be at its maximum, while the velocity component of the wave (the portion that measures the kinetic energy of atmospheric molecular motion) will be at a minimum. At the mouth of the bass trap (which is at a one-fourth wavelength distance from this rear boundary), the overall acoustic pressure is at its lowest, while its velocity component is at its highest potential. Because the wave's motion (force) is greatest at the opening, much of the signal can be absorbed by placing an absorptive material at the trap's opening. Low-density fiberglass lining can also be placed inside the trap to increase the amount of absorption (especially at harmonic intervals of the calculated fundamental).

Pressure Zone Trap

The pressure zone bass trap absorber (Figure 3.31) operates on the principle that sound pressure is doubled at physical boundary points (such as walls and ceilings). By rigidly supporting a number of medium-density fiberglass boards (commonly referred to as mdf 703) directly onto a surface boundary, the built-up pressure can be partially absorbed. Wood slats or other nonresonant reflective structures can also be built out from the fiberglass boards, so as to reflect mid

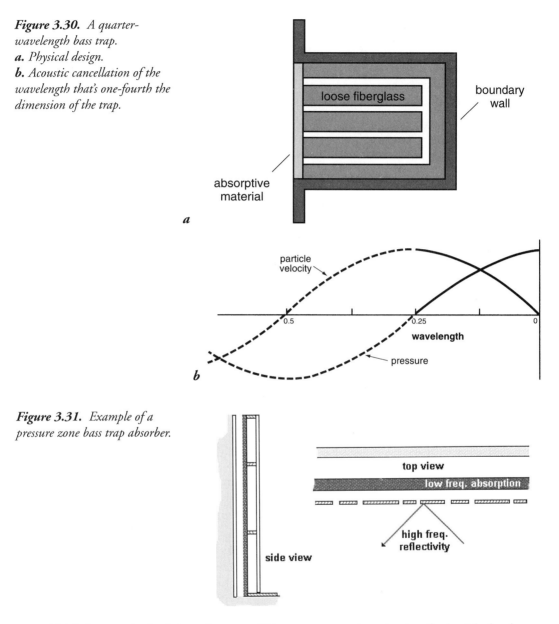

Figure 3.30. *A quarter-wavelength bass trap.*
a. *Physical design.*
b. *Acoustic cancellation of the wavelength that's one-fourth the dimension of the trap.*

loose fiberglass

boundary wall

absorptive material

a

particle velocity

0.5   0.25   0

wavelength

pressure

b

Figure 3.31. *Example of a pressure zone bass trap absorber.*

top view

low freq. absorption

high freq. reflectivity

side view

and high frequencies back into the room. Take care not to place the slats flush with the absorption material, as this would choke off the bass trap's ability to "breathe."

Functional Trap

Originally created in the 1950s by Harry F. Olson (former Director of RCA Labs), the functional bass trap (Figure 3.32) uses a material that's generally formed into a tube or half-tube structure, which is rigidly supported so as to reduce structural vibrations. By placing these devices into corners, room boundaries or in free-standing spots, a large portion of the undesired bass buildup frequencies can be absorbed. By placing a reflective membrane over the portion of the trap that faces into the room, frequencies above 400 Hz can be dispersed back into the listening environment. Figure 3.33 shows how these traps can be used in the studio to break up reflections and reduce bass buildup.

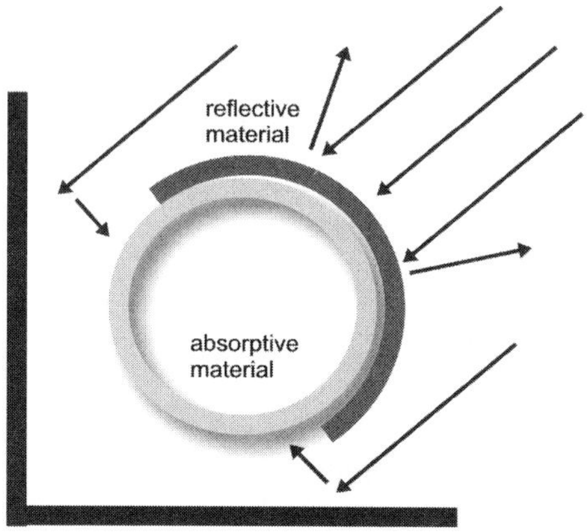

Figure 3.32. *A functional bass trap.*

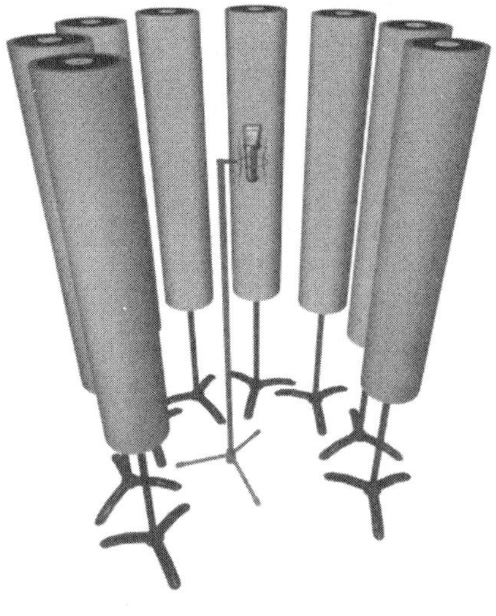

Figure 3.33. *The Quick Sound Field. (Courtesy of Acoustic Sciences Corporation, www.tubetrap.com)*

Reverberation

Another criterion for studio design is the need for a desirable room ambience and intelligibility, combined with the need for good acoustic separation between instruments and other instrument mic pickups. Each of these factors is governed by the careful control and tuning of the reverberation constants within the studio over the frequency spectrum.

Reverberation (reverb) is the persistence of a signal (in the form of reflected waves within an acoustic space) that continues after the original sound has ceased. The effect of these closely spaced and random multiple echoes gives us perceptible cues as to the size, density and nature of a space. Reverb also adds to the perceived warmth and depth of recorded sound and plays an extremely important role in the enhancement of our perception of music.

The reverberated signal itself (Figure 3.34) can be broken down into three components:

- The direct signal
- Early reflection
- Reverberation

Figure 3.34. The three components of reverberation.

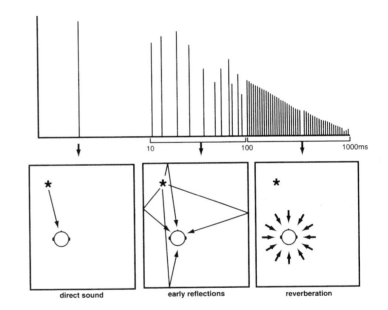

The direct signal is made up of the original sound that travels from the source to the listener. Early reflections consist of the first few reflections that are projected to the listener off of major boundaries within an acoustic space. These reflections generally give the listener subconscious cues as to the size of the room. The last set of signal reflections make up the actual reverberation characteristic. These signals are composed of random reflections that travel from boundary to boundary in a room and are so closely spaced that the brain can't discern the individual reflections. When combined, they are perceived as a single decaying signal.

Technically, reverb is considered to be the time that's required for a sound to die away to a millionth of its original intensity (resulting in a decrease over time of 60 dB), as shown by the following formula:

$$RT_{60} = V \times 0.049/AS$$

where:

RT is the reverberation time in seconds

V is the volume of the enclosure in cubic feet

A is the average absorption coefficient of the enclosure

S is the total surface area in square feet

As you can see from this equation, *reverberation time* is directly proportional to two major factors: the volume of the room and the absorption coefficients of the studio surfaces. A large environment with a relatively low absorption coefficient (such as a concert hall) will have a relatively long RT_{60} decay time, whereas a small studio (which might incorporate a heavy amount of absorption) will have a very short RT_{60}.

The style of music and the room application will often determine the optimum RT_{60} for an acoustical environment. Figure 3.35 shows a basic guide to reverb times for different applications and musical styles. Reverberation times can range from 0.25 sec in a smaller absorptive recording studio environment to 1.6 sec or more in a larger music or scoring studio. In certain designs, the RT_{60} of a room can be altered to fit the desired application by using movable panels or louvers (Figure 3.36) or by placing carpets in a room. Other designs may have sections of the studio environment that exhibit different reverb constants. One side of the studio (or separate iso-room) might be relatively nonreflective or "dead," while another section or room could be much more acoustically "live." The more reflective, live section is often used to bring certain instruments to life, such as strings, which rely heavily on room reflections and reverb. The recording of any number of instruments (including drums and percussion) can also greatly benefit from an acoustically "live" environment.

Figure 3.35. Reverberation times for various types of production studios at 512 Hz.

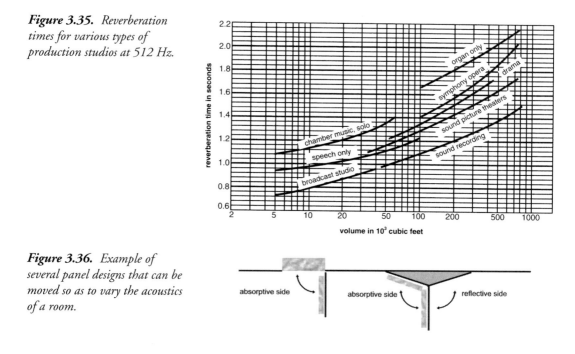

Figure 3.36. Example of several panel designs that can be moved so as to vary the acoustics of a room.

Isolation between different instruments and microphone channels is extremely important in the studio environment. If leakage isn't controlled, the room's effectiveness becomes severely limited over a range of applications. The studio designs of the 1960s and 1970s brought about the rise of the "sound sucker" era in studio design, whereby the absorption coefficient of many rooms was raised almost to an anechoic (no reverb) condition. With the advent of the music styles of the 1980s and a return to the respectability of live studio acoustics, modern studio and control room designs have begun to increase in "liveness" and size (with a corresponding increase in the studio's RT_{60}). This has reintroduced the buying public to the thick, live-sounding music production of earlier decades . . . when studios were larger, more live, acoustic structures.

Acoustic Echo Chambers

One other studio design that's been used extensively in the past (before the invention of artificial effects devices) for re-creating room reverberation . . . is the acoustic echo chamber. The echo chamber is often an isolated room that has highly reflective surfaces in which speakers and microphones are placed. The speakers are fed the signal to be reverberated, while the mics pick up a combination of sound from the speaker and reflections off the walls, ceiling and floor. By using one or more directional mics that have been pointed away from the room speakers, the pickup of direct sound can be minimized. Movable partitions also can be used to vary the room's decay time.

When properly designed, acoustic echo chambers have a very natural sound quality to them. The disadvantage is that they take up space (typically 18' × 15' × 12' or more) and require isolation from external sounds. Smaller rooms, hallways and stairwells can also be used for this purpose. Size and cost often make it unfeasible to build a new echo chamber, especially ones that can match the caliber and quality of the electronic reverb devices that are available today. However, this shouldn't discourage you from experimenting with the placement of mics (in the studio, iso-room, garage or other area in your facility or home). In short, you can try building a temporary echo chamber that can add an interesting degree of acoustic "spice" to your next project (Figure 3.37).

Figure 3.37. Example of how a room or studio space can be used as a temporary echo chamber.

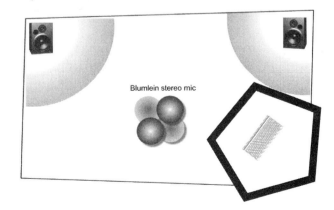

Blumlein stereo mic

Power Conditioning

Although this falls outside the scope of acoustic design, I'd like to touch upon one last topic before we conclude this chapter . . . *power conditioning*. Whether your facility is a project studio or full-sized professional facility, it's often a wise idea to regulate and/or isolate the voltage supply that's feeding one of your studio's most precious investments (besides you and your staff) . . . the equipment! The discussion of power conditioning can basically be broken down into three topics:

- Voltage regulation
- Keeping the lines quiet
- Eliminating power interruptions

In an ideal world, the power that's being fed to your studio outlets should be very close to the standard reference voltage of the country you live and/or work in (i.e., 120V, 220V, 240V, etc.). The real fact of the matter is that these "line" voltages regularly fluctuate from this standard level, causing voltage sags (a condition that can seriously underpower your equipment), surges (rises in voltage that can harm or reduce the working life of your equipment), transient spikes (sharp, high-level energy surges from lightning and other sources that can do serious damage) and brown-outs (long-term sags in the voltage lines).

Through the use of a power conditioning device (Figure 3.38), high-level, short-term spikes and surge conditions can be "clamped," thereby reducing or eliminating the possibility of equipment damage. Devices that are equipped with voltage regulation abilities are often able to deal with sags, long-term surges and brown-outs by electronically switching between the multiple voltage level taps of a transformer, so as to match the output voltage to the ideal mains level (or as close to it as is possible).

Figure 3.38. Furman IT-1220 balanced-output power conditioner. (Courtesy of Furman Sound, Inc., www.furmansound.com)

Just as it's a good idea to use balanced audio lines (whenever possible) to reduce noise and interference, it's equally wise to use a power conditioner that's balanced as well. In this case, the reason for balancing the line is to isolate the incoming power line from the ground line as much as possible. This reduces unwanted voltage potentials that might exist between the audio signal lines and chassis grounds within the facility (a basic cause of those pesky hums and buzzes). A balanced circuit might be more expensive, but as they say . . . silence is golden!

Finally, it's often a good idea to place a UPS (uninterruptible power supply) into your supply system. Just as the name implies, should the power be momentarily interrupted or give out altogether, a UPS can draw upon its battery supply to see you through or to give you enough time to safely shut your system down without losing data. Although a UPS might be unnecessary and overly expensive for a larger facility, such a device is often a good idea for computer-based project studios (particularly those in areas that are prone to having power problems). The biggest concern here is to make sure that you buy a UPS that has enough battery power to see you through a power interruption or that'll give you enough time to save your session and/or file data and safely shut the system down. In short—if you're looking to buy a UPS, make sure that it has a power rating that's high enough for your supply needs.

CHAPTER

Microphones: Design and Application

The Microphone: An Introduction

A *microphone* (which also goes by the name of *mic*) is often the first device in a recording chain. Essentially, a mic is a transducer that changes one form of energy (sound waves) into another corresponding form of energy (electrical signals). The quality of its pickup will often depend upon external variables (such as placement, distance and the acoustic environment), as well as design variables (such as microphone's operating type, overall design and quality). These interrelated elements tend to work together to affect the mic's overall sound quality.

In order to deal with the wide range of musical, acoustic and situational circumstances that might come your way (not to mention personal taste), a large number of mic types, styles and designs are available for use as "sonic tools." Because the particular sound characteristics of a mic might be best suited to a specific range of applications, an engineer can use his/her artistic talents to get the best possible sound from an acoustic source by carefully choosing a mic that fits the specific pickup application at hand.

When choosing the best microphone placement, we can start on the road to understanding mic techniques by considering a few simple rules:

Rule 1: *There are no rules, only guidelines.*
Although guidelines can help you achieve a good pickup, don't hesitate to experiment in order to get a sound that best suits your personal tastes.

Rule 2: *The overall sound of an audio signal is no better than the weakest link in the signal path.*

If a mic or its placement doesn't sound as good as it could, make the changes to improve it *before* you commit it to tape, disc or whatever. More often than not, the concept of "fixing it later in the mix" will often put you in the unfortunate position of having to correct a situation after the fact, rather than getting the best sound and/or performance onto tape the first time.

Rule 3: Whenever possible, use the "Good Rule" . . . which states: *Good musician + good acoustics + good mike + good placement = a good sound.*

This refers to the fact that a music track will only be as good as the performer, the instrument, the mike placement and the mike. If any of these fall short of their potential, the track will suffer accordingly. However, if all of these links are the best that they can be, the recording will often be something that you can be proud of!

The miking of vocals and instruments (both in the studio and onstage) is an art form. It's often a balancing act to get the most out of the Good Rule. Sometimes you'll have the best of all of the elements—sometimes you'll have to work hard to make lemonade out of a situational lemon. The best rule of all is to use common sense and to trust your instincts.

Beyond giving you a basic foundation into how microphones (and their characteristics) work, the goal of this chapter is to give you a set of insights and placement guidelines that can help you get the most out of your recordings.

Microphone Design

A microphone is a device that converts acoustic energy into corresponding electrical voltages in a number of ways; however, in recording and project studios, three transducer types are used: dynamic mics, ribbon mics, and condenser mics.

The Dynamic Microphone

In principle, the *dynamic* pickup system (Figure 4.1) operates by using electromagnetic induction to generate an output signal. The theory of electromagnetic induction states that *whenever an electrically conductive metal cuts across the flux lines of a magnetic field, a current of a specific magnitude and direction will be generated within that metal.*

Figure 4.1. *The Shure Beta 58A dynamic mic. (Courtesy of Shure Incorporated, www.shure.com)*

Dynamic mic designs (Figure 4.2) generally consist of a stiff Mylar diaphragm of roughly 0.35-mil thickness. Attached to the diaphragm is a finely wrapped core of wire (called a *voice coil*) that's precisely suspended within a high-level magnetic field. Whenever an acoustic pressure wave hits the face of this diaphragm (A), the attached voice coil (B) is displaced in proportion to the amplitude and frequency of the wave, causing the coil to cut across the lines of magnetic flux supplied by a permanent magnet (C). In doing so, an analogous electrical signal (of a specific magnitude and direction) is induced into the coil and across the output leads.

Figure 4.2. *Working properties of a dynamic microphone's diaphragm.*

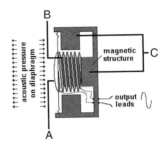

The Ribbon Microphone

Like the moving-coil microphone, the ribbon mic (Figure 4.3) also works on the principle of electromagnetic induction. The older ribbon design types, however, use a diaphragm of extremely thin aluminum ribbon (2 micrometers). Often, this diaphragm is corrugated along its width and is suspended within a strong field of magnetic flux. Sound pressure variations between the front and the back of the diaphragm cause it to move and cut across these flux lines, inducing a current into the ribbon that's proportional to the amplitude and frequency of the acoustic waveform.

Figure 4.3. *Cutaway detail of a ribbon microphone.*

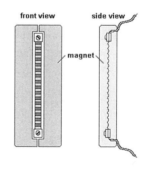

Because of the light mass of the ribbon diaphragm (compared to the moving coil), its electrical resistance is often on the order of 0.2 ohm (Ω). This impedance rating causes the output signal to be too low to drive a microphone input stage directly, so a step-up transformer must be used to bring the output impedance to the acceptable 150 Ω to 600 Ω range.

Recent Developments in Ribbon Technology

Over the past several decades, certain microphone manufacturers have made strides toward miniaturizing and improving the operating characteristics of ribbon mics. Beyerdynamic, for instance, designed the Beyerdynamic M160 (Figure 4.4a) and M260 systems. In the case of the M260, Beyer uses a rare-earth magnet to produce a magnetic structure that's small enough to fit into a 2" grill ball (much smaller than a traditional ribbon styled mic, such as the AEA-44 in Figure 4.4b). The ribbon (which is corrugated along its length to give it added strength and at each end to give it flexibility) is 3 microns thick, about 0.08" wide, 0.85" long, and weighs only 0.000011 ounce. A plastic throat is fitted above the ribbon, which houses a pop-blast filter. Two additional filters and the grill ball greatly reduces the ribbon's potential for blast and wind damage, which finally makes modern mics like these suitable for outdoor and handheld use.

Figure 4.4. Newer and older ribbon mic designs.
a. The Beyerdynamic M160 ribbon mic. (Courtesy of Beyerdynamic, www.beyerdynamic.com)
b. The AEA R44 ribbon mic. (Courtesy of Audio Engineering Associates, www.wesdooley.com)

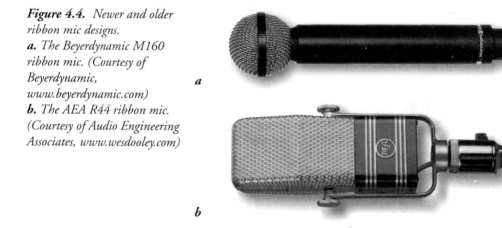

Another relatively recent advance in ribbon technology has been the development of the *printed ribbon mic*. In principle, the printed ribbon operates in precisely the same manner as the conventional ribbon pickup. However, a much more rugged diaphragm is made of a polyester film that has a spiral aluminum ribbon printed on it. The magnetic structure is produced by placing two ring magnets at the diaphragm's front and two in back, thereby creating a wash of magnetic flux that makes the electromagnetic induction process possible.

The Condenser Microphone

Condenser mics (like the ones shown in Figures 4.5 and 4.6) operate on an electrostatic principle rather than the electromagnetic principle that's used by the dynamic and ribbon mics. The basic head, or capsule, of the condenser mic consists of two very thin plates—one movable and one fixed. These two plates form a capacitor (or condenser, as it's still called in the UK and in many parts of the world). A capacitor is an electrical device that's capable of storing an electrical charge. The amount of charge that a capacitor can store is determined by its capacitance value and the voltage that's applied to it, according to the formula

$$Q = CV$$

Figure 4.5. *Exposed example of a condenser diaphragm. (Courtesy of ADK, www.a-dk.com, Photo by K. Bujack)*

Figure 4.6. *Inner detail of an ADK A-51s condenser microphone. (Courtesy of ADK, www.a-dk.com, Photo by K. Bujack)*

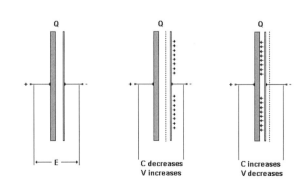

where

    Q is the charge, in coulombs

    C is the capacitance, in farads

    V is the voltage, in volts

The capacitance of the capsule is determined by the composition and surface area of the plates (which are fixed in value), the dielectric or substance between the plates (which is air and also fixed) and the distance between the plates (which proportionately varies with sound pressure). From this, it's fairly easy to see that plates of a condenser mic capsule form a sound pressure sensitive capacitor (Figure 4.7).

Figure 4.7. *Output and potential relationships as a result of changing capacitance.*

In the design used by most manufacturers, the plates are connected to opposite sides of a DC power supply, which provides a polarizing voltage to the capacitor (Figure 4.8). Electrons are drawn from the plate that's connected to the positive side of the power supply and are forced through a high-value resistor onto the plate connected to the negative supply side. This process continues until the charge on the capsule (that is, the difference between the number of electrons on the positive and negative plates) is equal to the capacitance of the capsule times the polarizing voltage. When this equilibrium is reached, no further appreciable current flows through the resistor. If the mic is fed a sound-pressure wave, the diaphragm's capacitance changes. When the distance between the plates decreases, the capacitance increases; when the distance increases, the capacitance decreases. According to the previous equation, Q, C and V are interrelated . . . therefore, if the charge (Q) is constant and sound pressure changes the diaphragm's capacitance (C), the voltage (V) must change in inverse proportion. Now that we have a voltage that proportionately changes with the sound wave, voilà . . . we have a condenser mic!

Figure 4.8. *As a sound wave decreases the condenser spacing by d, the capacitance will increase, causing the voltage to proportionately fall (and vice versa).*

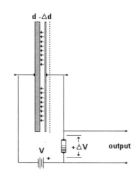

Along with our variable capacitor, a high-value resistor is placed into the circuit that produces a circuit time constant that's longer than a single audio cycle. This time constant is a measure of the time needed for a capacitor to charge or discharge. Because the resistor prevents the capacitor's charge from varying with the rapid changes in capacitance caused by the applied sound pressure, the voltage across the capacitor changes according to V = Q/C. Since the resistor and capacitor are in series with the power supply, the sum of the voltage dropped across them will equal the supply voltage. When the voltage across the capacitor changes, the voltage across the resistor will proportionately change—but in the opposite direction. The voltage across this resistor will then become the mic's output signal.

Because the diaphragm's output signal has an extremely high impedance, it's fed through an impedance conversion amplifier, which is placed into the circuit at a very short distance (often 2" or less) from the diaphragm. This amplifier is placed within the microphone's body in order to prevent hum, noise pickup and signal-level losses that would otherwise occur. It's also another reason why a condenser microphone requires a power supply voltage in order to operate.

Although most modern mics use some form of a FET (field effect transistor) to reduce the capsule impedance, certain highly prized older designs (and many newer "revival" models) use a vacuum tube that's housed within the mic case itself (Figure 4.9). These mics are generally valued by studios and collectors alike for their "tube-like sound," and they often exhibit a sonically favorable coloration that results from the physical design, as well as from the even-harmonic distortion and other phenomena that occur when tubes are used.

Figure 4.9. *Inner detail of a Neumann U-67 condenser microphone. (Courtesy of Neumann USA, www.neumannusa.com)*

The Electret-Condenser Microphone

Electret-condenser mics work on the same operating principles as an externally polarized condenser, except that a static polarizing charge is permanently stored within the diaphragm or on the mic's backplate. Because of this electrostatic charge, no external powering is required to charge the diaphragm. However, the capsule's high-output impedance will still require that an impedance-changing amplifier be used, which must be powered from a stable current source (such as an internal battery or phantom power supply).

Microphone Characteristics

In order to handle the wide range of applications that are encountered in studio and on-location recording, microphones often differ in their overall physical and electrical characteristics. The following highlights many of these characteristics in order to help you choose the best mic for an application.

Directional Response

The directional response of a mic refers to its sensitivity (output level) at various angles of incidence with respect to the on-axis (front) side of the microphone (Figure 4.10). This chart (known as the *polar response* or *polar pattern* of a microphone) is used to graphically plot a microphone's sensitivity with respect to direction and frequency over 360°.

Figure 4.10. *Directional axis of a microphone.*

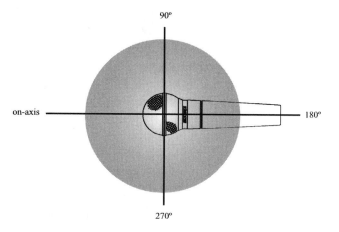

Microphone directionality can be classified into two categories:

*   Omnidirectional polar response
*   Directional polar response

The *omnidirectional mic* (Figure 4.11) is a pressure-operated device that's responsive to nondirectional acoustic sound pressure. In other words, the diaphragm reacts equally to all sound pressure fluctuations at its surface, regardless of the source's location. Pickups that display directional properties are pressure-gradient devices. This means that the system is responsive to differences in pressure between the front, back and sides of a diaphragm. For example, a purely pressure-gradient mic will exhibit a *bidirectional* polar pattern (cosine or more commonly figure 8), as shown in Figure 4.12.

Figure 4.11. *Typical pickup pattern of an omnidirectional microphone.*

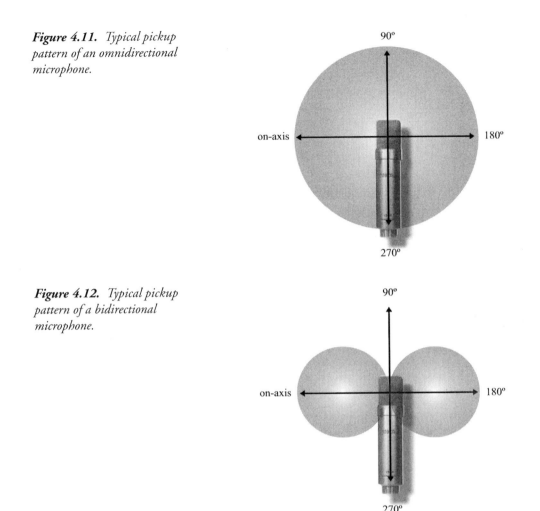

Figure 4.12. *Typical pickup pattern of a bidirectional microphone.*

Many ribbon mics exhibit a bidirectional pattern. Because the ribbon's diaphragm is often exposed to sound waves from both the front and rear axes, it's equally sensitive to sounds that emanate from either direction. Sounds from the rear will produce a voltage that's 180° out-of-phase with an equivalent on-axis signal (Figure 4.13a). Sound waves arriving 90° off-axis produce equal but opposite pressures at the front and rear of the ribbon (Figure 4.13b) that cancel at the diaphragm, resulting in no output signal.

Figure 4.14 graphically illustrates how the acoustical combination (as well as electrical and mathematical combination, for that matter) of a bidirectional (pressure-gradient) and omnidirectional (pressure) pickup can be combined to obtain other directional pattern types. In effect, an infinite number of directional patterns can be obtained from this mixture. The most widely known patterns resulting from these combinations are the *cardioid, supercardioid* and *hypercardioid* polar patterns (Figure 4.15).

Figure 4.13. *Sound sources on-axis and 90° off-axis at the ribbon's diaphragm.*
a. *The ribbon is sensitive to sounds at front and rear.*
b. *Sound waves from 90° and 270° off-axis are cancelled.*

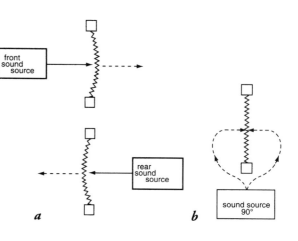

a   *b*

Figure 4.14. *Directional combinations of various bidirectional and nondirectional pickup patterns.*

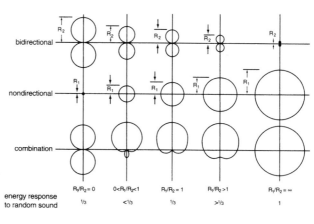

Figure 4.15. *Various polar patterns with output sensitivity plotted versus angle of incidence.*

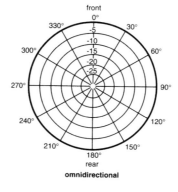

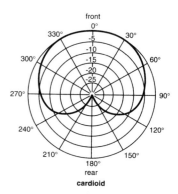

Figure 4.15. *(continued)*

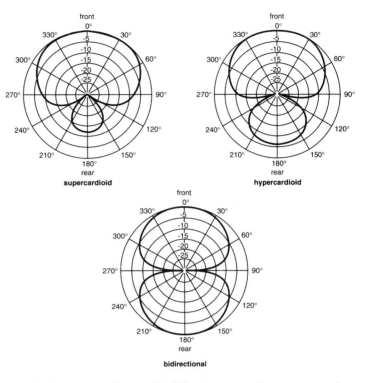

Often dynamic mics achieve a cardioid response (Figure 4.16) by incorporating a rear port into their design. This rear port serves as an acoustic labyrinth that creates an acoustic delay or resistance. A light felt or nylon screen is often used to dampen the diaphragm resonances over the entire frequency range.

Figure 4.16. *Typical pickup pattern of a cardioid microphone.*

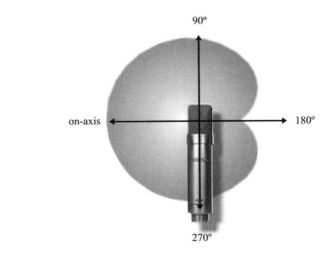

In Figure 4.17a, a cardioid dynamic pickup having a cardioid polar response is shown receiving a sound signal on-axis (at 0°). The diaphragm receives two signals: the incident signal that arrives from the front and the acoustically delayed rear signal. In this instance, the on-axis signal exerts a positive pressure on the diaphragm and also travels 90° to the side port where it is again delayed by 90° (totaling 180° at the rear of the diaphragm). During this short delay period, the on-axis signal has now begun to exert a negative pressure on the diaphragm (pulling it outwards, away from the pickup). Since the on-axis signal pressure is pulling the diaphragm outwards and the delayed rear signal is also pushing the diaphragm outwards (at that specific point in time), an output signal is produced.

Figure 4.17. *The directional properties of a cardioid mic.*
a. *Signals arriving at the front (on-axis) of the diaphragm will produce a full output level.*
b. *Signals arrive at the rear of the diaphragm (180°) will cancel each other out, resulting in a reduced output.*

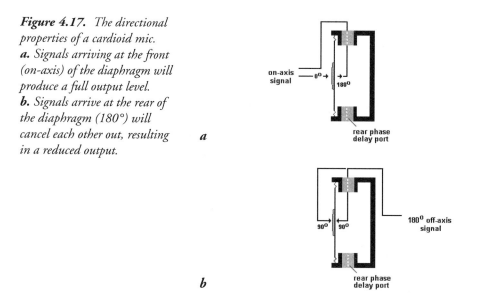

Whenever sounds arrive at the rear of the mic (180° off-axis), the signal travels to the side of the mic where it enters the delay labyrinth and is delayed by 90° (Figure 4.17b). The time that it takes for sounds emanating from the rear to travel to the front of the pickup is also equal to a phase shift of 90°. Since the delayed on-axis signal pressure is pushing the diaphragm inwards and the rear signal is at the same time pushing the diaphragm outwards, very little output is produced.

The attenuation of such an off-axis signal, with respect to an equal on-axis signal, is known as the *front-to-back discrimination* of a microphone and is rated in decibels.

Certain condenser mics can be electrically switched from one pattern to another by using a second capsule that's mounted around a central backplate. Configuring these dual-capsule systems electrically in-phase will result in the creation of an omnidirectional pattern; configuring them out-of-phase results in a bidirectional pattern. Varying them electrically (in either continuous or stepped degrees) between these two polar states yields a number of patterns, such as hyper-cardioid and cardioid (as can be seen in Figure 4.14).

Frequency Response

The on-axis frequency-response curve of a microphone is the measurement of its output over the audible frequency range when driven by a constant, on-axis input signal. This response curve (which is generally plotted in output level [dB] over the 20–20,000 Hz frequency range) often yields valuable information and can give clues as to how a microphone will react at specific frequencies.

A mic can be designed to respond equally to all frequencies (said to exhibit a flat frequency response, shown in Figure 4.18a), or it can be made to emphasize or de-emphasize the high-, middle-, or low-end response of the audio spectrum (Figure 4.18b), so as to give it a particular sonic character.

Figure 4.18. *Frequency-response curves. (Courtesy of AKG Acoustics, Inc., www.akg-acoustics.com)* ***a.*** *Response curve of the AKG C-460B/CK61 ULS.* ***b.*** *Response curve of the AKG D321.*

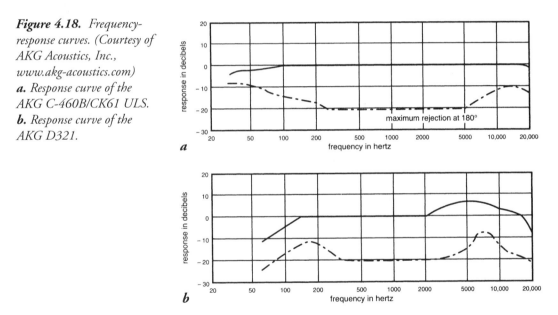

The solid frequency-response curve lines (shown in Figure 4.18) were measured on-axis and exhibit totally acceptable responses. However, certain designs may have a "peaky" or erratic curve when measured off-axis. These signal colorations could become evident when the mic is operating in an area where off-axis sounds (in the form of leakage) arrive at the pickup, often resulting in a change in tone quality when the leaked signal is mixed in with another properly miked signal. This off-axis frequency response can be charted along with the on-axis curve (shown as dotted 180° curves in Figure 4.18).

Low-Frequency Response Characteristics

At low frequencies, rumble (high-level vibrations that occur in the 3- to 25-Hz region) can be easily introduced into a studio or hall or along the surface of a large unsupported floor space, from any number of sources (such as passing trucks, air conditioners, subways, or fans). They can be reduced or eliminated in any of the following ways, among others:

- Use a shock mount to isolate the mic from the vibrating surface and floor stand.
- Choose a mic that displays a restricted low-frequency response.
- Restrict the response of a wide-range mic by using a low-frequency rolloff filter.

Another low-frequency phenomenon that occurs in most directional mics is known as *proximity effect*. This effect causes an increase in bass response whenever a directional mic is brought within 1' of the sound source. This bass boost (which is often most noticeable with vocals) will proportionately increase as the distance decreases. To compensate for this effect (which is somewhat greater for a bidirectional mics than for cardioids), a low-frequency rolloff filter switch is often located on the microphone body. If none exists, an equalizer can be used to roll off the low end. Either of these can be used to help restore the bass response to a flat and natural-sounding balance.

Another way to greatly reduce or eliminate proximity effect and the associated "popping" of the letters *p* and *b* is to replace the directional microphone with an omnidirectional mic when working at close distances.

On a more positive note, this increase in bass response has long been appreciated by vocalists for giving a full, "larger-than-life" quality to an otherwise thin voice. In many cases, the directional mic has become an important part of an engineer's, producer's and vocalist's toolbox.

Transient Response

A significant piece of data (which presently has no accepted standard of measure) is the *transient response* of a microphone (Figures 4.19). Transient response is the measure of how quickly a mic's diaphragm will react when it is hit by an acoustic wavefront.

Figure 4.19. Transient response characteristics of the various microphone types.

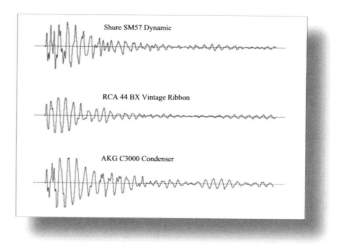

This figure varies wildly among microphones and is a major reason for the difference in sound quality among the three pickup types. For example, the diaphragm of a dynamic mic can be quite large (up to 2½"). With the additional weight of the coil of wire and its core, this combination can make for a large mass when compared to the power of the sound wave that drives it. Because of this, a dynamic mic can be very slow in reacting to a waveform, giving a rugged, gutsy and less accurate sound.

By comparison, the diaphragm of a ribbon mic is much lighter, so its diaphragm can react more quickly to a sound waveform, resulting in a clearer sound. The condenser pickup has an extremely light diaphragm, which varies in diameter from 2½" to ¼" and has a thickness of about 0.0015". This means that the diaphragm offers very little mechanical resistance to a sound-pressure wave, allowing it to accurately track the wave over the entire frequency range.

Output Characteristics

A microphone's output characteristics refer to its measured sensitivity, equivalent noise, overload characteristics, impedance and other output responses.

Sensitivity Rating

A mic's *sensitivity rating* is the output level (in volts) that a microphone will produce, given a specific and standardized input signal (rated in dB SPL). This figure specifies the amount of amplification that's required to raise the mic's signal to line level (–10 dBV or +4 dBm) and allows us to judge the relative output levels between any two mics. A microphone with a higher sensitivity rating will produce a stronger output-signal voltage than one with a lower sensitivity, given that both are driven by the same source and/or sound-pressure level.

Equivalent Noise Rating

The *equivalent noise rating* of a microphone can be viewed as the device's electrical self-noise. It expresses the equivalent dB SPL that would produce a voltage that's equal to the mic's self-noise voltage.

As a general rule, the mic itself doesn't contribute much noise to a system when compared to the mixer's amplification stages and/or the recording system or media (whether analog or digital). However, with recent advances in mic preamp/mixer technologies and overall reductions in system noise (including digital recording media), these noise ratings are becoming increasingly important.

When using a dynamic or ribbon pickup, the noise is generated by electrons moving within the coil or ribbon itself. Most of the noise that's produced by a condenser mic occurs in the built-in preamplifier. It almost goes without saying that certain microphone designs display a greater degree of self-noise than do other designs; thus, care must be taken in the choice of microphones for critical applications (such as with distant classical recording techniques).

Overload Characteristics

As the use of a microphone is limited at low levels by its inherent self-noise, it's also limited at high SPL levels by *overload distortion*.

In terms of distortion, the dynamic microphone is an extremely rugged pickup, often capable of an overall dynamic range of 140 dB. The typical condenser microphone often won't distort—except under the most severe sound-pressure levels. However, the condenser system differs from that of the dynamic in that at high acoustic levels the capsule's output might be high enough to overload the mic's preamplifier. To prevent this, many condenser mics offer a switchable attenuation pad that immediately follows the capsule output and serves to reduce the signal level and, therefore, reduce or eliminate overload distortion. When inserting such an attenuation pad in the microphone circuit, keep in mind that the mic's signal-to-noise ratio is degraded by the amount of attenuation. Therefore, it's wise to remove the inserted pad when using the microphone under normal SPL conditions.

Microphone Impedance

Microphones are available with different output impedances. *Output impedance* is a rating that's used to match the signal-providing capability of one device to the signal-drawing (input impedance) requirements of another device. Impedance is measured in ohms (with its symbol being Ω or Z). The most commonly used microphone output impedances are 50Ω, 150 to 250Ω (low), and 20 to 50 kΩ (high). Each impedance range has its advantage. In the past, high-impedance mics were less expensive to use because the input impedance of tube-type amplifiers was high. To be used with low-impedance mics, tube-type amplifiers required expensive input transformers. All dynamic mics, however, are low-impedance devices and use a built-in impedance step-up transformer to achieve a high-impedance output. A major disadvantage to using high-impedance mics is the susceptibility of their cables to the pickup of electrostatic noise (such as that caused by motors and fluorescent lights). In order to reduce such interference, a shielded cable is necessary that begins to act as a capacitor at lengths that are greater than 20–25', which serves to short out much of the high-frequency information that's picked up by the mic. For these reasons, high-impedance microphone pickups are rarely used in the professional recording process.

High-impedance mics and most line-level instrument lines use *unbalanced* signal lines (Figure 4.20), whereby one signal lead carries a positive current potential to a device, while a second ground shield is used to complete the signal's returning circuit path. When working at low signal levels (especially at mic levels), any noises, hums, buzzes or other induced interferences will be amplified along with the input signal.

Very-low-impedance mics (50 Ω) have the advantage that their lines are fairly insensitive to electrostatic pickup. They're sensitive, however, to induced hum pickup from electromagnetic fields (such as those that are generated by AC power lines). This extraneous pickup can be eliminated by using a twisted-pair cable, whereby the currents that are magnetically induced into

Figure 4.20. *Unbalanced microphone circuit. (Courtesy of Mackie Designs, www.mackie.com)*
a. *Basic wiring diagram.*
b. *Diagram for wiring an unbalanced mic (or line source) to a balanced XLR connector.*

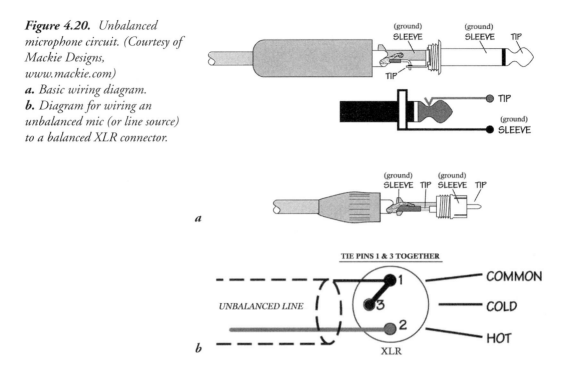

the cable will flow in opposite directions and be canceled out at the console or mixer's balanced microphone input stage.

Mic lines of 150 Ω to 250 Ω have low signal losses and can be used with cable lengths of up to several thousand feet. They're less susceptible to electromagnetic pickup than 50-Ω lines, but are more susceptible to electrostatic pickup. As a result, a shielded twisted-pair cable is used and the lowest noise is attained through the use of a *balanced* signal line. Within such a line, two wires carry the signal voltage, while a third lead and/or shield is used as a neutral ground wire. Neither of the two signal conductors is directly connected to the signal ground.

Balanced lines operate on the principle that the alternating current of an audio signal will be presented in opposite polarity potential between the two conductors (as occurs in any AC audio circuit), whereas any electrostatic or electromagnetic pickup will be simultaneously induced into both leads at equally polarities and level (Figure 4.21). The input transformer or balancing amplifier of the receiving device will only respond to the difference in voltage between the two leads. As a result, the unwanted noise signals (which are equal in phase and amplitude) will cancel, while the audio signal will be unaffected. The mic line type that's most used in recording, project and audio production is balanced with an impedance of 200 Ω.

The standard that has been widely adopted for the proper polarity of two-conductor, balanced XLR connector cables specifies pin #2 as being positive (+) or "hot" and pin #3 as being negative (–) or "neutral," with the outer shield and cable ground being connected to pin #1.

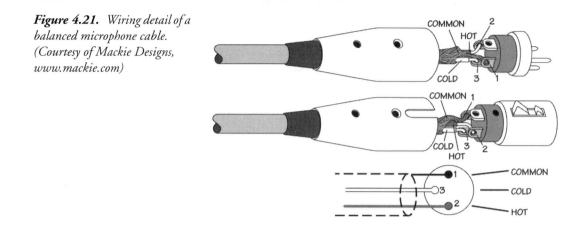

Figure 4.21. *Wiring detail of a balanced microphone cable. (Courtesy of Mackie Designs, www.mackie.com)*

If the hot and neutral pins of balanced mic cables are haphazardly pinned in a music or production studio, it's possible that any number of mics (and other equipment, for that matter) could be wired in opposite polarities. For example, if a single instrument is picked up by two mics that are improperly phased, the instrument might end up being either totally or partially canceled when mixed to mono.

Microphone Preamps

Since the output signals of most microphones are at levels far too low to drive the line level input stage of most recording systems, a mic preamplifier (preamp) must be used to boost it signal to acceptable levels (often by 30–60 dB).

With the advent of improved technologies in analog console design (along with similar improvements in reduced noise and distortion that have come with digital recorders, digital audio workstations, processors and consoles), many of the noise and spec figures that are outlined above have become more important than ever. To many professionals, microphone preamps that have been designed into modern and earlier-era mixer and console designs haven't kept up with the improvements in microphone technology . . . and don't have that special "sound" or simply aren't high enough in quality to be used in critical applications. For those who subscribe to these beliefs, the outboard mic preamp was developed (Figures 4.22 through 4.24). Alternatively, mic preamps are perfect for those who are using a digital audio workstation (or other line-level devices) who don't have a need for a mixer or console simply to boost one or more mics to a level that's acceptable for their computer's sound card.

The range of features, quality levels and particular sounds among mic pres (pronounced "preeze") are as varied as the basic, in-board mic preamps that they were designed to replace. Often offering low-noise, low-distortion specs, these devices may make use of tube, FET (field effect transistor) and/or integrated circuit technology and offer basic features such as variable input gain, phantom power and high-pass filtering. As with most recording tools, the desired type, sound and features are up to the individual, the producer and/or the artist and are totally a matter of personal style and taste.

Figure 4.22. *The Audio Buddy phantom powered dual mic preamp and direct box. (Courtesy of MIDIman, www.midiman.com)*

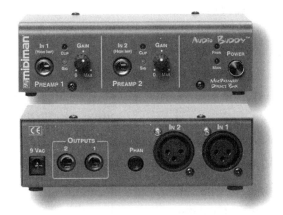

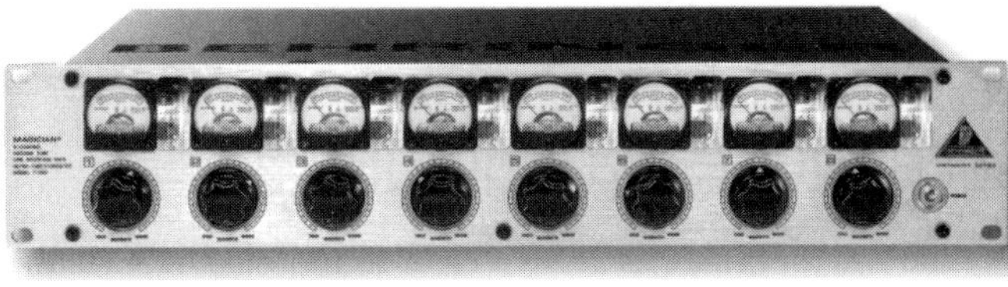

Figure 4.23. *Behringer's Magician 8 channel tube mic preamp. (Courtesy of Behringer International GMBH, www.behringer.de)*

Figure 4.24. *Focusrite ISA 110 Mic pre and equalizer. (Courtesy of Focusrite, www.focusrite.com)*

Phantom Power

Most modern professional condenser mics don't require internal batteries, external battery packs or individual AC power supplies in order to operate. These are designed to be powered directly from the console through the use of a *phantom power* supply.

Phantom power works by supplying a positive DC supply voltage of +48 V to both conductors (pins 2 and 3) of a balanced mic line. This voltage is equally distributed through identical value resistors, so that no differential exists between the two leads. This positive DC voltage is therefore not electrically "visible" to the input stage of a balanced mic preamp. Instead, only the alternating audio signal that's being simultaneously carried on the two audio leads is detected (Figure 4.25). The DC powering circuit is completed by supplying the negative side of the supply to the audio cable's grounding wire (pin 1). The ±1 percent tolerance, ¼W resistor values for R2 at various supply voltages (as some mics can also be designed to work at lower voltages than 48 V) are 6800 Ω for 48V, 1200 Ω for 24 V or 680 Ω for a 12-V supply.

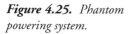

Figure 4.25. *Phantom powering system.*

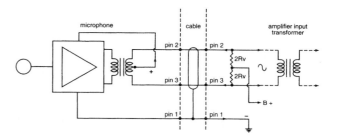

The resistors used in distributing power to the signal leads can also be used to provide a degree of power isolation between other mic inputs on a console. If a signal lead were accidentally shorted to ground (as could happen if defective cables or unbalanced XLR cables are used), the power supply should still be able to deliver power to other mics in the system. If two or more inputs were accidentally shorted, however, the phantom voltage could drop to levels that are too low to be usable.

Microphone Techniques

Each microphone has a distinctive sound character that's based on its specific type and design. A large number of types and models can be used for a variety of applications, and it's up to the engineer to choose the right one for the job. Choosing the appropriate mic, however, is only half the story. The placement of the microphone (either at a distance within a room or closely miked near an instrument) can play just as important a role. . . . It's definitely an art form and one of an engineer's most valued tools.

Because mic placement is an art form, there is no right or wrong. Placement techniques that are currently considered "bad" might easily be the accepted standard five years from now. As new musical styles develop, new recording techniques also tend to evolve. This helps to breathe new life into musical sound and production. The craft of recording should always be open to change and experimentation, two of the strongest factors that help keep music and the biz of music alive and fresh.

Pickup Characteristics as a Function of Working Distance

In studio and sound-stage recording, four fundamental styles of microphone placement are directly related to the working distance of a microphone from its sound source. These important pickup styles (which are described in the following sections) include distant miking, close miking, accent miking, and ambient miking.

Distant Microphone Placement

With *distant microphone placement* (Figure 4.26), one or more mics are positioned at a distance of 3' or more from the intended signal source. This technique will often yield the following results:

- It can pick up a large portion of a musical instrument or ensemble, thereby preserving the overall tonal balance of that instrument or ensemble. Often, a natural tone balance can be achieved by placing the mic at a distance that's roughly equal to the size of the instrument or sound source.

- It allows the room's acoustic environment to be picked up (and naturally mixed in) with the direct sound signal.

Figure 4.26. Example of an overall distant pickup.

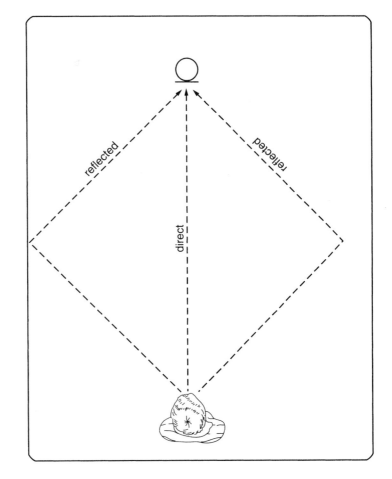

Distant miking is often used to pickup large instrumental ensembles (such as a symphony orchestra or choral ensemble). In this application, the pickup will largely rely upon the acoustic environment to help achieve a natural, ambient sound. The mic is placed at a distance so as to strike an overall balance between ensemble's direct sound and environmental acoustics. This balance is determined by a number of factors, including the sound source's size and overall volume level and mic distance and placement, as well as the reverberant characteristics of the room.

Distant miking techniques tend to add a live, open feeling to a recorded sound. However, this technique could put you at a disadvantage if the acoustics of a hall, church, or studio aren't particularly good. Improper or bad room reflections often create a muddy or poorly defined recording. To avoid this, the engineer can take one of the following actions:

• Temporarily correct for bad or excessive room reflections by using absorptive and/or offset reflective panels
• Place the mic closer to its source and add artificial ambience

If a distant mic is used to pick up a portion of the room sound, placing it at a random height can result in a hollow sound due to phase cancellations of the direct sound with the sounds that are reflected from the floor (Figure 4.27). If these delayed reflections arrive at the mic at a time equal to one-half a wavelength (or an odd integer multiple thereof), the signal will be 180° out-of-phase with the direct sound, thereby producing dips in the signal's pickup response. Because the reflected sound is at a lower level than the direct sound (as a result of traveling farther and losing energy when it hits the floor), the cancellation is usually only partially complete.

Figure 4.27. *Resulting frequency response from a mic that receives a direct and delayed sound from one source.*

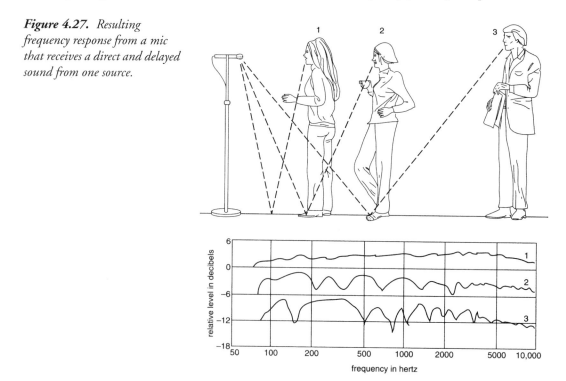

Moving the mic closer to the floor reduces the path length and raises the possibility of frequency cancellation. In practice, a height of ⅛" to 1/16" keeps the lowest cancellation above 10 kHz. One such microphone design type, known as a *boundary microphone* (Figures 4.28 and 4.29), places an electret-condenser or condenser diaphragm well within these height restrictions. For this reason, this mic type may be a good choice for use as an overall distant pickup, when the pickups need to be out of sight (i.e., placed on a floor, wall or large boundary).

Figure 4.28. *The boundary microphone system.*

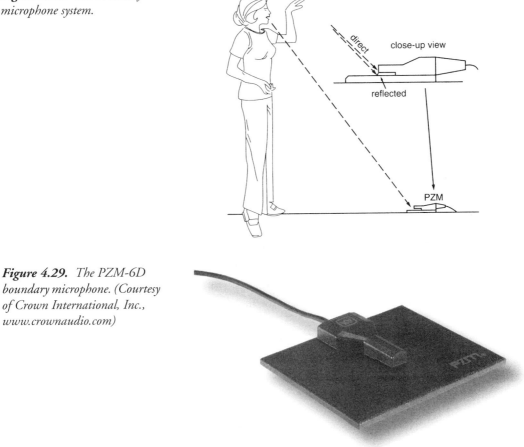

Figure 4.29. *The PZM-6D boundary microphone. (Courtesy of Crown International, Inc., www.crownaudio.com)*

Close Microphone Placement

When *close microphone placement* is used, the mic is often positioned about 1" to 3' from a sound source. This commonly used technique often yields two results:

- Creates a tight, present sound quality
- Effectively excludes the acoustic environment

Because sound diminishes with the square of its distance from the sound source, a sound originating 6' from a mic is generally insignificant in level when compared to a sound that originates 3" from the pickup (Figure 4.30). As a result, only the desired on-axis sound is recorded onto tape, while extraneous sounds (for all practical purposes) aren't picked up.

Figure 4.30. *Close miking reduces the effects of the acoustic environment.*

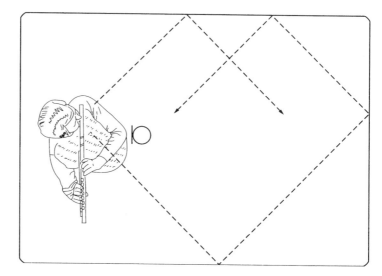

When sounds from one instrument are picked up by a nearby microphone (which is being used to pick up a different instrument), a condition known as *leakage* occurs. Because the microphones could contain both the direct and "leaked" sound (Figure 4.31), control over individual tracks in mixdown might be difficult without affecting the level and sound character of other tracks. Whenever possible, try to avoid this condition in the studio.

Figure 4.31. *Leakage due to indirect signal pickup.*

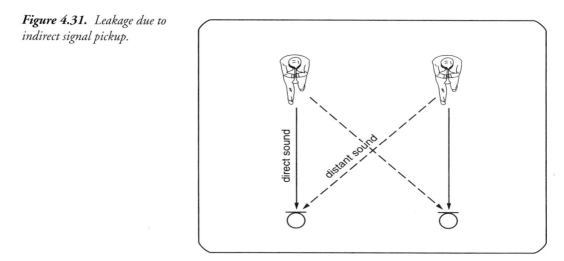

To avoid the problems of leakage, any or all of the following methods can be tried:

- Bring the mics closer to their respective instruments (Figure 4.32a)
- Place an acoustic barrier (known as a flat, gobo or divider) between the two instruments, as shown in Figure 4.32b
- Use directional mics
- Spread the instruments farther apart

Figure 4.32. Two methods to reduce leakage.
a. Mics placed closer to their sources.
b. An acoustic barrier used to reduce leakage.

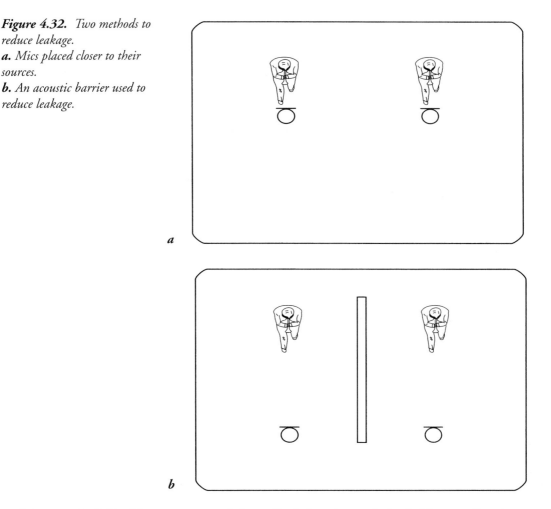

Whenever individual instruments are being miked close (or semi-close), it's generally wise to follow the *3:1 distance rule*. This principle states that in order to maintain phase integrity . . . for every unit of distance between a mic and its source, a nearby mic (or mics) should be separated by at least three times that distance (Figure 4.33).

Figure 4.33. Example of the
3:1 microphone distance rule:
For every unit of distance
between a mic and its source, a
nearby mic (or mics) should be
separated by at least three times
that distance.

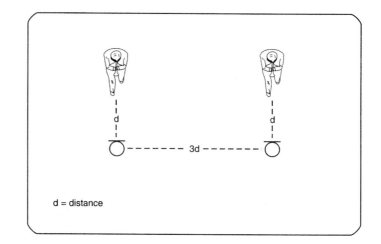

Although the close miking of a sound source offers several advantages, a mic should be placed only as close to the source as is necessary, *not* as close as possible. Miking too close can color the recorded tone quality of a source.

Do It Yourself Tutorial: Close Miking

Mic an acoustic instrument (such as a guitar, piano or violin) at a distance of 3".

Move (or have someone move) the mic over the instrument's body while listening to variations in the sound. Does the sound change? What's your favorite and least favorite position?

Because such techniques commonly involve distances of 1" to 6", the tonal balance (timbre) of an entire sound source often can't be picked up. Rather, the mic may be so close to the source that only a small portion of the surface is actually picked up, giving it a tonal balance that's very area-specific. At these close distances, moving a mic by only a few inches can easily change the instrument's overall tonal balance. If this occurs, try using one of the following remedies:

> Move the microphone along the surface of the sound source until the desired balance is achieved.
>
> Place the mic farther back from the sound source to allow for a wider angle (thereby picking up more of the overall instrument's sound).
>
> Change the mic.
>
> Equalize the signal until the desired balance is achieved.

Accent Microphone Placement

The overall pickup and tonal qualities of distant and close mic techniques often sound very different. Under certain circumstances, it's difficult to obtain a naturally recorded balance when mixing the two together. For example, if a solo instrument in an orchestra needs an extra mic for more volume and presence, placing the mic too close would result in a pickup that sounds overly present, unnatural and out of context with the distant, overall orchestral mics. To avoid this pitfall, a compromise in distance should be struck. A microphone that has been placed close (but not so close as to have an unnatural sound) to an instrument or section within a larger ensemble is known as an *accent microphone* (Figure 4.34).

Figure 4.34. *Accent microphone placed at proper compromise distance.*

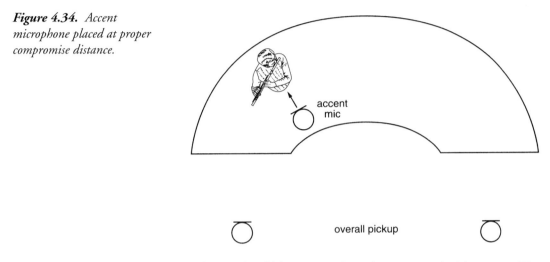

Whenever an accent mic is used, care should be exercised in placement and pickup type. The amount of accent signal that's introduced into the mix shouldn't change the balance and spatial relationship of the soloist to the overall pickup. A good accent pickup should only add presence to a solo passage and not stick out as separate, identifiable pickup.

Ambient Microphone Placement

An *ambient mic* is one that's placed at such a distance that the reverberant or room sound is more prominent than the direct signal. The ambient pickup is often a stereo cardioid pair or crossed figure-of-eight (Blumlein) pair that might or might not be mixed in with closely placed microphones.

To enhance recording, you can use ambient mic pickups in the following ways:

- In a live concert recording, ambient microphones can be placed in a hall to restore the natural reverberation that might be lost with close miking techniques.
- In a live concert recording, ambient microphones can be placed over the audience to pick up their reaction and applause.
- In a studio recording, ambient microphones can be used in the studio to add a sense of natural acoustics back into the sound.

Do It Yourself Tutorial: Ambient Miking

1. Mic an instrument or its amp (such as an acoustic or electric guitar) at a distance of 6" to 1'.

2. Place a stereo mic pair (in an XY and/or spaced configuration) in the room, away from the instrument.

3. Mix the two pickup types together. Does it "open" the sound up and give it more space? Does it muddy the sound up or breath new life into it?

Stereo Miking Techniques

For the purpose of this discussion, *stereo miking techniques* refer to the use of two microphones in order to obtain a coherent stereo image. These techniques can be used in either the close or distant miking of background vocals, large or small ensembles, single instruments, on location or studio applications . . . the only limitation is your imagination. The three fundamental stereo miking techniques are spaced pair, the XY technique, and the M-S method.

Spaced microphones (Figure 4.35) can be placed in front of an instrument or ensemble (in a left/right fashion) to obtain an overall stereo image. This technique places the two mics (of the same type, manufacturer and model) anywhere from only a few feet to more than 30 feet apart (depending on the size of the instrument or ensemble) and uses time and amplitude cues in order to create a stereo image.

Figure 4.35. *Spaced stereo miking technique.*

The primary drawback to this technique is the strong potential for the introduction of phase discrepancies between the two channels due to differences in a sound's arrival time at one mic relative to the other. When mixed to mono, these phase discrepancies could result in variations in frequency response and even the partial cancellation of various instruments or sound components in the pickup field.

The *XY technique* is an intensity-dependent system that uses only the cue of amplitude to discriminate direction. With the XY coincident-pair technique, two directional microphones of the exact same type, manufacture and model are placed with their grills as close together as possible (without touching) and facing at angles to each other (generally between 90° and 135°). The midpoint between the two mics is faced toward the source and the mic outputs are equally panned left and right.

Even though the two mics are placed together, the stereo imaging is excellent—often better than that of a spaced pair. Because of the close proximity of the mics, an added advantage is that there are no appreciable phase problems. The generally accepted polar pattern is cardioid (Figure 4.36a), although two crossed bidirectional mics (known as the Blumlein technique and named after the unheralded inventor, Alan Dower Blumlein), offset by 90° to each other, often yield excellent ambient results (Figure 4.36b).

Figure 4.36. *XY stereo miking technique.*
a. XY crossed cardioid pair.
b. Blumlein crossed bidirectional pair.

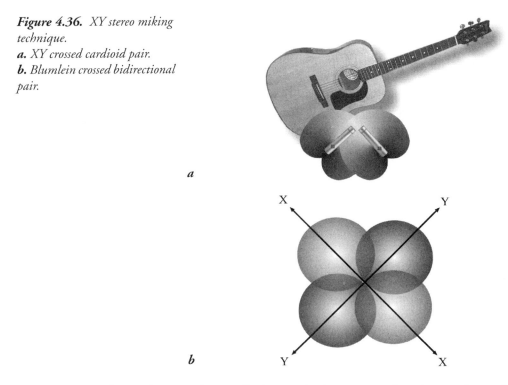

Stereo microphones that contain two diaphragms in the same case housing are also available on the used and new market. These mics will either be fixed (generally in a 90° or variable XY pattern) or be designed so that the top diaphragm can be rotated by 180°, so that they can be adjusted to various coincident XY angles.

Another coincident-pair method, known as the *M/S* (or *mid-side*) *technique* (Figure 4.37), is similar to XY in that it uses two closely spaced, matched pickups. The mid-side method differs from XY in that it requires the use of an external transformer, active matrix or software plug-in in order to work.

In the classic M/S configuration, one of the microphone capsules is designated to be the M (mid) position pickup, which is generally selected as having a cardioid pickup pattern that's oriented toward the sound source. The S (side) capsule is generally chosen to be a figure-8 pattern that's oriented sideways (with the 90° null side facing the cardioid's main axis). In this way, the direct sound is picked up by the mid capsule, while ambient and reverberant sound is picked

Figure 4.37. *M/S stereo mic technique.*

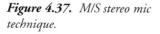

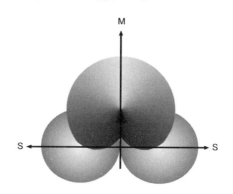

up by the side figure-8 capsule. These outputs are then electrically or mathematically (in the case of a digital M/S processor) combined through a sum-and-difference decoder matrix, which resolves them into a conventional XY stereo signal: (M + S = X) and (M–S = Y).

One advantage of this technique is its absolute monaural compatibility. When the left and right signals are combined, the sum the output will be ([M + S] + [M–S] = 2M). That's to say, that the side (ambient) signal will be cancelled out, while the mid (direct) signal will be accentuated. Since it's widely accepted that a mono signal looses it's intelligibility with added reverb, this tends to work to our advantage.

One of the amazing side-benefits of using M/S is the fact that it lets us continuously vary the mix of mid (direct) to side (ambient) sound that's being picked up either during the recoding (from the console location) . . . and even during mixdown, after it's been recorded. These are both possible by simply mixing the ratio of Mid to Side that's being sent to the decoder matrix (Figure 4.38). In a mixdown scenario, all that's needed is to record the Mid on one track and the Side on another (it's often best to use a digital recorder, as phase delays associated with analog recording can interfere with the decoding process). Upon mixdown, routing the M/S tracks to the decoder matrix allows you to make important decisions regarding stereo width and depth at a later, more controlled date.

Figure 4.38. *M/S decoder matrix.*
a. AEA MS-38 Lite active M/S matrix. (Courtesy of Audio Engineering Associates, www.wesdooley.com)

b. Waves M/S matrix plug in for digital audio workstations. (Courtesy of Waves, www.waves.com)

Recording Direct

As an alternative, the signal of an electric instrument (guitar, keyboard and so on) can be injected into a console and recorded directly to tape without the use of a microphone. This option can produce a cleaner, more present sound by bypassing the distorted components of a head/amp combination. It also reduces leakage into other mics by eliminating room sounds from the amplifier.

In the project or recording studio, the direct injection (D.I.) box (Figure 4.39) serves to interface an electric instrument to the audio console in the following ways:

- By reducing an instrument's line-level output to mic level for direct insertion into the console
- By changing an instrument's unbalanced, high-source impedance line to a balanced, low-source impedance signal that's needed by the console's input stage
- By electrically isolating audio signal paths (thereby reducing the potential for a ground loop hum)

Figure 4.39. The BSS Audio AR-133 direct injection (D.I.) box. (Courtesy of BSS Audio USA, www.bss.co.uk)

Most commonly, the instrument's output is plugged directly into the D.I. box (where it's stepped down in level and impedance), and the box's output is then fed into a console or mixer. If a "dirtier" sound is desired, certain boxes will allow high-level signal to be taken directly from the amp's external speaker jack.

It's also not uncommon for an engineer, producer and/or artist to combine the punchy, full sound of a mic with the present, crispness of a direct sound. These signals can then be combined onto a single tape track or recorded to separate tracks (thereby giving more flexibility in the mixdown stage).

Microphone Placement Techniques

This section is meant to be used as a general guide to mic placement for many of the more popular acoustic and electric instruments. It's important to keep in mind that these are only guidelines. Several general application and characteristic notes are detailed in Table 4.1, and descriptions of several popular mics are outlined in the "Microphone Selection" section to help give insights into what mic placement and technique will often work in a particular application.

As a general rule, choosing the best mic for an instrument or vocal will depend on the basic character of the "sound" that you're after. For example, a dynamic microphone often gives the sound a "rugged" or "punchy" character (which is often accentuated by a close proximity bass boost that's associated with most directional mics). A ribbon mic will often yield a mellow, slightly "croony" sound when used at close distances, whereas condenser mics will generally have a clear, present and full-range sound.

Before jumping into this section, I'd like to again take the time to point out "*The Good Principle*" to anyone who wants to be a better engineer, producer and/or musician:

> Good musician + good acoustics + good mike + good placement = a good sound.

As a rule, starting with an experienced musician who has a quality instrument that's well tuned is the best insurance toward getting the best possible sound.

Let's think about this for a moment. Say that we have a live rhythm session that involves drums, piano, bass guitar and scratch vocals. All of the players are the best around, except for the drummer, who is new to the studio process. . . . Unfortunately for all involved (unless you're very, very lucky), you've just signed on to teaching the drummer the ropes of proper drum tuning (if the instrument's any good), playing technique, etc. Both the engineer and the producer's jobs will be much harder, as you'll both have to take the time to work with him/her to tune the drums and set up the mics in order to get the best possible sound. Once tape's rolling, it'll be up to the producer to pull a professional performance out of someone who's new to the field.

Don't get me wrong, musicians have to start somewhere . . . but an experienced studio musician who comes into the studio with a great instrument that's tuned and ready to go (he/she might even clue you in on some sure-fire mic and placement techniques for their instrument) is simply a joy from a sound, performance and time/budget-saving standpoint. Simply put, if you and/or the project's producer have prepared enough to get all your "goods" lined up, the track will have a much better chance of being something that everyone can be proud of.

In addition to the art of playing an instrument, the art of mic choice, placement and style is one that's subjective and is often one of the calling cards of a good engineer. Experience simply comes with time and the willingness to experiment.

Table 4.1. Microphone selection chart.

Application	Microphone Characteristic/Choice
Natural, smooth tone quality	Flat frequency response
Bright, present tone quality	Rising frequency response
Extended lows	Dynamic or condenser with extended low-frequency response
Extended highs (detailed sound)	Condenser
Reduced "edge" or detail	Dynamic
Boosted bass at close working distances	Directional microphone
Flat bass response up close	Omnidirectional
Reduced pickup of leakage, feedback and room acoustics	Directional microphone, or omnidirectional microphone at close working distances
Enhanced pickup of room acoustics	Omnidirectional microphone, or directional microphone at greater working distances
Extra ruggedness	Moving-coil microphone
Reduced handling noise	Omnidirectional, or directional microphone with shock mount
Reduced breath popping	Omnidirectional, or directional microphone with pop filter
Distortion-free pickup of very loud sounds	Condenser with high maximum-SPL spec or dynamic
Noise-free pickup of quiet sounds	Low self-noise, high sensitivity

Brass Instruments

The following section describes many of the sound characteristics and miking techniques that are encountered in the brass family of instruments.

Trumpet

The fundamental frequency of a *trumpet* ranges from E3 to D6 (165 Hz to 1175 Hz) and contains overtones that stretch upwards to 15 kHz. Below 500 Hz, the sounds emanating from the trumpet project uniformly in all directions; above 1500 Hz, the projected sound becomes much more directional; and above 5 kHz, the dispersion emanates at a tight 30° angle from in front of the bell.

The formants of a trumpet (the relative harmonic and resonance frequencies that give an instrument its specific character) lie at around 1 to 1.5 kHz and at 2 to 3 kHz. Its tone can be radically changed by using a mute (a cup-shaped dome that fits directly over the bell), which serves to dampen frequencies above 2.5 kHz. A conical mute (a metal mute that fits inside the bell) tends to cut back on frequencies below 1.5 kHz while encouraging frequencies above 4 kHz.

Because of the high sound pressure levels that a trumpet can produce (up to 130 dB SPL), it's best to place a mic slightly off center of the bell at a distance of 1' or more (Figure 4.40). When closer placements are needed, a −10 to −20dB pad can help prevent input overload at the mic or console preamp input. Under such close working conditions, a windscreen often helps to protect the diaphragm from windblasts.

Figure 4.40. *Typical mic placement for a single trumpet.*

Trombone

Trombones come in a number of sizes; however, the most commonly used "bone" is the tenor that has a fundamental note range spanning from E2 to C5 (82 Hz to 520 Hz) and produces a series of complex overtones that range from 5 kHz (when played medium loud) to 10 kHz (when overblown). The trombone's polar pattern is nearly as symmetrical as the trumpet's: frequencies below 400 Hz are distributed evenly, whereas its dispersion angle is down to 45° from the bell at 2 kHz and above.

The trombone most often appears in jazz and classical music. The *Mass in C Minor* by Mozart, for example, has parts for soprano, alto, tenor and bass trombones. This style obviously lends itself to the spacious blending that can be achieved by distant pickups within a large hall or studio. On the other hand, jazz music often calls for closer miking distances. At 2" to 12", for example, the trombonist should play slightly to the side of the mic to reduce the chance of overload and wind blasts. In the miking of a trombone section, a single mic might be placed between two players, acoustically combining them onto a single channel and/or track.

Tuba

The *bass* and *double-bass tubas* are the lowest pitched of the brass/wind instruments. Although the bass tuba's range is actually a fifth higher than that of the double bass, it's still possible to obtain a low fundamental of B (29 Hz). A tuba's overtone structure is limited—with a top response ranging from 1.5 kHz to 2 kHz. The lower frequencies of the tuba (around 75 Hz) are dispersed evenly; however, as frequencies rise, the sonic distribution angle is reduced.

Under normal conditions, this class of instruments isn't miked at close distances. A working range of 2' or more, slightly off-axis to the bell, will yield the best results.

French Horn

The fundamental tones of the *French horn* range from B1 to B5 (65 Hz to 700 Hz). Its "oo" formant gives it the round, broad quality that can be found at about 340 Hz with other frequencies falling between 750 Hz and 3.5 kHz. French horn players often place their hands inside the bell to mute the sound and promote the formant at about 3 kHz.

A French horn player or section is traditionally placed at the rear of an ensemble, just in front of a rear, reflective stage wall. This wall serves to reflect the sound back toward the listener's position (which tends to create a fuller, more defined sound). An effective pickup of this instrument can be achieved by placing an omni- or bidirectional pickup between the rear, reflecting wall and the instrument bells, thereby receiving both the direct and reflected sound. Alternatively, the pickups can be placed in front of the players, thereby receiving only the sound that's being reflected from the rear wall.

Guitar

The following sections describe the various sound characteristics and miking techniques that are encountered for the guitar.

Acoustic Guitar

The popular steel-strung, *acoustic guitar* has a bright, rich set of overtones (especially when played with a pick). Mic placement and distance will often vary from instrument to instrument and may require experimentation to pick up the best tonal balance.

A balanced pickup can often be achieved by placing the mic (or an XY stereo pair) at a point slightly off-axis and above or below the sound hole at a distance of between 6" and 1' (Figure 4.41). Condenser mics are often preferred for their smooth, extended frequency response and excellent transient response.

Figure 4.41. *Typical mic placement for a guitar.*

The smaller-bodied classical guitar is normally strung with nylon or gut and is played with the fingertips, giving it a warmer, mellower sound than its steel-strung counterpart. To make sure that the instrument's full range is picked up, place the mic closer to the center of the bridge, at a distance of between 6" and 1'.

Miking Near the Sound Hole

The sound hole (located at the front face of a guitar) serves as a bass port, which resonates at lower frequencies (around 80–100 Hz). Placing a mic too close to the front of this port might give a sound that's boomy and unnatural. However, miking close to the sound hole is often popular on stage or around high acoustic levels because the guitar's output is highest at this position. To achieve a more natural pickup under these conditions, the microphone's output can be rolled off at the lower frequencies (5–10 dB at 100 Hz).

The Electric Guitar

The fundamentals of the average 22-fret guitar extend from E2 to D6 (82 Hz to 1174 Hz), with overtones that extend much higher. All of these frequencies might not be amplified, as the guitar chord tends to attenuate frequencies above 5 kHz (unless the guitar has a built-in low impedance converter or low-impedance pickups). The frequency limitations of the average guitar loudspeaker often add to this effect, as their upper limit is restricted to below 5 or 6 kHz.

Miking the Guitar Amp

The most popular guitar amplifier used for recording is a small practice-type amp/speaker system. This amp type is designed to help the suffering high end by incorporating a sharp rise in the response range at 4–5 kHz, thus helping to give it a clean, open sound.

By far, the most popular mic type for picking up an electric guitar amp is the cardioid dynamic. A dynamic mic tends to give the sound a full-bodied character without picking up extraneous amplifier noises. Often guitar mics will have a pronounced presence peak in the upper frequency range, giving the pickup an added clarity.

For increased separation, a microphone can be placed at a working distance of 2" to 1'. When miking at a distance of less than 4", mic/speaker placement becomes slightly more critical (Figure 4.42). For a brighter sound, the mic should face directly into the center of the speaker's cone. Placing it off-center to the cone produces a more mellow sound while reducing amplifier noise.

Figure 4.42. *Miking an electric guitar cabinet directly in front of and off-center to the cone.*

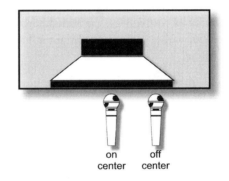

Recording Direct

A direct (D.I.) box is often used to feed the output signal of an electric guitar directly into the mic input stage of a recording console or mixer. By routing the direct output signal to a tape track, a cleaner, more present sound can be recorded (Figure 4.43a). This technique also reduces the leakage that results from having a guitar amp in the studio and even makes it possible for the guitar to be played in the control room or project studio.

A combination of direct and miked signals often results in a sound that has the "bite" of a D.I., but the characteristic richness and fullness of a miked amp. These may be combined onto one tape track; or whenever spare, open tracks exist, they can be assigned to their own, separate tracks, allowing for greater control during mixdown (Figure 4.43b).

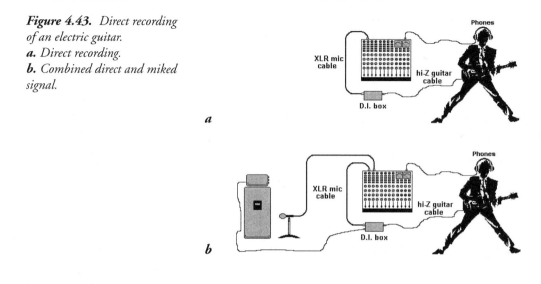

Figure 4.43. *Direct recording of an electric guitar.*
a. *Direct recording.*
b. *Combined direct and miked signal.*

The Electric Bass Guitar

The fundamentals of an *electric bass guitar* range from about E1 to F4 (41.2–343.2 Hz). If it's played loudly or with a pick, the added harmonics can range upwards to 4 kHz. Playing in the "slap" style or with a pick gives a brighter, harder attack, while a "fingered" style will produce a mellower tone.

In modern music production, the bass guitar is often recorded direct for the cleanest possible sound. As with the electric guitar, the electric bass can be either miked at the amplifier or picked up through a D.I. box. If the amp is miked, dynamic mics usually are chosen for their deep, rugged tones. The new, large-diaphragm dynamic designs tend to subdue the high-frequency transients. When combined with a boosted response at around 100 Hz, these large diaphragm dynamics give a warm, mellow tone that adds power to the lower register. Equalizing a bass can sometimes increase its clarity, with the fundamental being affected from 125–400 Hz and the harmonic punch being from 1.5–2 kHz.

One other tool that is commonly used for the electric and acoustic bass is the compressor. It's a basic fact that the signal output from the instrument's notes often varies in level, causing some notes to stand out while others dip in volume. A compressor that uses a smooth input/output ratio of roughly 4:1, a fast attack (8–20 milliseconds), and a slower release time (¼–½ second) can often smooth out these levels, giving the instrument a strong, present and smooth bass line.

Keyboard Instruments

The following section describes the various sound characteristics and miking techniques that are encountered for several keyboard instruments.

Grand Piano

The *grand piano* is an acoustically complex instrument that can be miked in a variety of ways, depending on the style and preferences of the artist, producer and/or engineer. The overall sound emanates from the instrument's strings, soundboard and mechanical hammer system. Because of its large surface area, a minimum miking distance of 4' to 6' is needed for the tonal balance to fully develop and be picked up. However, leakage from other instruments often means that these distances aren't practical or possible. As a result, pianos are often miked at distances that favor such parts of the instrument as these:

- *Strings and soundboard.* Often yields a bright and relatively natural tone.
- *Hammers.* Generally yields a sharp, percussive tone.
- *Soundboard holes alone.* Often yields a sharp, full-bodied sound.

In modern music production, two basic styles of grand pianos can be found in the recording studio: concert grands that traditionally have a rich and full-bodied tone (often used for classical music and range up to 9' in length) and studio grands that are more suited for modern music production and are designed to have a sharper, more percussive edge to their tone (often being about 7' in length).

Figure 4.44 shows a number of possible microphone positions that are acceptable for picking up of the grand piano. Although numbered mic positions are included, it's important to keep in mind that these are only guidelines from which to begin. Your own personal sound can be achieved through personal mic choice and experimentation with mic placement.

Figure 4.44. Possible miking combinations for the grand piano.

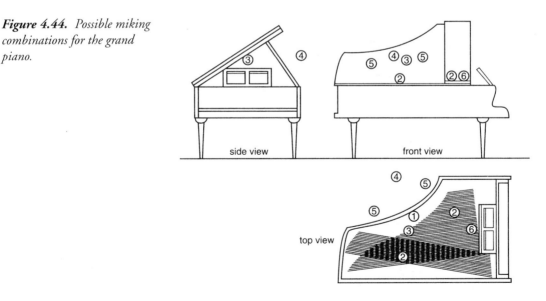

The following list explains the numbered miking positions shown in Figure 4.44:

- *Position 1.* The microphone is attached to the partially or entirely open lid of the piano. The most appropriate pickup type is the boundary mic, which can be permanently attached or temporarily taped to the lid. This method uses the lid as a collective reflector and provides excellent pickup under restrictive conditions (such as on stage and during a live video shoot).
- *Position 2.* Two mics are placed in a spaced stereo configuration at a working distance of 6" to 1'. One mic is positioned over the low strings and one is placed over the high strings.
- *Position 3.* A single mic or coincident stereo pair is placed just inside the piano between the soundboard and its fully or partially open lid.
- *Position 4.* A single mic or stereo coincident pair is placed outside the piano, facing into the open lid. This is most appropriate for the solo or accent miking of the instrument.
- *Position 5.* A spaced stereo pair is placed outside the lid, facing into the instrument.
- *Position 6.* A single mic or stereo coincident pair is placed just over the piano hammers at a working distance of 4" to 8" for a driving popular or rock sound.

A condenser or extended range dynamic mic is most often the preferred choice when miking an acoustic grand piano, as they tend to accurately represent the transient and complex nature of the instrument. Should excessive leakage be a problem, a cardioid or tighter polar pattern can be used. However, if leakage isn't a problem, an omnidirectional mic can help to capture the instrument's overall tonal balance.

Separation

Separation is often a problem that's associated with the grand piano whenever it is placed next to raucous neighbors. Separation, when miking a piano, can be achieved in the following ways:

- Place the piano inside a separate isolation room.
- Place a flat (acoustic separator) between the piano and its louder neighbor.
- Place the mics inside the piano and lower the lid onto its short stick. A heavy moving or other type of blanket can be placed over the lid to further reduce leakage.
- Overdub the instrument at a later time. In this situation, the lid can be removed or propped up by the long stick, allowing the mics to be placed at a more natural-sounding, distant position.

Upright Piano

You would expect the techniques for this seemingly harmless type of piano to be similar to those for its larger brother. This is partially true. However, because this instrument was designed for home enjoyment and not performance, the miking techniques are often different.

Since it's often more difficult to achieve a respectable tone quality when using an upright, you might want to try the following methods (Figure 4.45):

- *Miking over the top.* Place two mics in a spaced fashion just over and in front of the piano's open top—with one over the bass strings and one over the high strings. If isolation isn't a factor, remove or open the front face that covers the strings in order to reduce reflections and, therefore, the instrument's characteristic "boxy" quality. Also, to reduce resonances you might want to angle the piano about 17° relative to the wall, or move it away from the wall.

- *Miking the kickboard area.* For a more natural sound, remove the kickboard at the lower front part of the piano to expose the strings. Place a stereo spaced pair over the strings (one each at a working distance of about 8" over the bass and high strings). If only one mic is used, place the mic over the high-end strings. You might want to use caution, however, as this placement can pick up excessive foot-pedal noise.

- *Miking the upper soundboard area.* In order to reduce excessive hammer attack, place a microphone pair at about 8" from the soundboard, above both the bass and high strings. In order to reduce muddiness, the soundboard should be facing into the room or be moved away from the wall.

Figure 4.45. *One possible miking combination (over the top) for an upright piano.*

Electronic Keyboard Instruments

The signals from most electronic instruments (such as synthesizers, samplers and drum machines) are often taken directly from the device's line level output(s) and inserted into a console—either through a D.I. box or directly into a channel's line level input. Alternatively, the keyboard's output can be plugged directly into the tape machine's line level input.

The approach to miking an electronic organ can be quite different from the techniques just mentioned. A good Hammond or other older organ can sound wonderfully "dirty" through miked loudspeakers. Such organs are often played through a Leslie cabinet, which adds a unique, Doppler-based vibrato. Inside the cabinet is a set of rotating speaker baffles that spin on a horizontal axis and, in turn, produce a pitch-based vibrato as the speakers accelerate toward and away from the listener and/or mic.

The upper high-frequency speakers can be miked by either one or two mics (each panned left and right), with the low-frequency driver often being picked up by one mic. Motor and baffle noises can produce quite a bit of wind, possibly creating the need for a windscreen and/or experimentation with placement (Figure 4.46).

Figure 4.46. Miked rotating speakers of a Leslie cabinet.

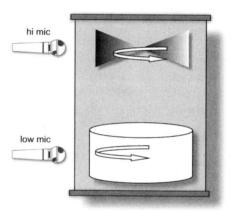

Percussion

The following section describes the different sound characteristics and miking techniques that are encountered for drums and other percussion instruments.

Drum Set

The standard drum kit (Figure 4.47) is often at the foundation of modern music; it provides the "heartbeat" of the basic rhythm track. Consequently, a proper drum sound is extremely important to the outcome of most music projects.

Generally, the drum kit is composed of the kick drum, snare drum, high-toms, low-tom (one or more), hi-hat and a variety of cymbals. A full drum kit is a series of interrelated and closely spaced percussion instruments, so it often takes real skill to translate the proper spatial and tonal balance onto tape. The larger-than-life driving sound of the acoustic rock drum set that we've all become familiar with is the result of an expert balance among playing techniques,

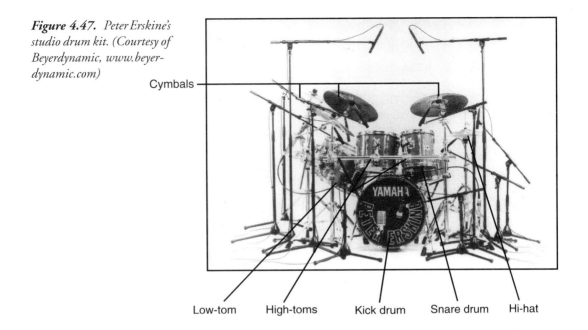

Figure 4.47. *Peter Erskine's studio drum kit. (Courtesy of Beyerdynamic, www.beyerdynamic.com)*

Cymbals

Low-tom   High-toms   Kick drum   Snare drum   Hi-hat

proper tuning and mic placement. Should any of these variables fall short, the search for that perfect drum sound could prove to be a long and hard one. As a general rule: a poorly tuned drum will sound just as out-of-tune through a good mic as it will through a bad one. Therefore, it's important to be sure that the individual drum components sound good to the ears, *before* attempting to place the mics.

Miking the Drum Set

After the drum set has been optimized for the best sound, the mics can be placed into their pickup positions (Figure 4.48). Because each part of the drum set is so different in sound and function, it's often best to treat each part as an individual instrument.

Figure 4.48. *Typical microphone placements for a drum set.*
a. Side view.
b. Front view.

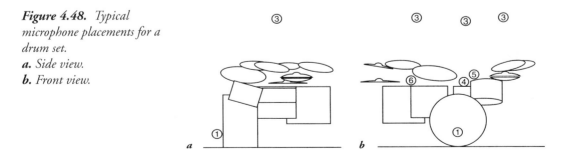

Figure 4.48. *(continued)*
c. Top view.

1. bass drum
2. mono top set
3. stereo top set
4. snare
5. high tom-tom
6. low tom-tom

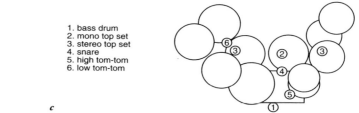

c

The following mic characteristics can be used to match a mic to the various parts of a drum: frequency response, polar response, proximity effect and transient response. Dynamic range is another important consideration when miking drums. Because a drum set is capable of generating extremes of volume and power (as well as softer, more subtle sounds), the chosen mics must be able to withstand strong peaks without distorting . . . yet still be able to capture the more delicate nuances of a sound.

Since the drum set usually is one of the loudest sound sources found in a studio setting, it's often placed on a rigidly supported 1½' riser. This reduces the amount of low-end "thud" that otherwise would be leaked through the floor into other parts of the studio. Depending on the construction, the following drum scenarios may occur:

- The drums are placed in their own room—isolated from other instruments.
- To achieve a bigger sound, the drums are placed in the large studio room while the other instruments are placed in smaller iso-rooms or are recorded direct.
- To reduce leakage, the drums are placed in the studio but are enclosed by 4' (or higher) divider flats.

Snare Drum

Generally, a snare mic is aimed just inside the top rim of the snare drum at a distance of about 1" (Figure 4.49). The mic should be angled for the best possible separation from other drums and cymbals. Its rejection angle should be aimed at either the hi-hat or rack-toms (depending on the leakage difficulties). Most often, the mic's polar response is cardioid, although a super-cardioid response offers a tighter pickup angle.

With certain musical styles (such as jazz), you might want a crisp or "bright" snare sound. This can be achieved by miking the snare's bottom head in addition to the top head and then combining the two mics onto a single track. As the bottom snare head is 180° out-of-phase with the top, generally it's wise to reverse the bottom mic's phase polarity.

When playing in styles where the snare springs are turned off, it's often wise to keep your ears open for snare rattles and buzzes that can easily leak into the snare mic (as well as most of the other drum mics and even those of other instruments).

Figure 4.49. *Mic positioning for the snare drum.*

Hi-Hat

The hi-hat usually produces a strong, sibilant energy in the high-frequency range, whereas the snare's frequencies often are more concentrated in the mid-range. Although moving the hat's mic won't change the overall sound as much as it would on a snare, you should still keep the following three points in mind:

Placing the mic above the top cymbal will help pickup the nuances of sharp stick attacks.

The open and closing motion of the hi-hat will often produce rushes of air. Consequently, when miking the hat's edge, angle the mic slightly above or below the point where the cymbals meet.

If only one mic is available or desired, both the snare and hi-hat can be picked up simultaneously by carefully placing the mic between the two, facing away from the rack toms as much as possible. Alternatively, a figure-8 mic can be placed between the two with the null axis facing toward the cymbals and the kick.

Rack-Toms

The upper rack-toms can be miked either individually (Figure 4.50) or with a single mic placed between the two at a short distance (Figure 4.51). When miked individually, a "dead" sound can be achieved by placing the mic close to the drum's top head (about 1" above and 1" to 2" in from the outer rim). A sound that's more "live" can be achieved by increasing the height above the head to about 3" to 6". If isolation or feedback is a consideration, a hypercardioid pickup pattern can be chosen.

Another way to reduce leakage and to get a deep, driving tone (with less attack) is to remove the tom's bottom head and place the mic inside, 1" to 6" away from the top head.

Figure 4.50. *Individual miking of a rack-tom.*

Figure 4.51. *Single mic placement for picking up two toms.*

Floor-Tom

Floor-toms can be miked in a manner similar to that for rack-toms (Figure 4.52). The mic can be placed 2" to 3" above the top head, or it can be placed inside 1" to 6" from the head. Again, a single mic can be used between two floor-toms, or each tom can have its own mic (which would yield a greater degree of control over panning and tonal color).

Kick Drum

The kick drum adds a low energy drive or "punch" to a rhythm groove. This drum has the capability to produce low frequencies at high sound pressure levels, so it's necessary to use a mic that can both handle and faithfully reproduce these signals. Often the best choice for the job is a large-diaphragm dynamic mic.

Figure 4.52. *Typical microphone placement for the floor-tom.*

Because of the extreme proximity effect (bass boost) that occurs when using a directional mic at close working distances and because the drum's harmonics vary over its large surface area, even a minor placement change can have a profound effect on the overall sound pickup. Moving the mic closer to the head (Figure 4.53) can add a degree of warmth and fullness, while moving it farther back often emphasizes the high-frequency "click." Placing the mic closer to the beater emphasizes the hard "thud" sound, whereas an off-center mic captures more of the drum's characteristic skin tone.

Figure 4.53. *Placing the mic at a distance just outside the kick drum head brings out the low-end and natural fullness.*

By placing a blanket or other damping material inside the drum shell firmly against the beater head, a dull and loose kick sound can be tightened to produce a sharper, more defined transient sound. Cutting back on the kick's equalization at 300–600 Hz can help reduce the dull "cardboard" sound, whereas boosting at from 2.5–5 kHz adds a sharper attack, "click" or "snap." It's also often a good idea to have a can of WD-40 or other light oil handy in case squeaks from some of the moving parts (most often the kick pedal) gets picked up by the mics.

Overheads

Overhead mics generally are used to pick up the high-frequency transients of cymbals with crisp, accurate detail, while also providing an overall blend of the entire drum kit. Because of the cymbals' transient nature, a condenser mic is often chosen for its accurate high-end response.

Overhead mic placement can be very subjective and personal. One type of placement is the spaced pair, in which two mics are suspended above the left and right sides of the kit. These mics are equally distributed, so as to pick up their respective cymbal and overall instrument components in a balanced fashion (Figure 4.54a). Another placement method is to suspend the mics closely together in a coincident fashion (Figure 4.54b). This often provides for an excellent overhead stereo image with a minimum of the phase cancellations that might otherwise result from the use of spaced overhead mics.

Figure 4.54. Typical stereo overhead pickup positions.
a. *Spaced pair technique.*
b. *X-Y coincident technique.*

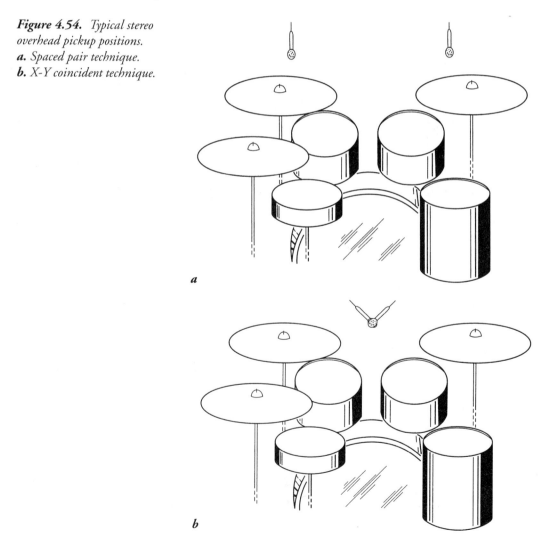

a

b

Again, it's important to remember that there are no rules for getting a good sound. If only one mic is available, place it at a central point over the drums; or, if you're close miking the individual drums, there may be times when you might not use overheads at all (the leakage spillover just might be enough to do the trick).

Tuned Percussion Instruments

The following section describes the various miking techniques that are encountered for tuned percussion instruments.

Congas and Hand Drums

Congas, tumbas and bongos are single-headed, low-pitched drums that can be individually miked at very close distances of 1" to 3" inches above the head and 2" in from the rim or the mics can be pulled back to a distance of 1' for a fuller, more "live" tone. Alternatively, a single mic or XY stereo pair can be used to pick up the drums (which are often played in pairs), at a placement point of about 1' above and between the two heads.

Another class of single-headed, low-pitched drums (known as hand drums) aren't necessarily played in pairs, but are often held in the lap or are strapped across the player's front. Although they can be as percussive as congas, these expressive instruments are often deeper in tone and will often require that the mic(s) be backed off in order to allow the sound to develop and/or fully interact with the room. In general, a good pickup can be achieved by placing a mic at a distance of 1' to 3' in front of the hand drum's head. Since a large part of the drum's sound (especially its low-end power) comes from its back hole, another mic can be placed in front of this port at a distance of 6" to 2'. Since the rear sound will be 180° out-of-phase from the front pickup, the mic's phase should be reversed whenever the two signals are combined.

Xylophone, Vibraphone and Marimba

The most common way to mic a tuned percussion instrument is to place two high-quality condenser or extended-range dynamic pickups above the playing bars and at a spaced distance that's appropriate to the instrument size (following the 3:1 general rule). A coincident stereo pair can help eliminate possible phase errors; however, a spaced pair will often yield a wider stereo image.

Stringed Instruments

Of all the instrumental families, the stringed instrument is perhaps the most diverse. Ethnic music often uses instruments that range from single-stringed to those that use highly complex and developed systems to produce rich and subtle tones. Western listeners have grown accustomed to hearing the violin, viola, cello and double bass (both as solo instruments and in an ensemble setting). Guitars abound as varieties of four-, six-, and twelve-stringed instruments.

Whatever the type, stringed instruments vary in their design type and in construction to enhance or cut back on certain harmonic frequencies. These variations are what give a particular stringed instrument its own characteristic sound.

Violin and Viola

The frequency range of the violin runs from 200 Hz to 10 kHz. For this reason, a good mic that displays a relatively flat frequency response should be used. The violin's fundamental range is from G3 to E6 (200 Hz to 1300 Hz), and it's particularly important to use a mic that is flat within the formant frequencies of 300 Hz, 1 kHz, and 1200 Hz. The fundamental range of the viola is tuned a fifth lower and contains fewer harmonic overtones.

In most situations, the violin or viola's mic should be placed on-axis to the instrument's front face. The distance will depend on the particular style of music and the room's acoustic condition. Miking at a distance will generally yield a mellow, well-rounded tone, whereas a closer position might pick up a scratchy, more nasal quality—the choice will depend on the instrument's tone quality.

For a solo instrument, a recommended miking distance is between 3' and 8', over and slightly in front of the player (Figure 4.55). Under studio conditions, a closer mic distance of between 2' and 3' is recommended. For a fiddle or jazz/rock playing style, the mic can be placed at a close working distance of 6" or less, as the increased overtones help the instrument to cut through an ensemble.

Figure 4.55. Example of a typical mic placement for the violin.

Under PA (public address) applications, distant working conditions are likely to produce feedback. In this situation, a clip-type microphone can be attached to the instrument's tailpiece. At such close working distances, less amplification is needed, thereby reducing the possibility of feedback.

Cello

The fundamental range of the cello is from C2 to CS (56 Hz to 520 Hz), with overtones up to 8 kHz. If the player's line of sight is taken to be 0°, then the main direction of sound radiation lies between 10° and 45° to the right. A quality mic can be placed level with the instrument and directed toward the sound holes. The chosen microphone should have a flat response and be placed at a working distance of between 6" and 3'.

Double Bass

The double bass is one of the orchestra's lowest-pitched instruments. The fundamentals of the four-string type reach down to E1 (41 Hz) and up to around middle C (260 Hz). The overtone spectrum generally reaches up to 7 kHz, with an overall angle of high-frequency dispersion being ±15° from the player's line of sight. Once again, a mic can be aimed at the "f" holes at a distance of between 6" and 1½'.

Voice

From a shout to a whisper, the human voice is a talented and versatile sound source, which displays a dynamic and timbral range that's matched by few other instruments. The male bass voice can ideally extend from E2 to D4 (82 Hz to 293 Hz) with sibilant harmonics extending to 12 kHz. The upper soprano voice can range upward to 1050 Hz with harmonics also climbing to 12 kHz.

The engineer/recordist should be aware of the following four possible traps that might be encountered when recording the human voice:

* *Excessive dynamic range.* This can be solved either by mic technique (physically moving away from the mic during loud passages) or by inserting a compressor into the signal path. Some vocalists might have dynamics that can range from a normal tone to practically screaming . . . all within a single passage. If you optimize your recording levels during a moderate-volume passage and the singer begins to belt his or her lines, the levels will become way too hot and will distort. Conversely, if you set your recording levels for the loudest passage, the moderate volumes will barely be heard at all and will be buried in the music. The solution is to place a compressor in the mic's signal path. The compressor automatically "rides" the signal's gain and reduces excessively loud passages to a level that the system can effectively handle. (See Chapter 12 for more information about compression and dynamic range altering devices.)

* *Sibilance.* This occurs when sounds such as "f," "s," and "sh" are overly accentuated. This often is a result of tape saturation and distortion at high levels or slow tape speeds.

Sibilance can be reduced when going to tape by inserting a frequency-selective compressor (known as a de-esser) into the chain.

- *Popping.* Explosive popping "P" sounds result from turbulent air puffs from the mouth striking the mic diaphragm. This problem can be avoided or reduced by placing a windscreen over the mic, by placing a pop filter windshield between the mic and the vocalist, or by using an omnidirectional mic (which is less sensitive to popping).

- *Excessive bass boost due to proximity effect.* This bass buildup often occurs when a directional mic is used at close working ranges. It can be reduced or compensated for in the following three ways: by increasing the working distance between the source and the mic, by using an omnidirectional mic (which doesn't display a proximity bass build up) or through equalization.

Woodwind Instruments

The flute, clarinet, oboe, saxophone and bassoon combine to make up the woodwind class of instruments. Not all modern woodwinds are made of wood, nor do they produce sound in the same way. The sound of the flute is generated by blowing across the hole in a tube, while others produce sound by using a reed to vibrate a column of air.

Historically, the woodwind's pitch was controlled by opening or covering finger holes along the sides of the instrument. This changed the length of the tube and, as such, the length of the vibrating air column. As instrumentation sophistication grew, the Boehm system of pads and levers further developed these instruments into their present forms.

It's a common misunderstanding that the natural sound of a woodwind instrument radiates entirely from its bell or mouthpiece. In reality, a large part of its sound often emanates from the fingerholes that span the instrument's entire length.

Clarinet

The clarinet comes in two pitches: the B clarinet with a lower limit of D3 (147 Hz), and the A clarinet with a lower limit of C3 (139 Hz). The highest fundamental is around G6 or 1570 Hz, whereas notes an octave above middle C contain frequencies of up to 1500 Hz when played softly. This spectrum can range upward to 12 kHz when played loudly.

The sound of this reeded woodwind radiates almost exclusively from the finger holes at frequencies between 800 Hz and 3 kHz; however, as the pitch rises, more of the sound emanates from the bell. Often, the best mic placement occurs when the pickup is aimed toward the lower finger holes at a distance of 6" to 1' (Figure 4.56).

Flute

The fundamental range of the flute extends from B3 to about C7 (247 Hz to 2100 Hz). For medium loud tones, the upper overtone limit ranges between 3 and 6 kHz. Commonly, the

Figure 4.56. *Typical mic position for the clarinet.*

instrument's sound radiates along the player's line of sight for frequencies up to 3 kHz. Above this frequency, however, the radiated direction often moves outward 90° to the player's right.

When miking a flute, placement depends on the type of music being played and the room's overall acoustics. When recording classical flute, the mic can be placed on-axis and slightly above the player at a distance of between 3' and 8'. When dealing with modern musical styles, the distance often ranges from 6" to 2'. In both circumstances, the microphone should be positioned at a central point between the mouthpiece and the instrument's footpiece. In this way, the instrument's overall sound and tone quality can be picked up with equal intensity (Figure 4.57). Placing the mic directly in front of the mouthpiece generally reduces feedback and leakage; however, breath noise is accentuated without getting the full overall body sound.

Figure 4.57. *Typical mic position for the flute.*

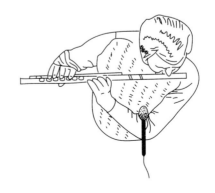

If mobility is important, a clip microphone can be secured near the mouthpiece or a specially designed contact pickup can be integrated into the instrument's headpiece.

Saxophone

Saxophones can vary greatly in size and shape. The most popular sax for rock and jazz is the S-curved B-flat tenor sax, whose fundamentals span from B2 to F5 (117 Hz to 725 Hz), and the E-flat alto, which spans from C3 to G5 (140 Hz to 784 Hz). Also within this family are the straight-tubed soprano and sopranino, as well as the S-shaped baritone and bass saxophones. The harmonic content of these instruments can range up to 8 kHz and can be extended by breath noises up to 13 kHz.

As with other woodwinds, the mic should be placed roughly in the middle of the instrument at the desired distance and pointed slightly toward the bell (Figure 4.58). Keypad noises are considered to be a part of the instrument's sound; however, even these can be reduced or eliminated by aiming the microphone closer to the bell's outer rim.

Figure 4.58. *Typical mic positions for the saxophone.* *a.* *Standard mic placement.*

a

b. Typical "clip-on" mic placement.

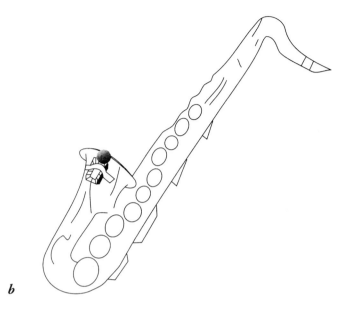

b

Harmonica

Harmonicas come in all shapes, sizes, and keys and are divided into two basic types: the diatonic and the chromatic. The actual pitch is determined purely by the length, width and thickness of the various vibrating metal reeds.

The "harp" player's habit of forming his or her hands around the instrument is a way to mold the tone by forming a resonant cavity. The tone can be deepened or a special "wahing" effect can be produced by opening and closing a cavity that's formed by the palms. Consequently, many harmonica players carry their preferred microphones with them (Figure 4.59) rather than being stuck in front of an unfamiliar microphone and stand.

Figure 4.59. The Shure 520DX "Green Bullet" microphone is a preferred harmonica pickup for many musicians. (Courtesy of Shure Brothers, Inc., www.shure.com)

Microphone Selection

The following list can be used to provide insights into a small number of professional mics that are used for music recording and pro sound applications. This list is by no means complete, as literally hundreds of mics are available, each with its own particular design, sonic character and application.

Shure SM-57

The SM-57 (Figure 4.60) is widely used by engineers, artists, touring sound companies, etc., for instrumental and remote recording applications. The SM-57's mid-range presence peak and good low-frequency response make it useful for use with vocals, snare drums, toms, kick drums, electric guitars, and keyboards.

Figure 4.60. Shure SM-57 dynamic mic. (Courtesy of Shure Brothers, Inc., www.shure.com)

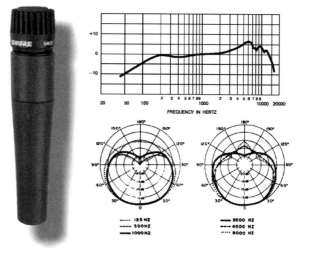

Specifications

Transducer type: Moving-coil dynamic

Polar response: Cardioid

Frequency response: 40–15,000 Hz

Equivalent noise rating: −7.75 dB (0 dB = 1 V/microbar)

Audix D2

This dynamic hypercardioid drum and instrument microphone (Figure 4.61) has warm contoured response for added bottom and punch on drums, instruments and brass. Features VLM noise rejection capsule and compact design for easy placement and is milled from a solid block of aluminum.

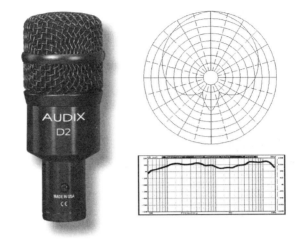

Figure 4.61. *Audix D2 dynamic mic. (Courtesy of Audix Corporation, www.audixusa.com)*

Specifications

Transducer type: Moving-coil dynamic

Polar response: Hypercardioid

Frequency response: 44–18,000 Hz

Maximum SPL rating: 144 dB

AKG D112

Large-diaphragm cardioid dynamic mics, such as the AKG D112 (Figure 4.62), are often used for picking up kick drums, bass guitar cabinets, and other low-frequency, high-output sources.

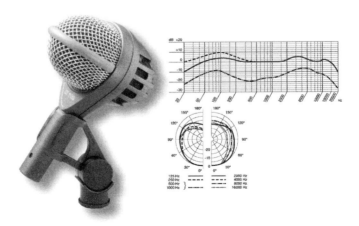

Figure 4.62. *AKG D112 dynamic mic. (Courtesy of AKG Acoustics, Inc., www.akg-acoustics.com)*

Specifications

> *Transducer type:* Moving-coil dynamic
> *Polar response:* Cardioid
> *Frequency response:* 30–17,000 Hz
> *Sensitivity:* –54 dB ± 3 dB re. 1 V/microbar

Beyerdynamic M-160

The Beyer M-160 ribbon microphone (Figure 4.63) is capable of handling high sound-pressure levels without sustaining damage, while providing the transparency that often is inherent in ribbon mics. Its hypercardioid response yields a wide-frequency response/low-feedback characteristic for both studio and stage.

Figure 4.63. *Beyerdynamic M-160 ribbon mic. (Courtesy of Beyerdynamic, www.beyerdynamic.com)*

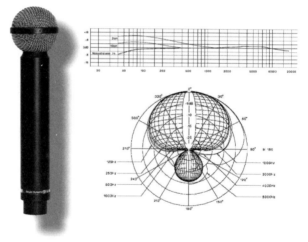

Specifications

> *Transducer type:* Ribbon dynamic
> *Polar response:* Hypercardioid
> *Frequency response:* 40–18,000 Hz
> *Sensitivity:* 52 dB (0 dB = 1 mW/Pa)
> *Equivalent noise rating:* –145 dB (0 dB = m W/2. 10–5 Pa)
> *Output impedance:* 200 Ω

Royer Labs R-121

The R-121 is a ribbon mic with a figure-8 pattern (Figure 4.64). Its sensitivity is roughly equal to that of a good dynamic mic, and it exhibits a warm, realistic tone and flat frequency response. Made using advanced materials and cutting-edge construction techniques, its response is flat and well balanced; low end is deep and full without getting boomy, mids are well defined and realistic, and the high end response is sweet and natural sounding.

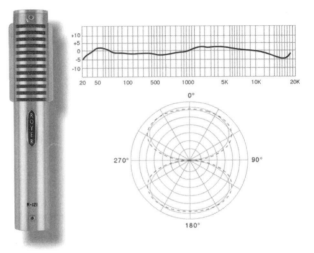

Figure 4.64. *Royer Labs R-121 ribbon mic. (Courtesy of Royer Labs, www.royerlabs.com)*

Specifications

Acoustic operating principle: Electrodynamic pressure gradient

Polar pattern: Figure-8

Generating element: 2.5-micron aluminum ribbon

Frequency response: 30–15,000 Hz ± 3 dB

Sensitivity: –54 dBV Re. 1 V/Pa ± 1 dB

Output impedance: 300 Ω at 1K (nominal), 200 Ω optional.

Maximum SPL: >135 dB

Output connector: Male XLR 3 pin (Pin 2 Hot)

Finish: Burnished satin nickel/matte black chrome optional.

AEA R44C

The AEA R44C ribbon mic (Figure 4.65) has the authentic sound, feel and look of the RCA 44B, and all parts interchange. The bidirectional ribbon element is only 1.8 microns thick, fabricated from NOS (New Old Stock) material that was originally manufactured for RCA. A black/bright chrome "radio" finish is standard. The microphone package includes a cushion mount and AEA custom vertical storage and shipping case. Each R44C is built to order and is individually numbered and signed.

Figure 4.65. AEA R44C ribbon mic. (Courtesy of Audio Engineering Associates, www.wesdooley.com)

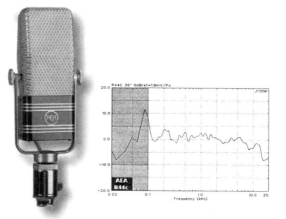

Neumann KM 180 Series

The "Series 180" consists of three compact miniature microphones (Figure 4.66): the KM 183 omnidirectional and KM 185 hypercardioid microphones and the successful KM 184 cardioid microphone. All "Series 180" microphones are available with either matte black or nickel finish and come in a folding box with a windshield and two stand mounts that permit connection to the microphone body, or the XLR-connector.

Figure 4.66. The Neumann KM 180 Series condenser mics. (Courtesy of Georg Neumann GMBH, www.neumann.com)

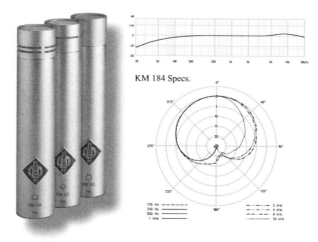

Specifications

Transducer type: Condenser

Polar response: Cardioid (183), cardioid (184) and hypercardioid (185)

Frequency response: 20 Hz–20 kHz

Sensitivity: 12/15/10 mV/Pa

Output impedance: 50 Ω

Equivalent noise level: 16/16/18 dB –A

AKG C3000B

The AKG C3000B (Figure 4.67) is a low-cost, large-diaphragm condenser mic. Its design incorporates a bass rolloff switch, a –10-dB pad and a highly effective internal windscreen. The mic's dual-diaphragm capsule design is floated in an elastic suspension for improved rejection of mechanical noise.

Figure 4.67. *The AKG C3000B condenser mic. (Courtesy of AKG Acoustics, Inc., www.akg-acoustics.com)*

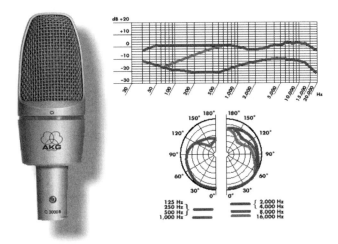

Specifications

Transducer type: Condenser

Polar response: Cardioid

Frequency response: 20–20,000 Hz

Sensitivity: 25 mV/Pa (–32 dBV)

Marshal MXL 2001

This low-cost, large capsule, gold diaphragm condenser (Figure 4.68) has all the fullness and warmth characteristic of the classic large capsule mics. It has a realistic, natural tone that makes it a good, all-around mic for both vocal and instrumental recording.

Figure 4.68. *The Marshal MXL 2001 condenser mic. (Courtesy of Marshall Electronics, Inc., www.mars-cam.com)*

Specifications

Transducer type: Condenser

Polar response: Cardioid

Frequency response: 30–20kHz

Sensitivity: 15 mV/Pa

Output impedance: 200 Ω

Equivalent noise level: –20 dB –A

Beyerdynamic MCD-100

The MCD 100 digital microphone (Figure 4.69) represents the optimum conversion of an analog signal into the digital domain. Because of the built-in 22-bit converter, it is possible to more accurately measure and reproduce the performance of the capsule. The microphone features a standard AES/EBU output via a standard 3-pin XLR connector. In order to provide Digital Phantom Powering (DPP: 6–10 V, 150 mA) and to facilitate connection to a digital console, there is a range of separate power supplies. For high SPL applications (140/150 dB SPL) a remote attenuator (–10/–20 dB) controlled from the DPP device is provided.

Figure 4.69. *Beyerdynamic MCD-100 digital Condenser mic. (Courtesy of Beyerdynamic, www.beyerdynamic.com)*

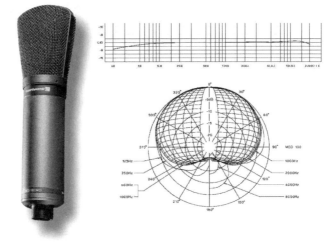

Specifications

Transducer type: Condenser

Polar response: Cardioid

Frequency response: 20–20,000 Hz

CHAPTER

The Analog Audio Tape Recorder

Over the course of the 1940s through the mid-1990s, the professional and project recording studio relied on magnetic media for the storage of analog sound onto reels of tape. Although analog recording is still highly regarded and even sought after by many studios, its use has steadily dwindled to the point that only a few new analog tape machine models are currently on the market.

With the ongoing debate of the merits of analog versus digital, I think it's fair to say that each has its own distinct type of sound and application in audio and music production. Although professional analog recorders are usually much more expensive than their cost-effective digital counterparts, as a general rule, properly aligned, professional analog "decks" will often have their own particular sound, often described as being "full," "punchy," "gutsy" and "raw" (when used on drums, vocals, entire mixes or almost anything that you want to throw at it). It's easy to see and hear why the analog tape recorder isn't dead yet . . . and probably won't be for some time.

Like its digital counterpart, the analog audio tape recorder (or ATR) can be thought of as a sound recording device that has the capacity to store audio information and, on request, play this information back using magnetic energy as the medium for recording analog signals onto magnetic tape. By definition, analog refers to something that's "analogous" or similar to and comparable in certain respects to something else. An analog ATR is called "analog" because of its ability to transform an electrical input signal into corresponding magnetic energy that can be stored onto tape in the form of magnetic remnants. On playback this magnetic energy can be reconverted back into a corresponding electrical signal that can be amplified, mixed and processed.

Magnetic Recording Media

In present-day production practices, audio and other forms of information can be recorded onto a wide range of magnetic medium types (including linear tape and hard disk). When speaking of analog recording, the tape itself is composed of several layers of material, each serving a specific function (Figure 5.1). The base material that makes up most of a tape's thickness is often composed of polyester or polyvinyl chloride (PVC), which is a durable polymer that's physically strong and can withstand a great deal of abuse before being damaged. Bonded to the PVC base is the all-important layer of magnetic oxide. The molecules of this oxide work together to create some of the smallest known permanent magnets, which are called domains (Figure 5.2). On an unmagnetized tape, these domains are oriented randomly over the entire surface of the tape. The net result of this random magnetization is a general cancellation of the north and south magnetic poles of each domain at the reproduce head, resulting in no signal at the recorder's output.

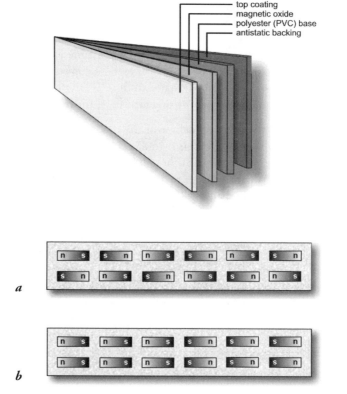

Figure 5.1. *Structural layers of magnetic tape.*

top coating
magnetic oxide
polyester (PVC) base
antistatic backing

Figure 5.2. *Orientation of magnetic domains on unmagnetized and magnetized recording tape.*
a. *The random orientation of an unmagnetized tape results in no output.*
b. *Magnetized domains result in an average flux output at the magnetic head.*

When a signal is recorded, the magnetization from the record head orients the individual domains (at varying degrees in positive and negative angular directions) in such a way that their average magnetism produces a much larger combined magnetic flux. When played back at the same, constant speed at which it was recorded, this alternating magnetic output can then be amplified and further processed for reproduction.

The Professional Analog ATR

Professional analog ATRs can be found in 2-, 4-, 8-, 16- and 24-track formats. Each configuration is generally best suited to a specific production and postproduction task. For example, 2-track ATRs are generally used to record the final, stereo mix of a project, whereas 8-, 16- and 24-track machines are usually used for multitrack recording. Although only a few professional analog machines are currently being manufactured (most notably, the Otari MTR-90 2" 24-track recorders), quite a few analog decks can be found on the used market, in varying degrees of working condition. Several examples of both new and used machines can be found in Figures 5.3 through 5.5.

Figure 5.3. Otari MX-5050 B3 2-channel recorder. (Courtesy of Otari Corporation, www.otari.com)

Figure 5.4. *Refurbished 1"
Ampex ATR-102 stereo
mastering recorder. (Courtesy of
ATR Service Company,
www.atrservice.com)*

Figure 5.5. *Otari MTR-90
MkIII 24-track master recorder.
(Courtesy of Otari Corporation,
www.otari.com)*

The Tape Transport

The process of recording audio onto magnetic tape depends on the transport's capability to pass the tape across a head path at a constant speed and with uniform tension. In other words, recorders must uniformly pass a precise length of tape over the record head within a specific period of time (Figure 5.6). During playback, the same time relationship is maintained by again moving the tape across the heads at the same speed, thereby preserving the original pitch, rhythm and duration.

This constant speed and tension movement of the tape across a head's path is initiated by simply pressing the Play button. The drive can be disengaged at any time by pressing the Stop button, which applies a simultaneous breaking force to both the left and right reels. The Fast Forward and Rewind buttons causes the tape to rapidly shuttle in their respective directions to

Figure 5.6. *Relationship of time to the physical length of recording tape.*

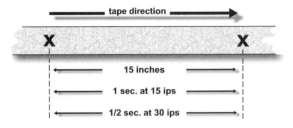

locate a specific point in the tape. Initiating either of these modes engages the tape lifters, which raise the tape away from the heads. Once the play mode has been engaged, pressing the Record button allows audio to be recorded onto any selected track or tracks.

Beyond these basic controls, you might expect to run into several differences between transports (often depending on the machine's age). For example, older recorders will often require that both the Record and Play buttons be simultaneously pressed in order to go into record; while others can begin recording when the Record button is pressed while already in the Play mode. Figure 5.7 shows the elements of the transport deck of a Fostex R8 ¼" multitrack recorder.

Figure 5.7. *Transport deck of Fostex R8 ¼" multitrack recorder. (Courtesy of Fostex Corp. of America, www.fostex.com)*

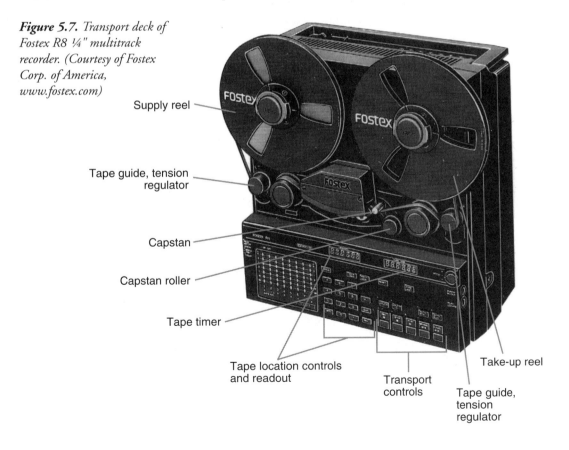

On certain older professional transports (particularly those wonderful Ampex decks from the 1950s and '60s), stopping a fast-moving tape by simply pressing the Stop button can stretch or ruin a master tape, because the inertia is simply too much for the mechanical brake design to deal with. A procedure known as "rocking" the tape is used to prevent such damage. A tape can be rocked to its stop position by engaging the fast-wind mode in a direction that's opposite to the current travel direction until the tape slows to a reasonable speed; then it's safe to press the Stop button.

In recent decades, tape transport designs have incorporated Total Transport Logic (TTL), which places monitoring and overall transport function under complete microprocessor control. This innovation has a number of distinct advantages. For example, TTL logic lets you push the Play or Stop buttons while the tape is in either fast-wind mode without fear of tape damage. With TTL, the recorder can sense the tape speed and direction and then automatically rock the transport until the tape can safely be stopped or slow the tape to a point where the deck can automatically slip into play or record mode.

Most modern ATRs are equipped with a shuttle control that allows the tape to be shuttled at various wind speeds in either direction. This allows a specific cue point to be located by listening to the tape at varying play speeds or to wind the tape onto its reel at a slower speed for long-term storage. The Edit button (which can be found on certain pro machines) often has two operating modes: stop-edit and dump-edit. If the Edit button is pressed while the transport is in the stop mode, the left and right tape reel brakes are released and the tape sensor is bypassed. This makes it possible for you to manually rock the tape back and forth until the edit point is found. Often, if the Edit button is pressed while in the play mode, only the take-up turntable is disengaged and the tape sensor is bypassed, allowing unwanted sections of a tape to be spooled off the machine (and into the trash can) while you listen to the material that's being dumped.

A safety switch, which is incorporated into all professional transports, initiates the stop mode when it senses the absence of tape along its guide path. Thus, the recorder stops automatically at the end of a reel or should the tape accidentally break. This switch can be built into the tape-tension sensor, or it might exist in the form of a light beam that's interrupted when tape is present.

Most professional ATRs are equipped with automatic tape counters that accurately readout time in hours, minutes and seconds (00:00:00). Many of these recorders have digital readout displays that can double as tape-speed indicators when in the "varispeed" mode. This function incorporates a continually variable control that lets you vary tape speed from the fixed industry standards. On many tape transports, this control lets you vary the speed over a ± 20 percent range from the 7½, 15 or 30 ips standard.

The Magnetic Tape Head

Most professional analog recorders use three magnetic tape heads, each of which perform a specialized task:

- Record
- Reproduce
- Erase

The function of a record head (Figure 5.8) is to electromagnetically translate the analog input signal into corresponding magnetic fields that can be permanently stored onto magnetic tape. The input current flows through coils of wire, which are wrapped around the head's magnetic pole pieces. This causes magnetic force to flow through the pole pieces and across the head gap. Like electricity, magnetism flows more easily through some media than through others. The head gap between poles creates a break in the magnetic field, thereby creating a physical resistance to the magnetic force that's been set up. Since the gap is in direct contact with the moving magnetic tape, the tape's magnetic oxide offers a lower resistance path to the field than does the nonmagnetic gap. Thus, the flux path travels from one pole piece, through the tape, to the other pole. The actual recorded signal occurs at the trailing edge of the record head (with respect to tape motion) because the magnetic domains retain the same polarity and magnetic intensity that they had on leaving the head gap.

Figure 5.8. *The record head.*

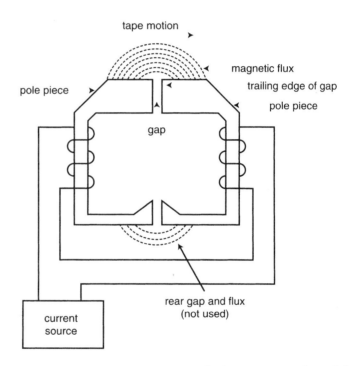

The reproduce or playback head (Figure 5.9) operates in a way that's opposite to that of the record head. When a recorded tape track passes across the reproduce head gap, a magnetic flux is induced into the pole pieces. This alternating field induces a current into the pickup coils that is then amplified and processed into a larger output signal.

The reproduce head's output is nonlinear because this signal is proportional to both the tape's average flux magnitude and the rate of change of this magnetic field. This rate of change increases as a direct function of and in direct proportion to the frequency of the recorded signal. Thus, the output level of a playback head effectively doubles for each doubling in frequency . . . resulting in a 6-dB increase in output voltage for each increased octave (Figure 5.10).

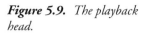

Figure 5.9. *The playback head.*

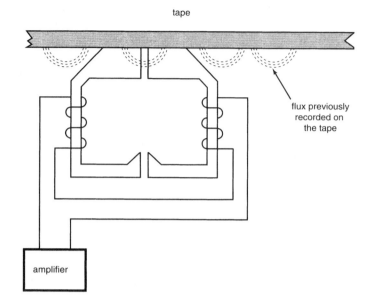

Figure 5.10. *The effects of increased frequency on the reproduced output at the magnetic head gap.*

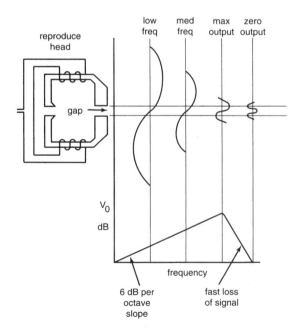

The tape speed and the length of the head gap work together to determine the reproduce head's upper-frequency limit, which in turn determines the system's overall bandwidth. The wavelength of a signal that's recorded onto tape is equal to the speed at which tape travels past the reproduce head, divided by the frequency of the signal. Therefore, the faster the tape speed, the higher the upper-frequency limit. Similarly, the smaller the head gap, the higher the upper-frequency limit. However, as the playback signal's frequency increases, more and more of the complete cycle will fall inside the head gap at any one point in time until the signal wavelength is equal to the gap width (Figure 5.11). At that point, the average output level will be zero. This reduced output (known as scan fling loss) is also a factor in determining the system's upper-frequency limit.

Figure 5.11. *Scanning loss occurs when the recorded wavelengths approach or exceed the width of the magnetic head gap.*

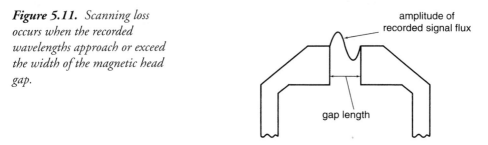

The function of the erase head is to reduce the average magnetization level of a recorded tape track to zero, thereby allowing the tape to be re-recorded. After the transport is placed into the record mode, a high-frequency and high-intensity sine wave signal is fed into the erase head (resulting in a tape that's alternately saturated in both the positive- and negative-polarity directions). Tape saturation is reached when all of the magnetic particles at the head gap are completely magnetized and an increase in magnetic force won't result in an increase in magnetism that's retained on tape. This alternating saturation serves to destroy any magnetic pattern that existed on the tape.

As the tape moves away from the erase head, the intensity of the magnetic field will decrease, leaving the domains in a random orientation, with a resulting average magnetization or output level that's as close to zero as tape noise will allow.

Equalization

Equalization (EQ) is the term used to denote an intentional change in relative amplitudes at different frequencies. Because the analog recording process isn't linear, equalization is needed to achieve a flat frequency-response curve when using magnetic tape. The 6-dB-per-octave boost that's inherent in the playback head's response curve makes it necessary to apply a complementary equalization cut of 6 dB per octave at the playback electronics (Figure 5.12).

Figure 5.12. *Flat frequency response curve due to complementary equalization.*

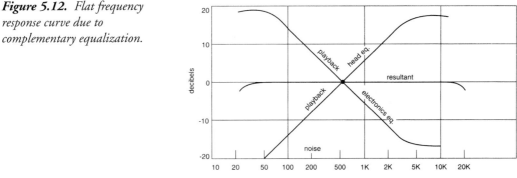

Bias Current

In addition to the nonlinear changes that occur in playback output level relative to frequency, there is another discrepancy in the recording process that exists between the amount of magnetic energy that's applied to the record head and the amount of magnetism that's retained by tape after the initial recording field has been removed.

As Figure 5.13a shows, the magnetization curve of magnetic tape is linear only between points A and B, and between points C and D. Signals greater than A (in the negative direction) and D (in the positive direction) have reached the saturation level and are subject to clipping distortion. Signals falling within the B to C range are too low in flux level to adequately affect the tape's magnetic particles. Thus, it's important to boost low-level signals at the recording head into the linear A–B and C–D ranges. This boost is generally created by mixing a bias current, or AC bias, in with the audio signal (Figure 5.13b).

Figure 5.13. *The effects of bias current on recorded linearity.*
a. *Magnetization curve showing distortion at lower levels.*
b. *After bias the signal is boosted back into the curve's linear regions.*

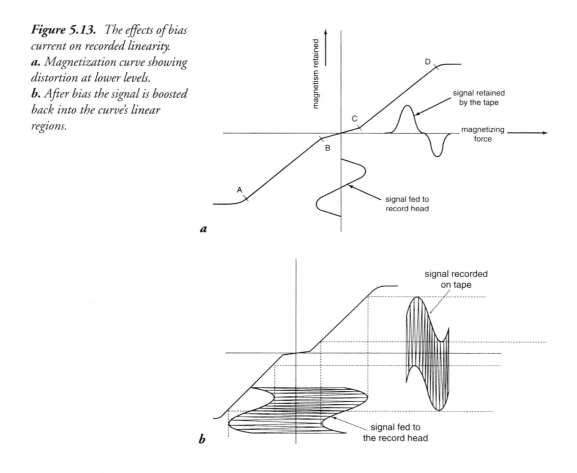

Bias current is applied by mixing the incoming audio signal to be recorded with an ultrasonic signal (often between 75 and 150 kHz). This has the effect of modulating the high-frequency bias signal with a lower-frequency carrier signal that represents the audible waveform. Since the combined energy (bias + audio signal) is much higher than the audio signal alone, the overall magnetic flux levels are given an extra "oomph," which effectively boosts the signal above the nonlinear zero-crossover range and into the linear portion of the curve. Since the reproduce head is incapable of playing back the high-frequency bias signal, only the modulated audio signal will be reproduced; the bias signal won't.

Recording Channels

No matter what the machine's track configuration, each recording channel of a modern ATR is designed to be electrically identical to the others; the channel circuitry is simply duplicated by the number of available channels.

Just as the magnetic tape head performs three functions, the electronics of an ATR must be specialized to perform the same processes: record, reproduce and erase. In older machines, each recording channel consists of a module that contains three electronics cards, with each one performing its own function. Newer ATRs now use input/output (I/O) modules, which incorporate all the adjustment controls and electronics of a recording channel on a single printed circuit board. I/O modules allow for greater channel interchangeability and service, and they give easy access to input, output, sync and equalization adjustments.

In the input (source) mode, the signal at the selected channel output is derived from its input signal. Thus, with the ATR transport in any mode (including stop), it's possible to meter and monitor all signals that are present at the input of the ATR.

In the reproduce mode, the output and metering signal is derived from the playback head. This mode can be useful in two ways: it allows previously recorded tapes to be played back, and it enables the monitoring of material off of tape while in the record mode. The latter provides an immediate quality check of the ATR's entire record and reproduce process.

The output signal of a professional ATR can be switched between three working modes: input, reproduce and sync. The sync mode is a required feature in multitrack ATRs because of the need to record material on one or more tracks while simultaneously listening to the playback of previously recorded tracks (a process known as overdubbing). Recording one or more new tracks while listening to previously recorded tracks through the reproduce head would actually cause the newly recorded track(s) to be out of sync with the others on final playback. To prevent such a time lag, the tracks to be played must be placed in the sync mode. This function actually plays the selected monitor tracks from their respective record head tracks. Since the record head is handling both the playback and recorded signals, there is no time lag and, thus, no signal delay (Figure 5.14).

In addition to control over input, reproduce and sync, the professional ATR has a selectable track function that's known as the record-enable switch. Activating this switch (which is usually labeled Safe or Ready) prevents the accidental erasure of a recorded track.

Figure 5.14. *The function of*
the sync mode.
a. *In the reproduce mode, the*
recorded signal lags behind the
monitored playback signal,
thereby creating an out-of-sync
condition.
b. *In the sync mode, the record*
head acts as both record and
playback head, bringing the
signals into sync.

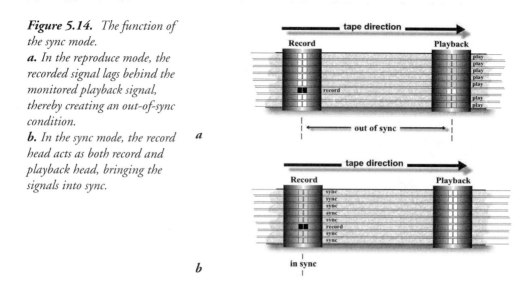

The Remote Control Unit and Autolocator

The remote control unit has evolved from being a simple control over deck transport functions to the present-day unit that contains all a recorder's transport-function and track-status controls. These units are usually located near the audio production console, which places the operating controls at the hand of the engineer.

One feature that's being built into many ATR remote control and computer control systems is the autolocator (Figure 5.15). This microprocessor-based device lets you enter important cue point locations into memory for recall at any time. When an engineer enters a cue point into a keypad or calls it from memory, the autolocator will go about the task of shuttling the tape to the desired position. At that point, the transport can be programmed to stop, place itself into the play mode or automatically audition (play–relocate–play) or punch-in (record–relocate–record) between two cue points.

Synchronizers (which are often used in high-level audio and video production) incorporate remote control and autolocator capabilities. Such a synchronization system (which is explained fully in Chapter 9) uses time code to exercise full, simultaneous control over the remote location functions of one or more transports within the production studio. Whenever multiple ATR, VTR (video tape recorders) and computer-based systems are synchronously interlocked, a single master is designated, and all other units will chase or follow the master tape position until the specified time code address has been reached.

MIDI Machine Control

The MIDI specification includes a protocol that allows the transport functions of suitably equipped analog, digital and computer-controlled media to be remotely controlled.

Figure 5.15. *Studer A827 analog multitrack recorder with autolocator. (Courtesy of Studer North America, www.studer.ch)*

Through the use of MIDI Machine Control (MMC), a wide range of transport commands can be communicated over standard MIDI lines from a central controller to a particular device or number of devices within a connected system. Basically, this means that remote transport control functions can now be executed cost-effectively from a MIDI-based central controller (such as a MIDI sequencer, digital audio workstation, mixing console or transport controller)—either manually or under automated computer control. As an example, the music production system in Figure 5.16 shows a computer-based MIDI sequencer that's been configured to be the master transport controller for an MMC-equipped 16-track analog tape recorder.

Figure 5.16. *Example of a music production system using MIDI machine control.*

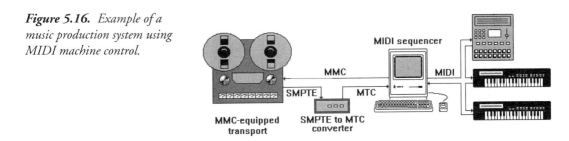

Since MMC is MIDI-based, its real beauty lies in the fact that it can be designed into software packages, controllers, audio and video recorders, mixing consoles and even musical instruments for little or no added cost to the manufacturer. As a result, this standardized protocol has grown in popularity with both professional and electronic music manufacturers alike. A more detailed description of MIDI and the MMC protocol can be found in Chapter 7.

Tape and Head Configurations

Professional analog ATRs are currently available in a wide range of track- and tape-width configurations. The most common analog configurations are 2-track mastering machines that use tape widths of ¼", ½" and even 1", as well as 16- and 24-track machines that use 2" tape. Figure 5.17 details many of the tape formats that can be currently found.

Figure 5.17. *Analog track configurations for various tape widths.*

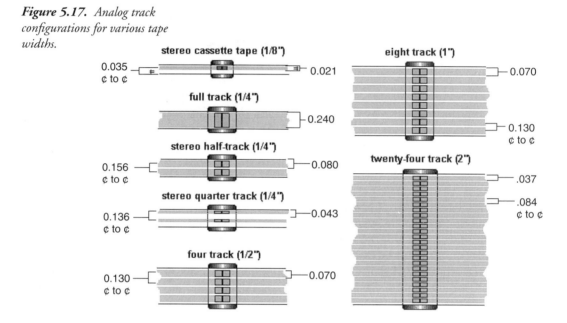

Optimal tape-to-head performance characteristics for an analog ATR are determined by several parameters: track width, head-gap width and tape speed. In general, track widths are on the order of 0.080" for a ¼" 2-track ATR; 0.070" for a ½" 4-track, 1" 8-track, and 2" 16-track formats; and 0.037" for the 2" 24-track format. These widths are large when compared to the 0.021" tracks that are commonly used in cassette recorder head design. With a greater recorded track width, an increased amount of magnetism can be retained by the magnetic tape, resulting in a higher output signal and an improved signal-to-noise ratio. The use of wider track widths also makes the recorded track less susceptible to signal-level dropouts. The guardband (an unrecorded area that exists between adjacent tracks) is used to help prevent crosstalk between channels.

Tape speed also has a direct bearing on the performance characteristics of an ATR, because it's directly related to the recorded signal's level and wavelength. At high tape speeds, the number of magnetic domains that pass over the tape head gap in a given period of time is greater than at slower speeds. Thus, the average magnetization that's received by the playback head is greater, produces a stronger output signal and requires less amplification. As a result, the resulting signal will have a greater dynamic range and less tape noise. Because higher tape speeds increase the recorded signal's wavelength, less high-frequency boosting by the record electronics will also be required, allowing higher recording levels before the tape will begin to saturate. Because of the response characteristics of the reproduce head, the recorded bandwidth will also increase at faster tape speeds.

The most common tape speeds used in audio production are 15 ips (inches per second), or 38 cm/sec, and 30 ips, or 76 cm/sec. Although 15 ips will "eat up" less tape, 30 ips has gained wide acceptance in recent years for its own particular sound (often having a tighter bottom end), as well as higher output and lower noise figures (which in certain cases has reduced or eliminated the need for noise reduction). On the other hand, 15 ips has a reputation for having a "gutsier," more rugged sound.

Print-Through

One form of deterioration in a recording's quality that can occur after a recording has been made is known as print-through. This effect is the result of the transfer of a recorded signal from one layer of tape to an adjacent track layer by means of magnetic induction, which gives rise to an audible false signal or pre-echo on playback. The effects of print-through are greatest when recording levels are very high, and the effect decreases by about 2 dB for every 1 dB reduction in signal level. The extent of this condition also depends on such factors as length of storage, storage temperature, and tape thickness (tapes with a thicker base material are less likely to have severe print-through problems).

Because of the directional properties of magnetic lines of flux, print-through has a greater effect on the outer-facing layers of tape (where magnetic induction is in-phase) than on the adjacent inner layers (where the magnetic induction is out-of-phase), as shown in Figure 5.18. Therefore, if a recorded tape is stored heads-out (with the tape being rewound onto the supply reel), a "ghost" signal will transfer to the outer layer and be heard as a pre-echo, before the original signal. For this reason, the standard means of professionally storing a recorded tape is in the tails-out position (with the tape always being wound onto the left-most take-up reel). When a tape is stored tails-out, the print-through echo will follow the original signal—a condition that allows the echo to be masked by the sound's natural decay, and that often is subconsciously perceived by the listener as reverb.

Figure 5.18. *Magnetic induction at the inner and outer layers of tape, causing print-through.*

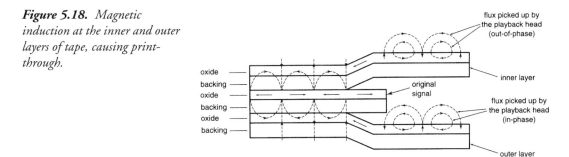

Cleanliness

It's very important that the magnetic recording heads and moving parts of an ATR transport deck be kept free from dirt and oxide shed. Oxide shed occurs when friction causes small particles of the magnetic oxide to flake off and accumulate on surface contacts. This accumulation is most critical at the surface of the magnetic recording heads, because even a minute separation between the magnetic tape and heads can cause separation loss. For example, a signal that's recorded at 15 ips, using a playback head that has an oxide shed buildup of 1 mil (0.001"), will be 55 dB below its standard level at 15 kHz. Denatured (isopropyl) alcohol or an appropriate cleaning solution should be used to clean transport tape heads and guides (with the exception of the machine's pinch roller and other rubber-like surfaces) at regular intervals.

Degaussing

Magnetic tape heads are made from a magnetically soft metal alloy, which means that the alloy is easily magnetized, but once the coil's current is removed, the core won't retain any of its magnetism. Small amounts of residual magnetism, however, will build up over time, which can act to partially erase high-frequency signals from a master tape. For this reason, it is a recommended practice that tape heads should be demagnetized after 10 hours of operation. A head demagnetizer works much like an erase head in that it saturates the magnetic head with a high-level alternating signal that randomizes the residual magnetic flux. Once a head has been demagnetized (after 5–10 seconds), it's important to move the tool away from the tape heads at a speed of less than 2" per second, so as to avoid inducing a larger magnetic flux back in the head. Before an ATR is aligned, the magnetic tape heads should always be cleaned and demagnetized in order to obtain accurate readings and to protect expensive alignment tapes.

Head Alignment

An important factor in the performance of an analog recorder is the physical alignment of each magnetic tape head. The erase, record, and playback heads will often have five adjustments: height, azimuth, zenith, wrap and rack.

The height determines the track's vertical positioning in relation to the tape path (Figure 5.19). If the track is recorded and reproduced on heads that have different height settings, not all of the recorded signal will be reproduced, resulting in a compromised signal-to-noise ratio and increased crosstalk between multitrack channels.

Figure 5.19. *The head gap's height must be centered on the track location.*

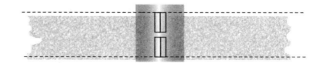

Azimuth refers to the head's tilt in the plane parallel to the tape (Figure 5.20). The head gap should be perpendicular (90°) to the tape, so that all track gaps are electrically in-phase with each other.

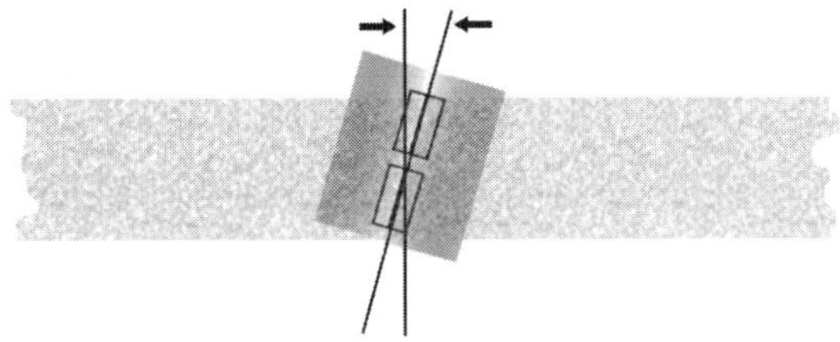

Figure 5.20. *The head gap must be perpendicular to tape travel.*

Zenith refers to the head's tilt toward or away from the tape. Zenith must be adjusted so that the tape contacts the top and bottom of the head with the same degree of force—otherwise the tape will tend to skew off its path. Skewing is bad news and occurs when tape rides up, down and even out of the tape's guide path.

Wrap refers to the angle at which the tape bends around the head. It also determines the degree of tape-to-head contact and thus controls the head's sensitivity to dropouts.

Rack determines the pressure of the tape against the head. The further forward the head is, the greater the pressure.

Although it's highly recommended that these adjustments be made by a qualified technician, height can be adjusted visually or by using a test tape in order to yield a maximum output at 1 or 3 kHz. Zenith can be tested by covering the pole pieces with a white grease pencil and playing a piece of scrap tape in order to observe the pattern that's formed as the grease pencil wears off. The edges of the wear pattern should be parallel. If they're not, use the screws on the head-block to adjust the zenith. The wrap angle can be checked at the same time as zenith by making sure that the wear pattern is centered on the gap. Rack adjustment may be needed if the wear pattern is wider on the record head than on the playback head, or vice versa.

Azimuth can be tested by deliberately skewing the tape across the heads (by pushing up and down on the edges of the tape near the front of the head), while playing back the 15-kHz tones of an older standard alignment tape. If the output increases, the azimuth will need to be adjusted in either of two ways (both of which use a full-track standard alignment tape). The first method for checking azimuth is to play the 15-kHz section of the tape and adjust the azimuth for the highest output on all channels. This will be a compromise because various channels on a multitrack machine will rise as others fall on either side of the proper setting. The peak at the correct setting will be fairly sharp while smaller, broader peaks occur on either side. To ensure that the proper peak is found, the head gap should be visually checked to make sure that it's perpendicular to the tape path before further adjustments are attempted. The second method uses the phase of a 12-kHz or 15-kHz test-tape signal in order to find the correct setting. After the peak is found (as in the first method), the output of the top channel is fed into the vertical input of an oscilloscope, while the bottom channel is fed into the horizontal input of the scope. The resulting pattern on the face of the scope represents the relative phase between the two channels. A straight line sloping up 45° to the right indicates that the two channels are in-phase, a circle indicates they are 90° out-of-phase, and a straight line sloping up 45° to the left indicates that the two channels are 180° out-of-phase. The azimuth should be adjusted so that the two channels are in-phase.

On a multitrack head, it's not possible to get all the gaps in-phase at one time because of gap scatter (variations from a vertical gap line), which occurs in manufacturing. Once the proper phase adjustment has been made for the outer tracks, the inner track phase will usually be within 60° of this figure. The record-head azimuth can be set in the same way as that of the playback-head: by playing the test tape while in the sync mode. The erase head's azimuth isn't nearly as critical, because its gap width is much wider. It will be correct as long as it's placed at the proper height and at right angles to the tape path.

Head adjustments don't have to be done frequently. Care and attention is generally indicated by the deterioration of the ATR's performance and should be performed by a qualified maintenance engineer.

Electronic Calibration

Since sensitivity, output level, bias requirements and frequency response vary considerably from one tape formulation to the next, ATRs have a number of adjustments that can be set for record/ playback level, equalization, bias current and so on. It's extremely important that a standard be adhered to in the adjustment of these level settings so that a recording made on one ATR will be compatible with another. The procedure used to set these controls to a standard level is called electronic calibration or electronic alignment.

Proper ATR alignment depends on the specific tape formulation and head alignment settings, so it's a good practice to recalibrate all production ATRs at regular or semiregular intervals. In major music recording studios, machine alignments are routinely made prior to each recording session, to ensure the standardization and performance reliability of each master tape.

Assuming that the record and reproduce heads are in proper alignment, the electronic calibration of the ATR can be performed. This procedure is carried out with reference to a standard set of equalization curves for each tape speed. Such curves have been established by the National Association of Broadcasters (NAB) in the United States, Japan and much of the world; the Deutsche Institute Norme (DIN) throughout Europe and the Audio Engineering Society (AES) is used at 30 ips. These standardized levels (Figure 5.21) are established through the use of a reproduce alignment tape, which is available in various tape speeds and track-width configurations and generally contain the following set of recorded tones:

- *Standard reference level.* A 1-kHz signal recorded at a standard reference flux level, 185 nWb/m-standard operating level, 250 nWb/m-elevated level or 320 nWb/m-DIN (European operating level).
- *Azimuth adjustment tone.* 10–16 kHz reference tone for a duration of 30 seconds.
- *Frequency response.* Tones at 16, 32, 63, 125, 250, 500, 1000, 2000, 4000, 8000, 10,000, 12,500, 16,000 and 20,000Hz.

Figure 5.21. *Pre- and postequalization curves for the NAB (top) and DIN (lower) standards.*

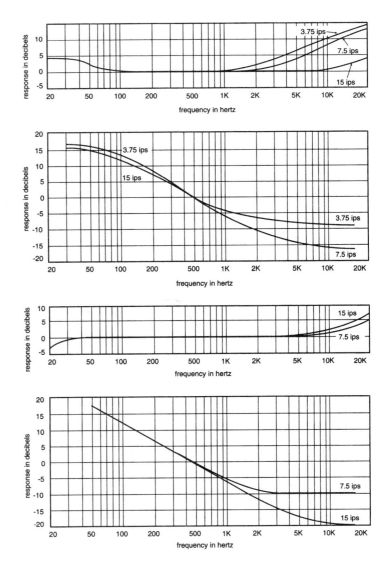

To anyone new to the alignment process, calibration might seem complicated; but with repetition, it will become second nature and usually takes less than 10 minutes to perform. Electronic calibration is carried out in two stages: playback alignment and record alignment.

Playback Alignment

To begin, place the alignment tape on the recorder in the tails-out position and rewind it onto the supply reel. In the playback mode, you can now follow the verbal instructions on the tape

and set the initial reference tone and subsequent frequencies to read at 0 VU (or other designated level) on all tracks.

Most alignment tapes are recorded in a full-track configuration; that's to say, the recorded signal
is laid down across the entire width of the tape. Because of an effect called "fringing," these test-
tape frequencies exhibit a bass boost, which ranges upward to +3 dB at 30 Hz. To avoid the
inaccuracies of fringing, the low-frequency EQ settings are postponed until after the record
alignment has been made.

The preceding process may now be duplicated with each output of the ATR being switched
into the sync mode. By adjusting the appropriate sync trim pot, all level and EQ controls can
likewise be aligned. After all the reproduce alignment adjustments have been made, rewind the
alignment test tape again and play or slow-wind it without interruption onto the take-up reel
for storage in the tails-out position.

Record Alignment

Setup for the record alignment stage consists of placing a fresh reel of unrecorded tape on the
machine. Ideally, this tape will be the same reel that is used in the actual recording session; if
not, it should at least have the same manufacturer and formulation.

The first setting to be made in the routine adjustment of an ATR's record electronics is the bias
adjustment control. This important adjustment determines the amount of AC-bias signal that's
to be mixed with the recorded signal at the record head, and has a direct relationship to the
machine's noise-to-distortion ratio. Too little bias signal will result in an increased noise floor,
increased distortion and an excessive rise in high frequencies. Too much bias results in a
reduced output at high frequencies. Thus, bias provides for a compromise among noise, distortion and frequency response . . . with the optimum setting being at some intermediate point.

To set the bias control, thread the proper tape on the machine and feed a 1-kHz signal to all
inputs of the ATR at operating level (0 VU on the console). With the track output selectors in
the reproduce mode, switch the transport to record, and use the appropriate adjustment tool to
locate the individual track controls marked Bias or Bias Adjust. The ATR's output level can
then be monitored from its VU meter bridge. Turn the bias trim counterclockwise until the test
signal drops to its lowest level. Slowly turn the trim pot clockwise until the signal rises to a peak
reading, after which the reading will begin to fall. Continue to increase the trim clockwise,
until the VU drops 1 dB in level. The bias level will now be optimized for the specific tape head
and formulation that's being used. Continue to adjust bias controls for all other channels. If the
recorded signal begins to peak off the VU scale, reduce the record input-level trim pot . . . not
the reproduce trim. This is not to be adjusted after it has been set to a standard operating level.

After the bias-adjust controls have been set, adjust the record-level and equalization controls.
With the tape rewound to the beginning of the reel, route a 1-kHz signal at 0 VU from the
console to all of the ATR's inputs. With all channel electronics switched to reproduce, place the
transport in the record mode and adjust the record level for all channels until each meter reads

0 VU. Place the ATR in the stop mode and, with all channel electronics switched to read the input signal, adjust each input level control to read 0 VU. The meter and output levels between the reproduce and input modes should now be matched. The final step is to adjust the record-equalization controls. In this adjustment, place the ATR back in the record mode. While monitoring the electronics, send tone signals of 100 Hz, 1 kHz and 10 kHz to all ATR inputs at 0 VU. The individual high-equalization (10 kHz) and low-equalization (100 Hz) trims are adjusted to 0 VU in order to attain the flattest possible frequency response over the audio spectrum. The step-by-step procedures for various speeds and tapes can be followed from Chart 5.1.

Chart 5.1 Playback and recording alignment procedures.

Playback

- Thread the playback alignment tape that has been produced especially for use at either 15 ips or 30 ips onto the machine. Always place the test tape on the take-up turntable (it's generally wise to store this tape tails-out for the purpose of an even wind tension) and rewind the tape onto the supply reel.
- Set repro level for 0 VU at 1000 Hz.
- Set high-frequency playback EQ for 0 VU at 10 kHz.

Record

- Thread the tape that's to be used during the session onto the machine. Set all output selectors to the bias position and set the machine into the record mode on all tracks.
- Adjust the erase peak control for maximum meter reading.
- Feed a 1000-Hz tone into the machine inputs.
- Set the output selector to playback (tape) and, beginning with a low bias setting, increase the amount of bias until the meter reading rises to a maximum. Continue increasing the bias level until the meter reading drops by 1 dB.
- Feed a 1000-Hz tone from the console into all machine inputs at a 0-VU (+4 dBm) level.
- Set all output selectors on the recorder to the input position and adjust the record calibrate controls for a 0-VU meter reading.
- Set the output selectors to playback (tape) and feed the console's 0-VU/1-kHz oscillator to all machine inputs. Adjust the playback levels to read as 0 VU.
- Set the console's oscillator to output a 0-VU tone at 10 kHz. Adjust the playback levels to read as 0 VU (at the peak of the meter swing, if the needle is unsteady).
- Feed a 100-Hz tone into the machine inputs at 0 VU level. Adjust the low-frequency playback EQ for a 0-VU meter reading.

CHAPTER

Digital Audio Technology

Over the past decade, digital audio has evolved from being an infant technology, available to only a few, to its present position as a primary driving force in audio production. In fact, digital audio (along with electronic music production) has had a direct and profound impact on the art, application and technology of every aspect of the audio industry.

At its most basic level, digital audio theory isn't difficult to understand. It's simply a means of encoding, processing and reproducing numeric representations of analog signal levels over time through the use of the binary number system.

Just as English-speaking humans communicate by combining any of 26 letters together into groups known as *words*, and by manipulating numbers using the decimal (base$_{10}$) system . . . the system of choice for a digital device is the binary (base$_2$) system. This numeric system provides a fast and efficient means for manipulating and storing digital data. By translating the alphabet, base$_{10}$ numbers or other form of information into binary form (represented as on/off, voltage/no voltage, magnetic flux/no flux or logical 1 and 0 conditions), a digital device (such as a computer or processor) can perform calculations and tasks that might otherwise be cumbersome, less cost-effective or downright impossible to perform in the analog domain. After a task has been performed, the results can be changed back into an analogous form that we humans can readily understand.

Before moving on, let's take another look at this concept. If you were to type the letters C, A and T into a word processor, the computer would translate the keystrokes into a series of 8-bit digital words that would be represented as [0100 0011], [0100 0001] and [0101 0100]. These "alpha-bits" don't have much meaning when examined individually. However, when placed together into a group, they represent a four-legged animal that's either seldom around or always underfoot (Figure 6.1). When these binary words are grouped together into a recognizable pattern, a meaningful message is conveyed. Similarly, a digital audio system works by sampling (measuring the instantaneous voltage levels of) an analog signal and converting these samples into a series of digitally encoded words. Upon reproduction, this stream of representative words is then converted back into a series of voltages that changes over time to represent the original analog signal.

Figure 6.1. Digital and analog equivalents for a strange four-legged animal (Conway, one of our cats).

The Basics of Digital Audio

In Chapter 2, we learned about the two most basic characteristics of sound: frequency (the component of time) and amplitude (the signal-level component). Digital audio can be likewise broken down into two analogous components: sampling (which represents time) and quantization (which represents level).

Sampling

In the world of analog audio, signals are passed, recorded, stored and reproduced as changes in voltage levels that continuously change over time (Figure 6.2). The digital recording process, on the other hand, doesn't operate in a continuous manner. Rather, digital recording takes periodic *samples* of a changing audio waveform (Figure 6.3) and transforms these sampled signal levels into a representative stream of binary words that can be manipulated or stored for later processing and/or reproduction.

Figure 6.2. An analog signal is continuous.

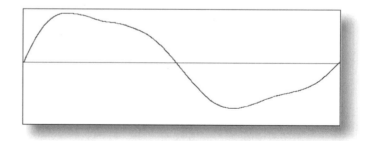

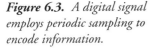

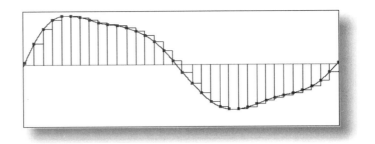

Figure 6.3. A digital signal employs periodic sampling to encode information.

Within a digital audio system, the *sampling rate* is defined as the number of measurements (samples) that are taken of an analog signal in one second. Its reciprocal, sampling time, is the elapsed time between each sampling period. For example, a sample rate of 48 kHz corresponds to a sample time of 1/48,000th of a second. Because sampling is tied directly to the component of time, the sampling rate of a system determines its overall bandwidth, meaning that a system with higher sample rates is capable of storing more frequencies at its upper limit.

During the sampling process (Figure 6.4), an incoming analog signal is sampled at discrete and precisely timed intervals (as determined by the sample rate). At each interval, this analog signal is momentarily "held" (frozen in time), while the converter goes about the process of determining what the voltage level actually is (with a degree of accuracy that's defined by the converter's circuitry and the chosen bitrate). The converter then generates a binary-encoded word that's numerically equivalent to the analog level that's currently being sampled. Once done, the converter can store the representative word into a memory medium (tape, disk, disc, etc.), release its hold and then go about the task of determining the level of the next sampled voltage.

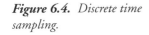

Figure 6.4. Discrete time sampling.

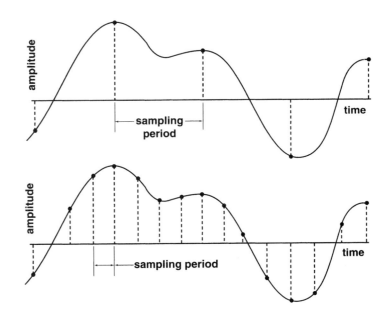

Figure 6.4. (continued)

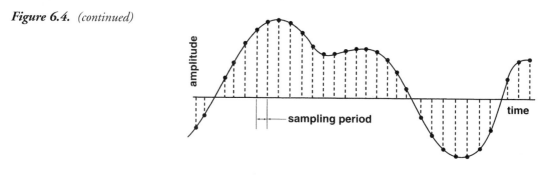

The Nyquist Theorem

According to the *Nyquist Theorem*, in order to digitally encode the desired frequency bandwidth, the selected sample rate must be at least twice as high as the highest frequency to be recorded (sample rate ≥ 2 × highest frequency). Thus, an audio signal with a bandwidth of 20 kHz would require a sampling rate of at least 40,000 samples/second. In addition, it's equally important that no audio signal greater than half the sampling frequency enter into the digitization process. If frequencies greater than one-half the sample rate are allowed to enter into the conversion process, erroneous frequencies—known as *alias frequencies* (Figure 6.5)—could enter into the audible audio signal band as false frequencies, which might be heard as audible harmonic distortion.

In order to eliminate the effects of aliasing, a low-pass filter is placed before the analog-to-digital (A/D) conversion stage. In theory, an ideal filter would pass all frequencies up to the Nyquist cutoff frequency and have an infinite attenuation thereafter (Figure 6.6a). However, in the real world, such a "brick wall" filter doesn't exist. For this reason, a slightly higher sample rate must be chosen in order to account for an attenuation slope that's required for the filter to be effective (Figure 6.6b). A sample rate of 44.1 kHz, for example, has been chosen in order to accurately encode an effective bandwidth up to 20 kHz.

Oversampling

Oversampling is a process that's commonly used in professional and consumer digital audio systems to improve anti-aliasing filter characteristics. Oversampling has the effect of further reducing intermodulation and other forms of distortion.

Whenever oversampling is used, the effective sampling rate within a converter's filtering block is multiplied by a specific factor (often ranging between 12 and 128 times the original rate). This significant increase in the sample rate is accomplished by interpolating sampled level points between the original sample times. In effect, this technique makes educated guesses as

Figure 6.5. *Alias frequencies introduced into the digital audio chain.*
a. *Sampled frequencies above the Nyquist half-sample frequency limit.*
b. *Alias frequencies introduced into the audio band.*

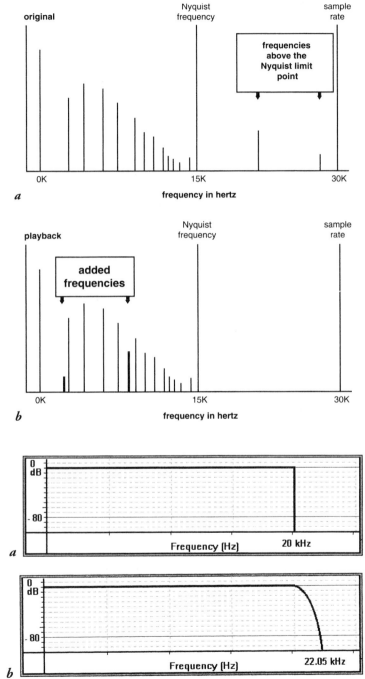

Figure 6.6. *Anti-alias filtering.*
a. *An ideal filter would have an infinite attenuation at the 20,000-Hz Nyquist cutoff frequency.*
b. *Real-world filters require an additional frequency "guardband" in order to fully attenuate unwanted frequencies above the half-bandwidth Nyquist limit.*

to what the sample levels would be and where they'd fall at the newly generated sample time, and then creates an equivalent digital word at that point. This increased sample rate likewise results in a much wider frequency bandwidth (so much so that a simple, less-expensive single-order filter can be used to cut off the frequencies above the Nyquist limit). By shifting the sample rate back to its original value after the filtering block, the bandwidth of the Nyquist filter is likewise narrowed to such a degree that it approximates a much more complex and expensive cutoff filter.

Quantization

Quantization represents the amplitude component of the digital sampling process. It's the technique of translating the instantaneous voltage levels of a continuous analog signal into discrete sets of binary digits (bits) for the purpose of manipulating and/or storing audio data in the digital domain. The amplitude of the incoming analog signal is broken down into a series of discrete voltage steps. Each step is then assigned an analogous set of binary numbers that are arranged together to form a binary word (Figure 6.7). This representative word encodes the signal level with as high a degree of accuracy as can be permitted by the word's bit length and the system's overall design.

Figure 6.7. *The instantaneous amplitude of the incoming analog signal is broken down into a series of discrete voltage steps, which are then assigned as an equivalent binary-encoded word.*

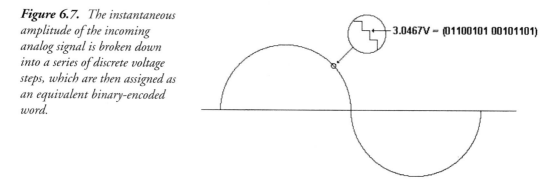

3.0467V = (01100101 00101101)

Currently, the most common binary word length for professional audio is 16-bit (for example, [0110010100101101]); however, systems having 20- and 24-bit capabilities are also currently available.

Digital signal processors (such as reverb, equalizers and dynamics processors) often perform their internal calculations using a 24- or 32-bit wordlength structure. This keeps calculation errors to a minimum by increasing the processor's overall headroom. Because errors are most likely to accumulate within the *least-significant bits* (LSB—the final and smallest numeric value within a digital word), these bits (which reside below the audio signal bitrate's LSB) can be dropped at the device's output. The final result is an n-bit data stream that's relatively free of errors. This leads to the conclusion that greater word lengths will often translate directly into an increased resolution (and thus higher quality) due to the added number of finite steps into which a signal can be digitally encoded. The following example details a number of encoding steps for the most commonly encountered bit lengths:

8-bit word = (nnnn nnnn) = 256 steps

16-bit word = (nnnnnnnn nnnnnnnn) = 65,536 steps

20-bit word = (nnnnnnnnnn nnnnnnnnnn) = 1,048,576 steps

24-bit word = (nnnnnnnnnnnn nnnnnnnnnnnn) = 16,777,216 steps

Signal-to-Error Ratio

Although analog signals are continuous, the process of quantizing a signal into an equivalent digital word isn't. Since the number of discrete steps that can be encoded within a digital word limits the accuracy of the quantization process, the representative digital word can only be a close approximation of the original analog signal level.

A digital system's *signal-to-error ratio* is closely akin (although not identical) to an analog system's signal-to-noise (S/N) ratio. Whereas an S/N ratio is used to indicate the overall dynamic range of an analog system, the signal-to-error ratio of a digital audio device indicates the degree of accuracy that's used when encoding a signal's dynamic range with regard to the step-related effects of quantization. Given a properly designed system, the signal-to-error ratio for a signal coded with n bits is:

$$\text{Signal-to-error ratio} = 6n + 1.8 \text{ (dB)}$$

For a 16-bit system, this would yield an error figure of 97.8 dB, a value that's about 30 dB below the noise figures of most conventional analog tape recorders.

Dither

Through the addition of small amounts of white noise (a signal that includes the random distribution of all frequencies at equal levels across the entire audio spectrum), it's possible to further reduce signal-to-error and distortion to figures that are below their standard levels. By adding such noise (known as *dither*), it's possible to encode signals that are less than the least-significant bit (that is, less than a single quantization step). It's interesting that introducing small amounts of noise into the conversion process yields results that are far preferable to the increased quantization distortion that would otherwise result.

The Digital Recording/Reproduction Process

The following sections provide a basic overview of the various stages that are encountered within the process of encoding analog signals into equivalent digital data (Figure 6.8a) and converting digital data back into its original analog form (Figure 6.8b).

Figure 6.8. *The digital audio chain.*
a. *Recording.*
b. *Reproduction.*

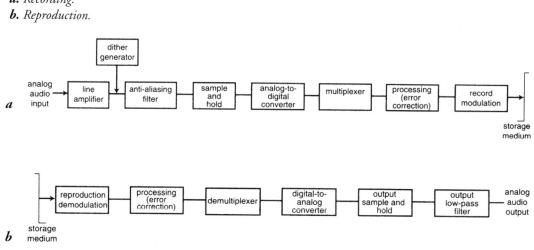

The Recording Process

In its most basic form, the digital recording chain includes a low-pass filter, a sample-and-hold circuit, an analog-to-digital converter and the circuitry for signal coding and error correction.

At the digital sampling system's input, the analog signal must be band-limited with a low-pass filter, so as not to allow frequencies above half the sample rate frequency to pass into the A/D conversion circuitry. Such a stop-band (anti-aliasing) filter generally makes use of a gradual roll-off slope at its high-frequency cutoff point (because an immediate or "brick wall" slope would introduce severe signal distortion and phase shifts). In order to accurately encode the full bandwidth, the Nyquist Theorem requires that a sampling rate be chosen that's higher than twice the highest frequency to be recorded. For example, a system with a bandwidth that reaches into the 20-kHz range is often sampled at a rate of 44.1K or 48K samples/second.

Following the low-pass filter, a sample-and-hold (S/H) circuit holds and measures the analog voltage level for the duration of a single sample period (which is determined by the sample rate, i.e., less than 1/44,100th of a second). At this point, computations must be performed to translate the sampled level into an equivalent binary word. This process of A/D conversion is a critical component of the digitization process, in that the sampled DC voltage level at the S/H circuit must be accurately quantized to the nearest step level in a very short period of time.

After the signal has been converted into a digital bit form, the data must be conditioned for further data processing and storage. This conditioning includes data coding, data modulation and error correction. In general, the binary digits aren't directly stored onto a recording medium as raw data. Rather, *data coding* is used to translate this data (along with synchronization and address information) into a form that allows the data to be most efficiently and accurately stored. The most common form of digital modulation is pulse-code modulation, or PCM (Figure 6.9).

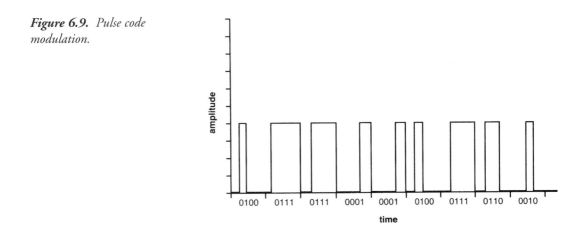

Figure 6.9. *Pulse code modulation.*

The density of the information in PCM recording and playback equipment is extremely high. So much so that any imperfections (such as dust or fingerprints that might adhere to the surface of any magnetic or optical recording medium) will readily generate error signals. To keep these errors within acceptable limits, a form of *error correction* is used. Several forms of error correction can be used, depending on the media type. One method uses redundant data in the form of parity bits and check codes; a second uses error correction that involves interleaving techniques (whereby data is deliberately scattered across the digital bit stream, according to a complex mathematical pattern) in order to reduce the effects of dropouts (Figure 6.10).

Figure 6.10. *An example of interleaved error correction.*

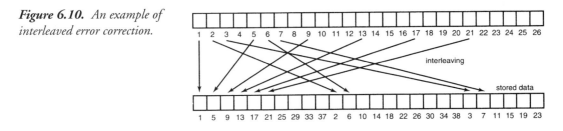

The Reproduction Process

The digital reproduction chain basically works in a manner that's complementary to the standard encoding process.

Because most digital media encodes data onto magnetic media in the form of highly saturated on/off transition states, the reproduced signal must be reconditioned so as to restore the digital bitstream back to its originally modulated binary state. Once this is done, the data is then de-interleaved back into its original form, where it can be easily converted back into PCM data. It's a simple fact that without error correction, the quality of most digital audio media would be greatly reduced or (in the case of the CD) almost useless.

After the signal has been reconstructed back into its original PCM form, the process of digital-to-analog (D/A) conversion can take place. Often, a stepped resistance network (sometimes called an R/2R network) is used to convert each word back into an analogous voltage level at each point in the sample playback phase. During a complementary sample-and-hold period, each bit in the word that's currently being converted is assigned to a leg in the network (moving from the most-significant to the least-significant bit). Each leg is designed to pass one-half the reference voltage level that can be passed by the previous step (Figure 6.11). The presence or absence of a logical "1" in each step can be used to turn on each voltage leg, and thus determine how much of the overall voltage source will be summed together and passed on to the converter's analog output.

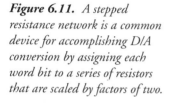

Figure 6.11. *A stepped resistance network is a common device for accomplishing D/A conversion by assigning each word bit to a series of resistors that are scaled by factors of two.*

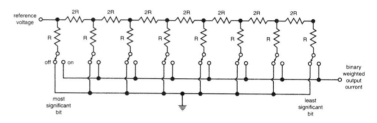

Following the conversion process, another complementary low-pass filter is inserted into the signal path. This filter smoothes out nonlinear steps that are introduced by the sampling process and results in a smoothed waveform that (when the circuit is properly designed) will faithfully represent the originally recorded analog waveform.

Digital Transmission

In this digital age, it's become increasingly common for audio data to be distributed from one device to another or throughout a connected production system in the digital domain. Using this approach, digital audio can be transmitted in its original numeric form and thus (in theory) won't degrade whenever copies are made.

When looking at the differences between the distribution of digital and analog audio, you should keep in mind that, unlike its counterpart, the transmitted bandwidth of digital audio data occurs in the megahertz range. Therefore, digital audio transmission has more in common with video signals than the lower bandwidth, analog audio range. This means that impedance must be much more closely matched and that quick-fix solutions (such as using a Y-cord to split a digital signal between two recorders) are major no-nos. Failure to follow these precautions could seriously deform the digital signal.

Because of these tight restrictions, several digital transmission standards have been adopted that allow digital audio data to be quickly and reliably transmitted between compliant devices. The two most commonly encountered formats are the AES/EBU and S/PDIF protocols.

AES/EBU

The *AES/EBU (Audio Engineering Society and the European Broadcast Union) protocol* has been adopted for the purpose of transmitting digital audio between professional digital audio devices. This standard (which is most often referred to as simply an "AES" digital connection) is used to convey two channels of interleaved digital audio through a single, three-pin XLR microphone cable. This balanced configuration connects pin 1 to the signal ground, while pins 2 and 3 are used to carry signal data. AES/EBU transmission data is low-impedance in nature (typically 110 Ω) with waveform amplitudes that range between 3 and 10 V. These factors allow a maximum cable length of up to 328 feet (100 meters) to be used at sample rates of less than 50 kHz without encountering undue signal degradation.

Digital audio channel data and subcode information is transmitted in blocks of 192 bits that are organized into 24 words, each 8 bits long. Within the confines of these data blocks, two subframes are transmitted during each sample period that convey information and digital synchronization codes for both channels in a L-R-L-R . . . fashion. Because the data is transmitted as a self-clocking biphase code (Figure 6.12), wire polarity can be ignored and when two devices are directly connected, the receiving device will usually derive its reference clock timing from the digital source device.

Figure 6.12. *AES/EBU subframe format.*

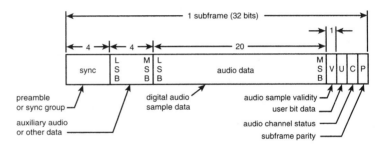

In the late 1990s, the AES protocol was amended to include the "stereo 96k dual AES signal" protocol. This was created to address signal degradations that can occur when running longer cable runs at sample rates that are above 50 kHz. In order to address the problem, the dual AES standard allows stereo rates above 50 kHz (such as 96/24) to be transmitted over two, synchronized AES cables (with one carrying the L and one the R information).

S/PDIF

The *S/PDIF (Sony/Phillips Digital Interface)* has been widely adopted for transmitting digital audio between consumer digital audio devices and is similar in data structure to its professional counterpart. Instead of using a balanced 3-pin XLR cable, however, the S/PDIF standard has adopted the single conductor, unbalanced phono (RCA) connector (Figure 6.13a), which conducts nominal peak-to-peak voltage levels of 0.5 volts with an impedance of 75 Ω. In addition to using RCA wire cable connections, S/PDIF can also be transmitted between devices using Toslink optical connection lines (Figure 6.13b), which are often referred to as "lightpipe" connections.

As with the AES/EBU protocol, S/PDIF channel data and subcode information is transmitted in blocks of 192 bits, consisting of 12 words 16 bits long. A portion of this information is reserved as a category code that provides the necessary setup information (sample rate, copy

Figure 6.13.
a. *RCA wire connection.*
b. *Toslink optical connection.*

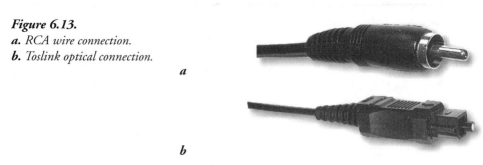

a

b

protection status and so on) to the copy device. Another portion is set aside for transmitting audio data that's used to relay track indexing information (such as start ID and program ID numbers), allowing this relevant information to be digitally transferred from the master to the copy. It should be noted that the professional AES/EBU protocol isn't capable of digitally transmitting these codes during a copy transfer.

SCMS

Initially, the DAT medium was intended to provide consumers with a way of making high-quality digital recordings for their own personal use. Soon after its inception, however, for better or for worse, the recording industry began to see this new medium as a potential source of lost royalties due to home copying and piracy practices. As a result, the RIAA (Recording Industry Association of America) and the former CBS Technology Center set out to create a "copy inhibitor." After certain failures and long industry deliberations, the result of these efforts was a process that has come to be known as the *Serial Copy Management System* or *SCMS*.

SCMS (pronounced scums) has been incorporated into many consumer digital devices to prohibit the unauthorized copying of digital audio at the 44.1-kHz sample rate (the standard CD rate). This copy inhibitor doesn't apply to the making of analog copies, to digital copies that are made using the AES/EBU protocol or to sample rates other than 44.1 kHz.

So what is SCMS? Technically, it's a digital protection flag that is encoded in byte 0 (bits 6 and 7) of the S/PDIF's subcode area. This flag can have only one of three possible states:

- Status 00: No copy protection, allowing unlimited copying and subsequent dubbing.
- Status 10: No more digital copies allowed.
- Status 11: A single copy can be made of this product, but that copy cannot be copied.

Suppose that we have two DAT machines that are equipped with SCMS (with one being used for playback and the other for recording). If we try to digitally copy a DAT that has a 10 SCMS status, you would simply be out of luck. But suppose that you found a DAT that has an 11 status flag. By definition, the bitstream data would inform the copy machine that it's OK to record the digital signal. However, the status flag on the subsequent copy tape would then be changed to a 10 flag. If at a later time you were to clone this DAT copy, the machine doing the second-generation copy couldn't be placed into Record. At that point, you have two possible choices: you could record the signal using the analog ports (often with minimum signal degradation), or you could use a digital format converter that (among other things) lets you strip the SCMS copy protection flags from the bitstream and continue to make multigenerational copies (it should be noted that some digital audio editors are able to strip out or reset the protection flag bits).

Signal Distribution

Both the AES/EBU and S/PDIF digital audio signals can be distributed from one digital audio device to the next in a daisy-chain fashion (Figure 6.14). This method works well only if a few devices are to be chained together. If a number of devices are connected together, time-base errors (known as jitter) might be introduced into the path, with possible side effects being added noise and a slightly "blurred" signal image. One way to reduce potential time-base errors is to use a digital distribution device that can route data from a single digital source to a number of individual device destinations (Figure 6.15).

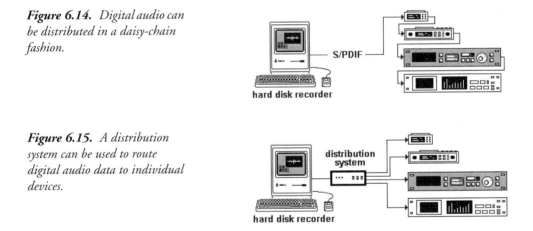

Figure 6.14. *Digital audio can be distributed in a daisy-chain fashion.*

Figure 6.15. *A distribution system can be used to route digital audio data to individual devices.*

What Is Jitter?

Jitter is a controversial and widely misunderstood phenomenon. It has been explained best by Bob Katz of Digital Domain (Orlando, Florida) in his article "Everything You Always Wanted to Know about Jitter but Were Afraid to Ask." The following is a brief excerpt . . . the full article (as well as those on other relevant topics) can be found at www.digido.com:

> Jitter is time-base error. It is caused by varying time delays in the circuit paths from component to component in the signal path. The two most common causes of jitter are poorly designed Phase Locked Loops (PLL's) and waveform distortion due to mismatched impedances and/or reflections in the signal path.
>
> Here is how waveform distortion can cause time-base distortion: The top waveform (*a* in Figure 6.16) represents a theoretically perfect digital signal. Its value is 101010, occurring at

equal slices of time, represented by the equally spaced dashed vertical lines. When the first waveform passes through long cables of incorrect impedance, or when a source impedance is incorrectly matched at the load, the square wave can become rounded, fast rise times become slow, also reflections in the cable can cause misinterpretation of the actual zero crossing point of the waveform. The second waveform (*b* in Figure 6.16) shows some of the ways the first might change; depending on the severity of the mismatch you might see a triangle wave, a squarewave with ringing, or simply rounded edges. Note that the new transitions (measured at the Zero Line) in the second waveform occur at unequal slices of time. Even so, the numeric interpretation of the second waveform is still 101010! There would have to be very severe waveform distortion for the value of the new waveform to be misinterpreted, which usually shows up as audible errors—clicks or tics in the sound. If you hear tics, then you really have something to worry about.

Figure 6.16. Example of time-base errors.
a. A theoretically perfect digital signal source.
b. The same signal with jitter errors.

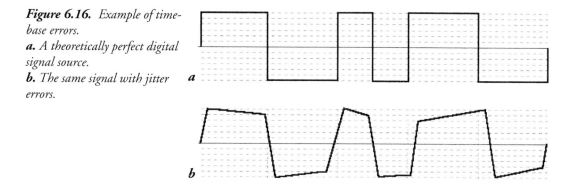

If the numeric value of the waveform is unchanged, why should we be concerned? Let's rephrase the question: "*when* (not *why*) should we become concerned?" The answer is "hardly ever". *The only effect of timebase distortion is in the listening; as far as it can be proved, it has no effect on the dubbing of tapes or any digital-to-digital transfer (as long as the jitter is low enough to permit the data to be read. High jitter may result in clicks or glitches as the circuit cuts in and out).* A typical D to A converter derives its system clock (the clock that controls the sample and hold circuit) from the incoming digital signal. If that clock is not stable, then the conversions from digital to analog will not occur at the correct moments in time. The audible effect of this jitter is a possible loss of low-level resolution caused by added noise, spurious (phantom) tones, or distortion added to the signal.

A properly dithered 16-bit recording can have over 120 dB of dynamic range; a D to A converter with a jittery clock can deteriorate the audible dynamic range to 100 dB or less, depending on the severity of the jitter. I have performed listening experiments on purist, audiophile-quality musical source material recorded with a 20-bit accurate A/D converter (dithered to 16 bits within the A/D). The sonic results of passing this signal through processors that truncate the signal at -110, -105, or -96 dB are: increased "grain" in the image, instruments losing their sharp edges and focus; reduced soundstage width; apparent loss of level causing the listener to want to turn up the monitor level, even though high-level signals are reproduced at unity gain. Contrary to intuition, you can hear these effects without having to turn up the listening volume beyond normal (illustrating that low-level ambience cues are very important to the quality of reproduction). Similar degradation has been observed when jitter is present. Nevertheless, the loss due to jitter is subtle, and primarily audible with the highest-grade audiophile D/A converters.

Multichannel Transmission of Digital Audio

In addition to the above standards for transmitting mono or stereo audio data, several standards exist for communicating eight or more digital audio channels from one device to another. These include the MADI, ADAT lightpipe and TDIF protocols.

MADI

The *MADI (Multi Channel Audio Digital Interface)* standard was jointly proposed as an AES standard by representatives of Neve, Sony and SSL as a straightforward, clutter-free interface for connecting multitrack tape recorders to mixing consoles (as well as to each other).

The format allows up to 56 channels of linearly encoded digital audio to be connected via a single 75-Ω, video-grade coaxial cable at distances of up to 120' (50 meters) or at greater distances when using fiber optic cable. MADI is a serial interface format that's compatible with the AES/EBU twin-channel format (whereby the data, Status, User and parity bit structure is preserved). Its channels are sequentially transmitted (starting with channel 0 and ending with channel 55) at a data transmission rate of 100 Mbit/sec. This results in an overall bandwidth that can handle audio data and numerous sync codes at various sample rate speeds (including allowances for changing pitch either up or down by 12.5 percent at rates between 32 and 48 kHz.

ADAT Lightpipe

A wide range of modular digital multitrack recorders, sound cards and hardware devices use the Alesis "*lightpipe*" system for transmitting audio via a standardized optical cable link. Lightpipe connections use standard Toslink connectors and cables to transmit up to eight digital audio channels over a sequential, optical digital bitstream. Although these connectors are identical to those that are used to optically transmit S/PDIF stereo digital audio, the datastreams are incompatible with each other.

Lightpipe data can only travel from a single source to a destination in one direction; thus, two cables are needed to distribute data both to and from a device. Only digital audio data is transmitted over the serial bitstream, meaning that sync data will not be passed. Should sync be needed to lock one or more devices together (as in a multiple ADAT system), separate sync cabling (often 9-pin serial connectors) will be needed.

TDIF

The *TDIF (Tascam Digital InterFace)* is a proprietary format that uses a 25-pin D-sub cable to transmit and/or receive up to eight channels of digital audio between compatible devices.

Unlike the lightpipe connection, TDIF is a bidirectional connection. This means that only one cable is needed to connect the ins and outs of a source to another device.

Although systems that support TDIF-1 cannot send and receive sync information (a separate wordclock connection is required for that), the newer TDIF-2 is capable of receiving and transmitting sync through the connection, without any additional cabling.

Wordclock

One aspect of digital audio recording that never seems to get enough attention is the need for sample rate synchronization within a connected system that uses multiple digital multitrack devices. In order to reduce such gremlins as clicks, pops and jitter (oh my!), it's sometimes necessary to lock the overall sample rate timing to a single master clock signal (so that the conversion sample and hold states for all digital audio channels and devices will occur at exactly the same point in time).

As an example, don't ya just hate it when you're in a room that has three or four clocks, and none of them read the same (or even roughly the same) time! In places like this you never quite know what the time really is. Similarly, it's likely that each digital system within a production system will have its own internal sample clock (in the form of a crystal timing reference that tells the device when to perform its sample and hold calculations). If these devices were to be connected together via digital audio line connections, the timing reference of one device wouldn't accurately match the timing reference of another that's receiving the data . . . and so on throughout the digital audio chain. Even though the devices are all running at the same sample rate, these timing mismatches often result in clicks, ticks and unwanted audible grunge. In order to correct for this, the internal clocks of all the digital devices within a connected chain must be referenced to a single timing reference, known as *wordclock*.

Similar to the distribution of time code, within a digital production network, there can only be one master wordclock reference (Figure 6.17). This reference source can be derived from a digital mixer, sound card, digital audio workstation . . . or any desired source that can transmit wordclock. Often, this reference pulse is chained between necessary devices through the use of BNC and/or RCA connectors, using low-capacitance cables (often 75-Ω video-grade coax cable is used, although, this cable grade isn't always necessary on shorter cable runs).

It's interesting to note that wordclock isn't generally needed when making a digital copy from one device to another via S/PDIF or AES cable connections. This is because the timing information is actually embedded within the data bitstream itself. Only when we begin to connect multichannel digital devices together will we see the immediate need for wordclock.

It almost goes without saying that there will often be differences in connections and parameter setups, from one system to the next. In addition to proper cabling, there'll often be a need to make specific hardware and software changes to get all the device blocks to communicate. In order to better understand your particular system during setup (and to keep frustration to a minimum), it's always a good idea to keep your device manuals close at hand.

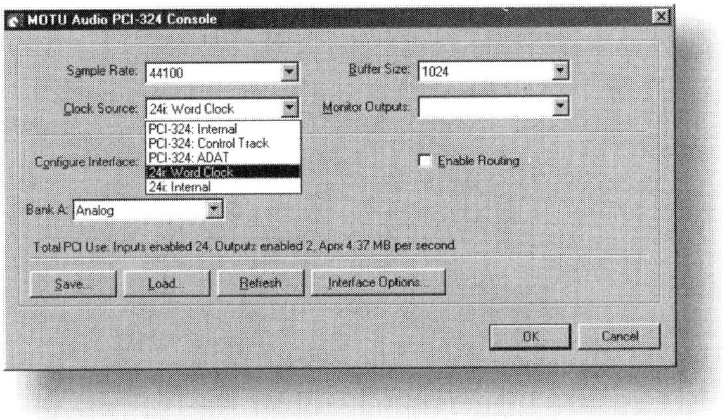

Figure 6.17. *Within a digital production network, there can only be one master wordclock reference.*

Digital Audio Recording Systems

For the remainder of this section, we'll be looking at the various types of digital audio recording devices which are currently available on the market. From my own personal viewpoint, I find the here-and-now of recording technology to be not only exciting and full of cost-effective possibilities . . . I also love the fact that there are lots of device-type options. In other words, a digital system that might work really well for *me* . . . might not be the best and easiest solution for *you*! In the earlier years of recording, there were basically only a few ways to successfully make a recording. Now, there are numerous technological options . . . it's up to you to find the one that best suits your needs, budget and personal working style.

As we take an in-depth look at many of these system choices, I hope that you'll take the time to learn about each one (and possibly even try your hand at listening to and/or working with each system type). This knowledge can also spur you on to learn more about and then test-drive a system that best fits your personal needs and preferred working style.

The Fixed-Head Digital Audio Recorder

The fixed-head digital audio recorder is a reel-to-reel system that often emulates its analog counterpart in form and function, although most other similarities end there. These recording systems use digital audio conversion and special data encoding structures to store digital audio data onto specially formulated digital audio tape, using state of-the-art, thin-film heads (Figure 6.18). Although two-channel fixed-head recorders do exist, the vast majority of these systems are multitrack. Of these, the most commonly found recorders are 24- and 48-track recorders that use the Digital Audio Stationary Head (DASH) format, which was ratified and jointly established by Sony, Willi Studer AG and Matsushita Electric Industries.

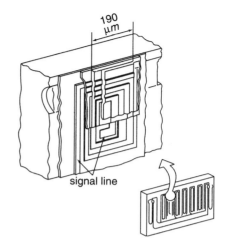

Figure 6.18. *Detail of digital thin-film head construction. (Courtesy of Sony Professional Audio, www.sony.com/proaudio)*

The DASH Format

DASH was developed to help ensure standardization between different generations of digital ATR and tape manufacture. The format consists of fast, medium and slow versions, the choice of which is determined by the tape speed of the recorder. Using this system, data isn't encoded onto a single track, but is actually spread over several interleaved data tracks in order to achieve the high degree of data densities that are needed to record digital audio onto longitudinal tape.

Error Correction

The operation of a DASH encoder is based on the Cross Interleave Code (CIC), with increased interleaving being made between the even- and odd-numbered words, which can correct up to a maximum of three words.

The interleaving of even and odd words make it possible for the tape to be spliced, thereby allowing physical tape edits to be made. The correctability of burst (large-scale) errors is determined by the encoders and is the same for all three speed versions. Error correction encoding and decoding are done independently for each track. For example, if excessive errors appear on one of the recorded tracks (as might occur with a dropout), the correction capabilities on other tracks won't be affected (a feature that safeguards the recording under adverse conditions).

In addition to allowing the tape to be spliced, cross fading, punch in/out and electronic editing operations are all feasible under this professionally styled system.

DASH Recording Systems

Recorders using the DASH format are commonly available in 24- and 48-track configurations. With the development of new large-scale integrated circuits (LSIs), smaller and lighter recorders are currently available that are lower in cost and power consumption than their first-generation models.

Both Sony and Studer have developed a series of popular multichannel DASH recorders. The Sony 24-track PCM-3324S and the Studer D 827 Mark II MCH use ½-inch tape that runs at a speed of 30 ips and operate using 16- or 24-bit wordlengths and sample rates of 48, 44.1 and 44.056 kHz. Using thin-film head technology (a technology borrowed from integrated circuit fabrication), the digital tracks are longitudinally spread across the width of the tape, along with two outside analog tracks and an additional control or external data track. The error detection circuitry of both of these recorders allows splice edits to be made by creating a seamless cross fade that interpolates the data before and after the splice. The headblock mechanism for the Sony 3324 is shown in Figure 6.19.

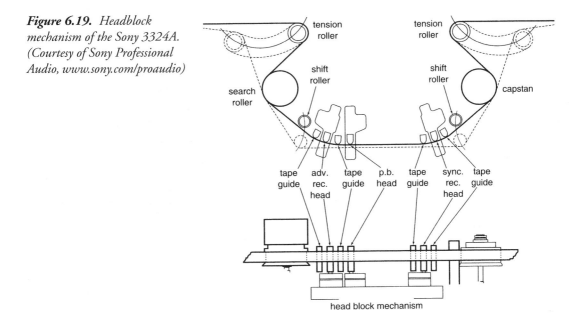

Figure 6.19. *Headblock mechanism of the Sony 3324A. (Courtesy of Sony Professional Audio, www.sony.com/proaudio)*

Both the Sony PCM-3348 (Figure 6.20) and the Studer D 827 Mark II MCH DASH recorders (Figure 6.21) have the added bonus of being able to record 48 tracks of digital audio that's fully compatible with the 24-track DASH format. Using a unique system, tapes that were previously recorded on a 24-track DASH machine can be recorded and reproduced with no signal degradation using tracks 1–24, while tracks 25–48 can continue to be recorded onto the same tape with no compatibility problems.

Figure 6.20. The Sony PCM3348 digital audio multitrack recorder. (Courtesy of Sony Professional Audio, www.sony.com/proaudio)

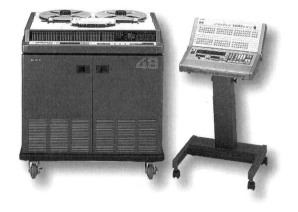

Figure 6.21. The Studer D 827 Mark II MCH 24/48 track digital audio multitrack recorder. (Courtesy of Studer Professional Audio AG, www.studer.ch)

The Rotating-Head Digital Audio Recorder

Rotating-head digital recorders fall into several categories, including DAT (digital audio tape) systems and MDM (modular digital multitrack) recording systems.

The Rotary Head

Because of the tremendous degree of data density that's required to record/reproduce PCM digital audio (approximately 2.77 million bits/second), the recording of data onto a single, linear tape track is virtually impossible. In order to record such a wide bandwidth, a rotating-head helical scan path is used to effectively increase the overall head-to-tape contact speed. This process allows the longitudinal tape speed to be substantially lowered, resulting in a vast reduction in tape consumption. An example of a transport that employs rotary head technology is shown in Figure 6.22.

Figure 6.22. *Helical scan path.*
a. Tape path around drum mechanism.
b. Recording of slanted tracks.

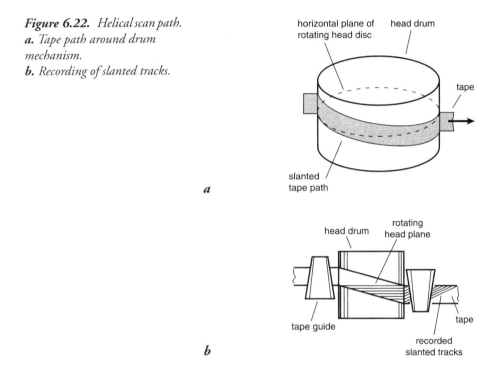

Digital Audio Tape (DAT) System

The digital audio tape or *DAT* format has allowed for the creation of a compact, dedicated PCM digital audio recorder (Figures 6.23 and 6.24) that displays a wide dynamic range, low distortion and an unmeasurable amount of wow and flutter. This recorder is designed to equal or exceed many standard digital specifications for pro audio.

Figure 6.23. *Tascam DA-P1 portable DAT recorder. (Courtesy of Tascam, www.tascam.com)*

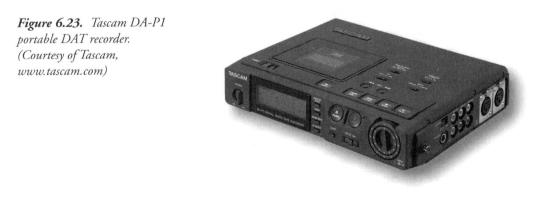

Figure 6.24. *Sony PCM 7040 Professional DAT recorder. (Courtesy of Sony Professional Audio, www.sony.com/proaudio)*

DAT technology makes use of an enclosed compact cassette that's even smaller than a compact audio cassette. Equipped with both analog and digital input/outputs, the DAT format can record and play back at three standard sampling frequencies: 32 kHz, 44.1 kHz, and 48 kHz (although sample rate capabilities and system features may vary from one recorder to the next).

On a consumer DAT machine, the 44.1 kHz sampling frequency is often reserved for prerecorded DAT tapes and is designed to discourage the unlawful copying of prerecorded program material through the implementation of the SCMS copy code system. Current DAT tapes offer running times of up to two hours when sampling at 44.1 kHz and 48 kHz, and the standard is capable of three record/reproduce modes for the 38-kHz (broadcast) sampling rate:

- Option 1 provides two hours of maximum recording time with 16-bit linear quantization.
- Option 2 provides up to four hours of recording time with 12-bit nonlinear quantization.
- Option 3 (which is rarely, if ever used) allows for the recording of four-channel, nonlinear 12-bit audio.

DAT Tape/Transport Format

The actual track width of the DAT format's helical scan can range downward to about 1/10th the thickness of a human hair (which allows for a recording density of 114 million bits per square inch). This is the first time such a density has been achieved using magnetic media. To assist with tape tracking and therefore to maintain the highest quality playback signal, a sophisticated tracking correction system is used to center the heads directly over the tape's scan path.

The head assembly of a DAT recorder uses a 90° half-wrap tape path. Figure 6.25 shows a two-head assembly in which each head is in direct contact with the tape 25 percent of the time. Such a discontinuous signal requires that the digital data be encoded within the slant track in digital "bursts," which necessitates the use of a digital buffer. On playback, these bursts are again smoothed out into a continuous data stream. These encoding methods have the following advantages:

- Only a short length of tape is in contact with the drum at one time. This reduces tape damage and allows high-speed searches to be performed while the tape is in contact with the head.
- A low tape tension can be used to ensure a longer head life.

Figure 6.25. *A half-wrap tape path, showing 90° tape contact.*

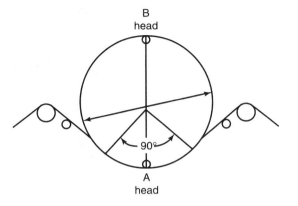

The high-speed search function (approaching 300 times normal speed) is a key feature of the DAT format. In addition to this function, the format makes provisions for non-audio information to be written into the digital stream's subcode area. This subcode area serves as a digital identifier (just as a compact disc uses subcodes for selection and timing information) and can be written as any one of three data types: start ID, which is used to indicate the beginning of a selection; skip ID, which indicates that a selection should be skipped over; and program number. These identifiers assist the user in finding selections when searching at high speeds.

Subcode information can be written or erased at any time without altering the audio program. In addition, the subcode area can be used to encode SMPTE time code for use with audio-for-video and synchronous music production.

Modular Digital Multitrack Systems

One of the most monumental developments in recent recording history has been the introduction of the *modular digital multitrack system*, or *MDM* (Figure 6.26). MDMs are compact multitrack digital audio recorders that are capable of recording eight tracks of digital audio onto standard videotape cassettes that can often be bought at your favorite neighborhood drugstore.

These recorders are said to be "modular" because several of them can be linked together (in groups of eight) in a synchronous fashion that allow them to become a large-scale multitrack recording system (having a theoretical maximum limit of 128 tracks! . . . although most systems max out in the 24- and 32-track range). If the capability to create and expand a digital audio recording system to suit your needs isn't enough to pique your interest, you might want to check out the price of an MDM. As of this writing, a basic 8-track digital MDM can be purchased new for less than $2000! Given the fact that a tape-based digital multitrack can easily sell for more than $60,000, it's not hard to understand why the MDM has become somewhat of a revolution in production technology . . . in both project and professional production studios.

Figure 6.26. *Tom Gioia of the Visionary Music Group with his DA-88 and DA-38 MDM recorders. (Courtesy of Tascan, www.tascam.com)*

ADAT MDM Format

The *ADAT* standard, which was created by the Alesis Corporation (Figures 6.27 and 6.28) is a family of rotary-head, 8-track modular digital multitrack recorders that use standard S-VHS videotape and includes such features as 16/20-bit wordlength capabilities, professional sample-rate standards (including varispeed over a wide range), autolocate functions, autoloop/rehearse functions that work in both the play and record modes and multiple cross fade times. When using a standard 120-minute S-VHS tape at the highest sampling rate, a recorder that conforms to the ADAT standard can yield recording times of slightly over 40 minutes, which increases to 53 minutes when a 160-minute tape is used.

Figure 6.27. *Alesis XT20 20-bit modular digital multitrack recorder. (Courtesy of Alesis Studio Electronics, Inc., www.alesis.com)*

Figure 6.28. *Alesis BRC (Big Remote Control) unit for the ADAT. (Courtesy of Alesis Corporation, www.alesis.com)*

Analog input/output connections vary between models and can include ¼-inch unbalanced/balanced jacks or RCA or XLR connections (in either –10 or +4 dBm configurations). Many models allow for multiple I/O connections to be made via a single, standard 56-pin Elco connector at professional +4 dBm reference levels.

Remote functions are carried out through the use of the LRC (Little Remote Control), BRC (Big Remote Control) or external computer control. The LRC is shipped with every ADAT and contains all the device's basic front panel transport controls on a single, palm-sized remote. The optional BRC (Figure 6.28) is capable of acting as a full-featured remote control, digital editor, expanded-feature autolocator and synchronizer on a single tabletop or free-standing surface. It's capable of controlling up to 16 remote ADAT units and can bounce data from one track to another in the digital domain (while also allowing tracks to be shifted in location). An example of track location shifting is its ability to copy a set of background vocals in a song's bridge to another point near the end of the song, without the use of a digital audio editor or manually having to copy tracks. All that's needed is to define a track and location source, as well as its destination, and the BRC will take care of all transport, record and audio functions.

Digital I/O is transmitted (in a single direction) through the use of a Toslink fiber-optic cable line that's capable of simultaneously carrying all eight channels. This link is used to connect an ADAT to a growing number of digital peripheral I/O devices (such as digital audio interfaces, Mixers, effects processors and even synthesizers), to connect multiple ADATs to a central BRC remote or to connect one ADAT to another in order to make direct digital-tape copies.

On the sync front, since only audio data is passed over an optical Toslink cable, it's necessary to interconnect all of the synchronous devices by chaining the sync ports together using a 9-pin sync cable. This port is used to transmit a proprietary code for interlocking the devices together, with near sample accuracy. From a user's perspective, this code will be displayed as SMPTE and through the use of a special conversion device or system that includes an ADAT sync interface, this code can often be converted to/from SMPTE and/or MIDI time code.

DTRS MDM Format

Although the rotary-head *Digital Tape Recording System (DTRS)* was created by Tascam (Figure 6.29), several manufacturers offer MDM recorder systems that adhere to this standard.

Figure 6.29. Tascam DA-38 modular digital multitrack recorder. (Courtesy of Tascam, www.tascam.com)

These 8-track modular recorders are capable of recording up to 108 minutes of digital audio onto a standard 120-minute Hi-8mm video tape, and (like ADATs) can be combined with other DRTS recorders to create a system that has 24, 32 or more tracks.

Digital I/O is made through the proprietary TDIF connection, which uses a special 25-pin D-sub connector to link to digital mixers, hard disk recorders or external accessories or to other, linked DRTS recorders. As of this writing, TDIF comes in two flavors: TDIF-1 is capable of transmitting and receiving all eight channels of digital audio (in a bidirectional I/O fashion) over a single cable. TDIF-2 is also an eight-channel I/O format; however, it's also capable of transmitting sync without the need for an external sync connection.

As with all MDMs, the sync capabilities of a DRTS recorder will vary between models and manufacturers. Like the ADAT, the various recorders may need to be synced together through the use of a 15-pin D-sub cable, which will need to be chained between the various synchronized devices. This proprietary sync code can also be converted to MTC or SMPTE time code, when using conversion boxes or other special interface options.

Autolocators (such as the Tascam RD-848) can be used to remote control many of the basic transport, track arming and autolocation functions, from a single desktop surface.

Third-Party Developments and Accessories for the MDM

An ever-growing number of accessories are available for both the ADAT and DTRS MDM formats. These accessories can be used to link both the digital audio and synchronization signals to an external device in order to edit, process, interface and/or synchronize an MDM system to a host of applications, digital mixers, workstations, etc.

A wide range of manufacturers and third-party companies continue to create new and innovative applications and accessories for integrating these devices together and into an existing system. The best way to keep on top of these developments is to contact the manufacturers, search the Web and keep reading the trade magazines.

Sampling Systems

One of the most important revolutions in audio history has been the merging of sound recording with computer technology to create a sample- and soundfile-based approach to production (Figure 6.30).

The encoding of audio data into digital memory or onto a digital storage medium provides us with a means for storing and/or manipulating defined blocks of digital data. This data can be stored as a sample file (data that can be imported, played, manipulated and exported to/from various instrument-based sampling devices) or as a soundfile (providing recorded "tracks" that can be recorded, processed and reproduced from disk or disc).

Figure 6.30. *Example of a computer-based music production system. (Courtesy of Mark of the Unicorn, Inc., www.motu.com)*

Perhaps the most important difference between a tape-based system (be it digital or analog) and a sample-based recording system . . . is the fact that the latter is *random access*. Random access production refers to digital audio data's ability to be stored within read-only memory (ROM), random-access memory (RAM), or a disk/disc-based memory medium in such a way that data can be instantaneously accessed, processed or reproduced in any order and at any time. A tape-based medium, however, differs greatly as its linear nature requires time to search and locate to the various points where audio is stored.

Neither medium is better than the other, as each has its strong points and inherent weaknesses. Most recording professionals can't deny that the two production styles complement each other in ways that can often immeasurably boost the effectiveness of a production system.

Samplers

A *sampler* (Figure 6.31) is a digital audio device/instrument that's capable of recording, musically transposing, processing and reproducing segments of digitized audio directly from RAM for playback in a polyphonic, musical fashion.

Figure 6.31. *Emu E5000 Ultra Sampler. (Courtesy of E-MU/Ensoniq, www.emu.com)*

Basically, a sampler can be thought of as a device that lets you record, edit and reload the samples into its internal memory. Once loaded, these sounds (whose length and complexity is limited only by RAM size and your imagination) can be looped, modulated, filtered and amplified according to user or factory setup parameters, so as to modify a sample's overall waveshape and envelope. Signal processing capabilities, such as basic editing, looping, gain changing, reverse, sample-rate conversion, pitch change and digital mixing capabilities, are also often designed into the system.

A sampler's design will often include a MIDI keyboard or set of trigger pads that lets you poly-phonically play samples as musical chords, sustain pads, triggered percussion sounds or sound effects. These samples can be played according to the standard Western musical scale (or any other scale, for that matter) by altering the playback sample rate. Pressing a low-pitched key on the keyboard will cause the sample to be played back at a low sample rate, while pressing a high-pitched one will cause the sample to be played back at a rate that would put Mickey Mouse to shame. Choosing the proper sample-rate ratios lets you play samples simultaneously at various pitches that correspond to standard music intervals.

Once a set of samples have been recorded or recalled from disk, they can be assigned to a spe-cific note on a MIDI keyboard, or mapped over a series of keys in a polyphonic fashion. A sam-pler (or synth) with a specific number of voices (i.e., 64 voices) simply means that up to 64 notes can be simultaneously played on a keyboard at any one time. Each sample in a multiple-voice system can be assigned (using a process known as "splitting" or "mapping") across a per-formance keyboard (Figure 6.32). In this way, a sound can be assigned to play across a "zone" or range of notes and/or velocity "layers" in various ways. For example, a single key on a loaded piano setup could be layered to trigger two samples (pressing the key at velocities ranging from 0 to 64 could sound a sample of a piano note that was played softly, while a velocity range of 65 to 127 might play a sample of a piano note that was struck hard). In this way, mapping can be used to create a more realistic instrument or wild set of soundscapes that change not only with the played keys, but with different velocities as well.

Figure 6.32. *A split keyboard setup.*

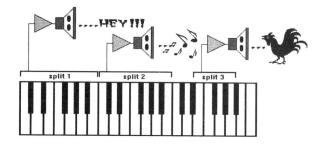

Most samplers have extensive edit capabilities that let you modify sounds in much the same way as a synthesizer would modify such electronically generated characteristics as LFO (low-fre-quency oscillation), velocity, panning, expression, ADSR (attack, delay, sustain and release) and other types of envelope processing parameters, keyboard scaling, aftertouch . . . etc. Many often include such features as integrated signal processing, multiple outputs (offering isolated chan-nel outputs for added mixing and signal processing power, or for recording individual voices to a multitrack recording system) and integrated MIDI sequencing capabilities.

Distributing data to and from a sampler can be done in many ways. Almost every sampler has a floppy or hard drive mechanism for saving waveform or system data and many are equipped with a port for adding a CD-ROM drive. Distribution between samplers or to/from a sample editing program can be handled through the use of the slower MIDI sample-dump standard, or via a high-speed SCSI (small computer system interface) port.

In this day and age, a wide range of previously recorded sample files can be purchased in the standard CD audio format (requiring the user to manually edit the sounds into the sampler's native format), or as soundfiles that are already edited in the sampler's native format. These commercially available libraries are often the mainstay of both electronic musicians and visual postproduction facilities.

Commercial sample libraries need not be the only option. Professional and nonprofessional artists will often record and edit their own samples for their own personal library. These sounds can be created from original acoustic or electronically generated sound sources, in addition to lifting soundclips from previously recorded source material (such as CD, TV, records and videotapes).

Software Samplers

In addition to hardware sampling systems, a growing number of software samplers exist that use a computer's existing memory, processing and sound output systems in order to polyphonically reproduce samples in real time.

Offering much of the same functionality as their hardware counterparts, these software systems (Figure 6.33) are capable of editing, mapping and splitting sounds across a MIDI keyboard, using on-screen graphic controls that have improved to the point of being competitive with hardware systems in cost effectiveness, accessibility and ease of use.

Figure 6.33. Gigastudio Hard Disk Sampling Workstation. (Courtesy of Nemesys Music Technology, Inc., www.nemesysmusic.com)

Sample Editing

When a recorded sound is transferred into a sampler, the original source material will often contain extraneous sounds, breathing and fidget noises or other music that occurs both before and after the desired sample (Figure 6.34a). Using the sample edit function in a sampler, these unwanted sounds can be deleted by trimming the in- and out-points to include only desired sounds. Trimming is accomplished by instructing the system's microprocessor to ignore (not access or reproduce) the samples that exist before a user-defined in-point and/or those samples following a desired out-point (Figure 6.34b). After trimming, the final sample can be played, processed (if a fade or other function needs to be performed) and then saved to floppy or hard disk for later recall.

Figure 6.34. *Sample editing.*
a. *Unedited sample.*
b. *Sample that has been trimmed and faded at its end.*

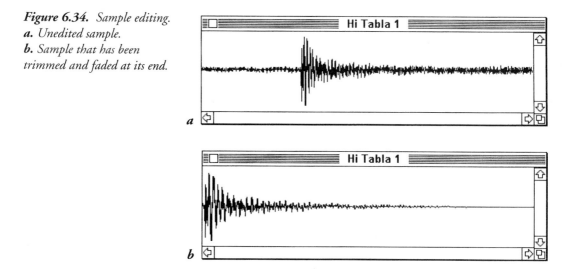

Looping

Another editing technique that's used regularly by sample artists to maximize the system's available RAM- and disk-based memory is a process known as *looping*. Through this technique, a sample that occupies a finite memory space in RAM can be sustained for long periods of time (well past the length of the original sample), thereby preventing the sound from abruptly ending while the key on a MIDI keyboard is still held down.

Such a loop is created by defining a segment of sound within a sample that doesn't significantly change in amplitude and composition over time and then repeatedly accessing this section from RAM (Figure 6.35). This loop can be created from waveform segments that are very short, or they can be longer.

Figure 6.35. *Example of a sample with a sustain loop. (Courtesy of Sonic Foundry, www.sonicfoundry.com)*

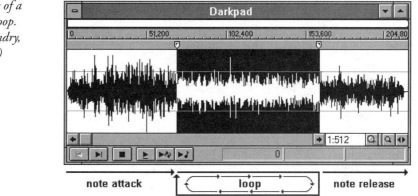

When creating a looped splice, life can often be made easier by following this simple rule: Match the waveform shape and amplitude at the beginning of the loop with the same waveform shape and amplitude at its end.

This simply means that the waveform begin/end amplitudes must match (Figure 6.36). If they don't, the signal levels will vary and an annoying "pop" or audible "tick" will result. Many samplers and sample-editing programs provide a way to automatically search out the closest level match or display the loop crossover points on a screen, so the amplitude levels can be manually matched.

Figure 6.36. *A loop waveform window allows the beginning and end levels of a loop to be manually matched. (Courtesy of Sonic Foundry, www.sonicfoundry.com)*

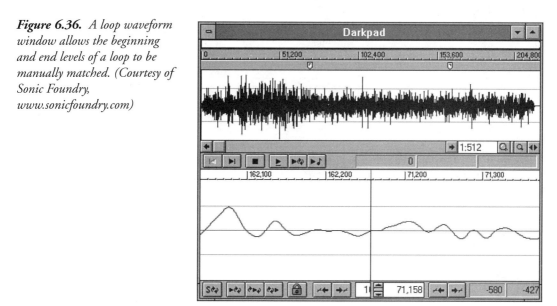

Certain samplers allow more than one loop to be programmed into a samplefile. This has the effect of making the sample sound less repetitive and more natural and adds to its overall expressiveness when played on a keyboard. In addition to having multiple sustain loops, a release loop can be programmed to decay the sample when the keyboard note is released.

The Sample Editor

Over the past few years, several MIDI sample-dump formats have been developed that allow samples to be transmitted and received in the digital domain. These sample dumps can be transferred between like samplers or between a personal computer and various samplers. To take full advantage of the latter, sample-editing software (Figures 6.37 and 6.38) was developed to perform such tasks as:

- Loading samples into a computer . . . where they can be stored to hard disk, arranged into a library that best suits the users needs and transmitted to any sampling device in the system (often to samplers with different sample rates and bit resolutions).

- Editing and arranging the sample before copying to disk using standard computer cut-and-paste edit tools. Because most samplers support multiple loops, segments of a sample can be repeated to saves valuable RAM memory.

- Digitally processing a signal to alter or mix a samplefile, using such functions as gain changing, mixing, equalization, inversion, reversal, muting, fading, crossfading and time compression.

Figure 6.37. *Example diagram of a sample-editor network.*

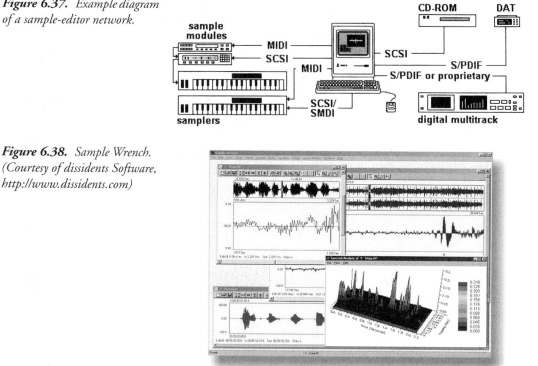

Figure 6.38. *Sample Wrench. (Courtesy of dissidents Software, http://www.dissidents.com)*

Distribution of Sampled Audio

Within a sample-based MIDI setup, it's important that sampled audio be distributed using a method that's as fast and as painless as possible. Therefore, standards have been adopted that allow samplers and related software programs to communicate and store sample-based digital audio. It should be noted that samplers might or might not support these protocols. As always, consulting the manuals for each device before attempting a data transfer will most likely reduce stress, frustration and wasted time.

MIDI Sample-Dump Standard

The *sample-dump standard (SDS)* was developed and proposed by the MIDI Manufacturers Association as a protocol for transmitting sampled digital audio and sustain-loop information between sampling devices. This data is transmitted over regular MIDI lines as a series of MIDI system-exclusive (Sys-Ex) messages, which are unspecified in length and data structure.

Although samplers of different manufacture and model type can be used to perform similar musical functions, the inner electronics and the way that data is internally structured often vary from device to device. As a result, most samplers communicate using their own unique system-exclusive data structure (as identified by a unique manufacturer and device ID number). In order for data to be successfully transmitted between samplers, they must be of the same or compatible manufacture and design. If this isn't the case, a computer-based program (such as a sample editor) must be used to translate from one sampler's data format and structure into one that can be understood by another make or model.

Samplers that communicate sample data via SDS have the distinct disadvantage of being rather slow, since the digital audio data is transmitted over standard MIDI lines at the 31.25 kbaud rate. When transmitting anything more than a short sample, be prepared to take a coffee break.

SCSI Sample Dump Formats

A number of computer-based digital audio systems and professional samplers are capable of transmitting and receiving sampled audio via *SCSI (small computer systems interface)*. SCSI is a bidirectional communications line that's commonly used by personal computers to exchange digital data at high speeds. When used in digital audio applications, it provides a direct parallel data link for transferring soundfiles at a rate of 16 Mbytes/sec or higher (literally hundreds of times faster than MIDI). Although the data format will change from one device to the next (meaning that data can only be transmitted between like devices or a device/PC combination that's installed with the proper device driver), SCSI still wins out as a fast and straightforward way to transfer data to and from an editing program, hard disk or CD-ROM sample library.

SMDI

The *SCSI Musical Data Interchange (SMDI)* was developed as a standardized, non-device-specific format for transferring digitally sampled audio between SCSI-equipped samplers and computers at speeds up to 300 times faster than MIDI's transmission rate of 31.25 kbytes per second. Using this format, all you need to transfer digital audio directly from one supporting sampler to another is to connect the SMDI ports by way of a standard SCSI cable and follow the steps for transferring the sample.

Although SMDI is loosely based on MIDI SDS, it has more advantages over its slower cousin than just speed. For example, SMDI can distribute stereo or multichannel samplefiles. Also, it isn't limited to files that are less than two megawords in length and can transmit associated file information (such as filename, pitch values, and sample number range). Sound patch and device-specific setup parameters can also be transmitted and received over SMDI lines via standard system-exclusive (Sys-Ex) messages.

Hard-Disk Recorders

Once developers began to design updated sample editors, it was discovered that through additional processing hardware, digital audio could be recorded and edited directly to a computer's hard disk. Thus, the concept of the *hard disk recorder* was born. These systems serve as computer-based hardware and software packages for the recording, manipulation and reproduction of digital audio data that resides on disk or disc. These systems are commonly (but not necessarily) designed around and are controlled by a standard personal computer and its associated hardware.

There are numerous advantages to using a hard-disk recording system in an audio production environment:

- *The ability to handle longer sample files.* Hard-disk recording time is often limited only by the size of the disk itself.
- *Random-access editing.* As audio is recorded onto disk, any point within the program can be accessed at any time, regardless of the order in which it was recorded.
- *Nondestructive editing.* This process allows audio segments (often called regions) to be placed in any context and/or order in a program without changing or effecting the originally recorded soundfile in any way. Once edited, these regions can be consecutively reproduced to create a single performance, or individually at specified SMPTE time-code addresses.
- *DSP.* Digital signal processing can be performed on a segment or entire sample file in either real time or non-real time (often in a nondestructive fashion).

Add to this the fact that computer-based digital audio devices can integrate many of the tasks that are related to both digital audio and MIDI production . . . and you have a system that offers the artist and engineer an unprecedented degree of control, editing, processing and connectivity.

The Driving Forces of Change

Over the history of disk- and disc-based systems, the style, form and function of hard disk recording has changed to meet the challenges of faster processors, bigger drives, improved hardware systems and the onward push of marketing forces to sell, sell, sell! As a result, there are numerous hard disk system types that are designed for various purposes, budgets and production styles. Many of the long-held limitations and distinctions between system types and operating principles have gone by the wayside or have been blurred, as new technologies and programming techniques continue to turn out new products on a staggeringly regular basis.

Although the field of hard disk recording has matured into a technology that's become pervasive in all forms of media production, it's still in a continual form of evolution as hardware, software and personal working tastes change. As with all evolutionary revolutions, it's always a good idea to keep abreast of these changes by reading the trade mags, searching the Web and keeping your eyes and ears open.

Hard Disk Editing: The Basic Tools of the Trade

In addition to recording and playing back larger soundfiles, one of the strongest features of a hard-disk recorder is how quickly and easily it can perform extensive edits on a soundfile, when compared to the time it would take to perform a similar function using an analog recorder. Furthermore, if you don't like the final results, these systems will easily let you "undo" an edit, save the resulting audio program, create a number of different edited versions and then choose between them . . . all without altering the originally recorded soundfiles.

Before taking a closer look at many of the system types that are available today, let's delve into a few of the basic production techniques and advantages of this technology.

Nondestructive Editing

Non-destructive editing refers to a disk- or disc-based recorder's ability to edit a soundfile without altering the data that was originally recorded to disk. This important capability allows any number of edits, alterations or program versions to be performed and saved to disk without changing or destroying any of the original soundfile data.

The nondestructive editing process is accomplished by accessing defined segments of a recorded soundfile (often called *regions*) and then outputting them in a user-defined order. In effect, when you're creating a specific region, you're telling the program to define a block of memory that begins at a specific memory address on the hard disk (or within RAM memory) and continue until the ending address has been reached (Figure 6.39). Once defined, these regions can be inserted into a list (often called a *playlist*) or multitrack session in such a way that they can be assembled and manipulated into a final edited program. Once done, the regions can then be saved and reproduced in any order . . . without affecting the original soundfile. For example, Figure 6.40 shows a snippet from *Gone with the Wind* that contains the immortal words

"Frankly, my dear, I don't give a damn." By segmenting it into three defined regions, we could use a hard disk recorder to change the word order in several ways.

Figure 6.39. *Example of a defined region.*

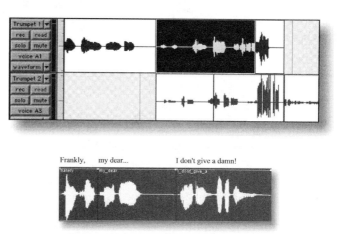

Figure 6.40. *Example of how a playlist could output Rhett's statement.*

Destructive Editing

Destructive editing, on the other hand, occurs whenever the recorded data is altered and rewritten to disk in such a way that it can't be recovered in its original form. Obviously, this edit form is often less desirable than its nondestructive counterpart, because edited data can't be easily recovered without saving each edited rendition.

Although destructive editing isn't the preferred method, there are times when it can be very useful. For example, you might want to save an edited file as a single, separate soundfile that can be easily retrieved and reproduced without additional playlist assembly or processing. You might also want to make edits to a song, process and then save it (under the same or possibly a different name) as a mastered single.

Basic Editing Techniques

One of the strongest assets of a hard disk recording system is its ability to edit segments of digital audio with speed and ease. The following sections offer a brief introduction to many of these valuable editing tools.

Graphic Editing

Most hard disk recording systems graphically display soundfile information on a computer or LCD screen as a series of vertical lines that represent the amplitude of a waveform over time in a WYSIWYG (what-you-see-is-what-you-get) fashion (Figure 6.41). Depending on the system type, soundfile length and the degree of zoom, the entire waveform can be shown on the screen, or only a portion can be shown with it continuing to scroll off one or both sides of the screen.

Figure 6.41. *Cool Edit 2000's 2-channel editing screen. (Courtesy of Syntrillium Software Corp., www.syntrillium.com)*

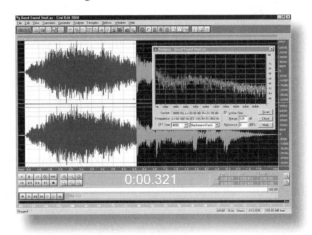

Graphic editing differs greatly from the "razor blade" approach that's used to cut analog tape in that it gives both visual and audible cues as to where a precise edit point should be. Using this common display technique, waveforms that have been cut, pasted, reversed and assembled are visually reflected on the screen. Often these edits are nondestructive, allowing the original soundfile to remain intact.

Only when a waveform is zoomed-in fully is it possible to see the individual waveshapes of a soundfile (Figure 6.42). Often when a soundfile display is "zoomed-in" to a level that shows individual sample points, the individual volumes can be redrawn to remove potential offenders (such as clicks and pops) or to smooth amplitude transitions between loops or adjacent regions.

Figure 6.42. *Often, when a soundfile is fully zoomed-in, the actual shape of a waveform will be displayed at the sample level. (Courtesy of Syntrillium Software Corp., www.syntrillium.com)*

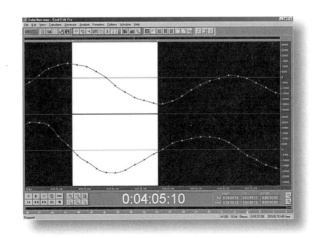

When working in a graphic editing environment, regions can be defined by positioning the cursor within the waveform, pressing and holding the mouse or trackball button, and then dragging the cursor to the left or right. Usually, the selected region is then highlighted for easy identification. After the region is defined, it can be edited, marked, named or otherwise processed.

Cut and Paste Edit Techniques

Once you've introduced yourself to the world of graphic editing, the next step is to begin to *cut and paste* a soundfile into a basic sequence that best fits the project. The basic cut and paste techniques used in hard disk recording are entirely analogous to those used in a word processor or other graphics-based program.

- *Cut.* Places the highlighted region into memory and deletes the selected data (Figure 6.43).
- *Copy.* Places the highlighted region into memory and doesn't alter the selected waveform in any way (Figure 6.44).
- *Paste.* Copies the waveform data that's within the system's clipboard memory into the soundfile beginning at the current cursor position (Figure 6.45).

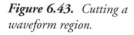

Figure 6.43. *Cutting a waveform region.*

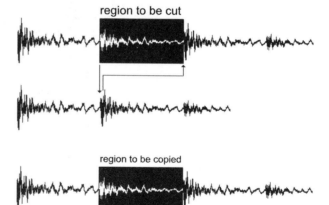

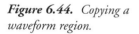

Figure 6.44. *Copying a waveform region.*

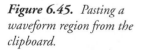

Figure 6.45. *Pasting a waveform region from the clipboard.*

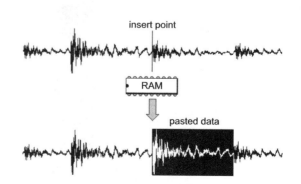

Do It Yourself Tutorial: Copy and Paste

1. Consult your editor's manual regarding basic cut and paste commands.

2. Open a soundfile and define a region that includes a musical phrase or sentence.

3. Cut the region and try to paste it into another point in the soundfile in a way that makes sense (musical or otherwise).

4. Feel free to cut, copy and paste to your heart's desire to create an interesting or totally wacky soundfile.

Digital Signal Processing

In addition to being able to cut, copy and paste regions within a soundfile, it's also possible to alter a soundfile or segment using *digital signal processing (DSP)* techniques. In short, DSP works by directly altering the samples of a soundfile or defined region according to a program algorithm (a set of programmed instructions) so as to achieve a desired result. These processing functions can be performed either in *non-real time* or *real time*.

- *Non-real-time DSP.* Using this method, signal processing (such as changes in level, EQ, dynamics or reverb) is too calculation intensive to be carried out during playback. Instead, the processor must perform the DSP calculations and write the new file to disk as a separate file. Upon playback, the processed region is then joined with the unprocessed part of the original file in a way that's seamless and inaudible.

- *Real-time DSP.* This process differs from its non-real-time counterpart in that the program is capable of using the computer's CPU or additional hardware to perform complex DSP calculations during actual playback. Because no calculations are written to disk in an off-line fashion, significant savings in time and disk space can be realized when working with productions that involve complex or long processing events. In addition, real-time processing systems often embed signal processing instructions within the session file that allows gain changes and effects to be recalled, automated or changed at any time.

Basic DSP Techniques

The following section details many of the basic DSP functions that can be found on a hard-disk editing system.

Amplitude

Besides basic cut and paste techniques, processing the amplitude of a signal is one of the most common types of changes that are likely to be encountered. These include such processes as gain changing, fading and normalization.

Gain changing relates to the altering of a region or track's overall amplitude level, such that a signal can be proportionally increased or reduced to a specified level (often in dB or percentage value).

In order to increase a soundfile or region's overall level or signal resolution, a function known as *normalization* can be used. Normalization (Figure 6.46) refers to changes in level whereby the file's greatest signal amplitude will be at 100 percent (full digital signal level), with all other levels in the soundfile or region being changed to proportionately match.

Figure 6.46. Original signal and normalized (full-gain) signal level.

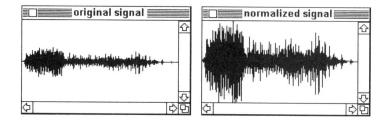

Fades and Crossfades

The *fading* (either in or out) of a region is accomplished by increasing or reducing a signal's relative amplitude over a defined duration. For example, fading in a file (Figure 6.47a) proportionately increases a region's gain from infinity (zero) to full gain. Likewise, a fade-out has the opposite effect of creating a transition from full gain to infinity (Figure 6.47b). These DSP functions have the advantage of creating a much smoother transition than would otherwise be humanly possible when performing a manual fade.

Figure 6.47. Examples of various fade curves.
a. Fade-in.
b. Fade-out.

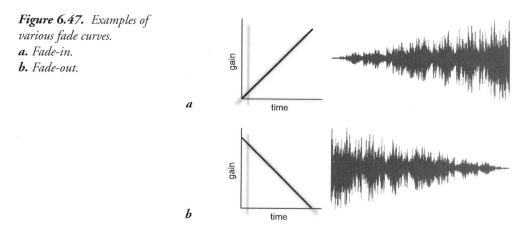

A *cross-fade* (or *X-fade*) is often used to smooth the transition between two audio segments that either are sonically dissimilar or don't match in amplitude at a particular edit point (a condition that would otherwise lead to an audible "click" or "pop").

A crossfade basically is a fade-in and fade-out that overlaps to create a smooth transition from one segment to the next (Figures 6.48 and 6.49). Technically, this process averages the amplitude of the signals over a user-definable length of time to mask the offending edit point.

Figure 6.48. Example of a cross-faded stereo soundfile. (Courtesy of Steinberg, www.steinberg.de)

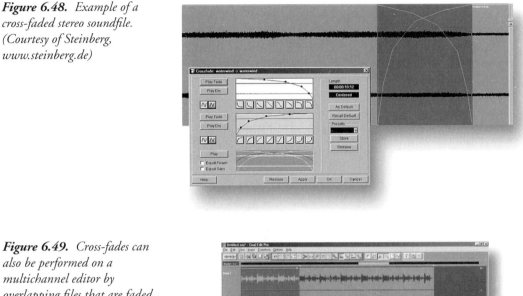

Figure 6.49. Cross-fades can also be performed on a multichannel editor by overlapping files that are faded in and out. (Courtesy of Syntrillium Software Corp., www.syntrillium.com)

Advanced DSP Editing Tools

Most hard disk recording systems offer editing functions that go beyond basic cut and paste and gain commands. However, the number of processing functions, the degree of complexity and their flexibility will vary from system to system. Depending on its capabilities, these functions can either be performed in real or non-real time.

Equalization

Digital EQ has become a common feature that varies between systems in flexibility, layout and musicality (Figures 6.50 and 6.51). Most systems can give full parametric control over the entire audible range, with a variable degree of bandwidth (Q), allowing the range of affected frequencies to be changed from being subtle and broadband . . . to tightly controlled and severe (i.e., a notch filter).

Figure 6.50. Waves Q10 paragraphic EQ screen. (Courtesy of Waves, www.waves.com)

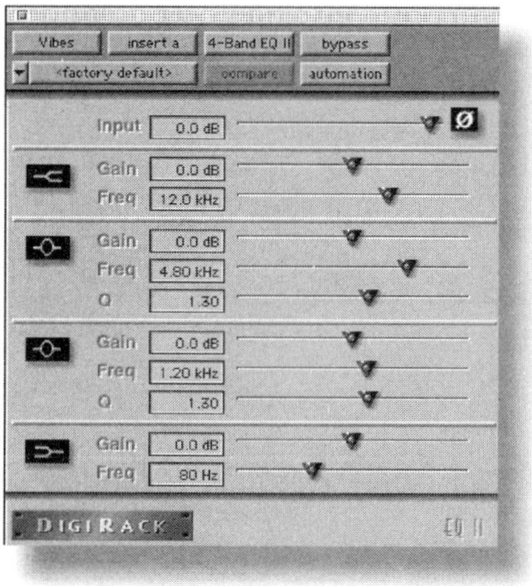

Figure 6.51. Pro Tools EQ screen. (Courtesy of Digidesign, www.digidesign.com)

Dynamic Range

Dynamic range processors can be used to change the signal level of a program. Using algorithms that emulate a compressor (a device that reduces gain by a ratio that's proportionate to the input signal), limiter (reduces gain at a fixed ration above a certain input threshold) or expander (increase the overall dynamic range of a program), the dynamics of a region, single track or entire program can be adjusted with complete flexibility and repeatability and often under the system's automation (Figure 6.52).

Figures 6.52. *Cool Edit Pro's dynamic processing screen. (Courtesy of Syntrillium Software Corp., www.syntrillium.com)*

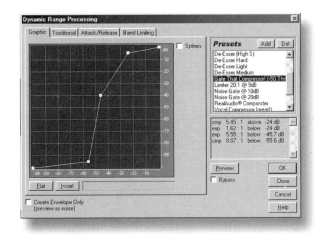

Pitch and Time Change

The *pitch change* function lets you shift the relative pitch of a defined region or entire soundfile either up or down by a specific percentage ratio or musical interval. Many systems will shift pitch by determining a ratio between the present and the desired pitch and then add (lower pitch) or drop (raise pitch) samples from the existing region or soundfile (Figure 6.53).

Figure 6.53. *Basic disk-based pitch shifting techniques.*
a. *Digital audio can be shifted downward by interpolating and adding data to the original segment and then reproducing the data at its original sample rate.*
b. *Digital audio can be shifted upward by dropping samples from the segment and then reproducing this data at its original sample rate.*

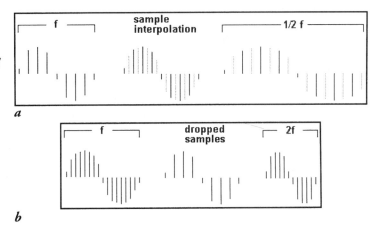

In addition to raising or lowering a soundfile's relative pitch, most systems can combine variable sample rate and pitch shift techniques to alter the *duration* of a region or soundfile (Figure 6.54). These systems can perform pitch- and time-related combinations such as the following:

- *Pitch shift only.* A program's pitch can be changed while recalculating the file so that its length remains the same.

- *Change duration only.* A program's length can be changed while shifting the pitch so that it matches that of the original program.

- *Change both pitch and duration.* A program's pitch can be changed while also having a corresponding change in length.

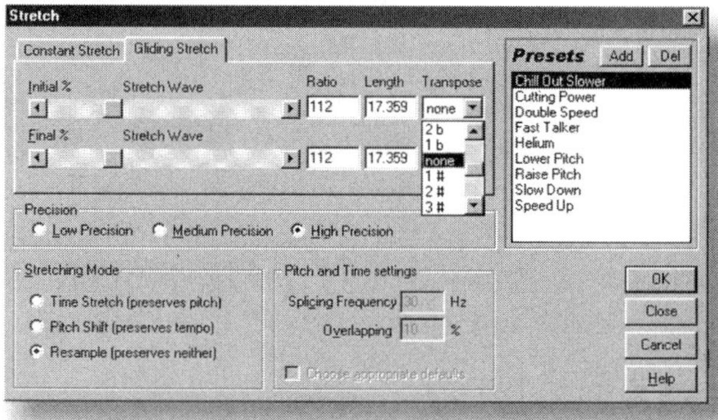

Figure 6.54. A time/Stretch function dialog box. (Courtesy of Syntrillium Software Corp., ww.syntrillium.com)

Hard Disk Editing/Recording Systems

For the remainder of this section, we'll be looking at the various types of hard disk system options that are currently available on the market.

Hard Disk Multitrack Recorders

With the advent of modular digital multitrack recorders, the more expensive analog multitrack and much more expensive digital multitrack recorders became less and less of a recording medium option for the average project and professional production studio. A certain number of working professionals and enthusiasts are opting not to go with tape-based recording systems in favor of the multitrack hard disk recorder (Figures 6.55 through 6.57).

Figure 6.55. Tascam MX-2424 PB multitrack hard disk recorder. (Courtesy of Tascam, www.tascam.com)

Figure 6.56. *Mackie HDR 24/ 96 multitrack hard disk recorder. (Courtesy of Mackie Designs, www.mackie.com)*

Figure 6.57. *Euphonix r-1 multitrack recorder. (Courtesy of Euphonix, www.euphonix.com)*

Unlike editors that are based around the personal computer, these dedicated hardware systems mimic the basic transport, operational and remote controls of a traditional multitrack recorder. In essence, the basic allure of such systems is the fact that they offer simple, dedicated multi-track controls, while offering the speed and flexibility benefits of random access audio.

Hard Disk Portastudios

Hard disk portastudios (Figure 6.58) combine the mixing and simple control surface of a tape-based portastudio system with the speed, quality and flexibility of a hard disk recorder. This type is often chosen by musicians, as it gives the user a portable mixing/control surface that can quickly and easily record and access digital audio files.

Figure 6.58. *Fostex VF-16 track digital multitracker. (Courtesy of Fostex Corp. of America, www.fostex.com)*

Digital Audio Editors

By far, the most commonly used hard disk recording systems for sound effects, music editing, broadcast, desktop/Internet audio and project studio recording are computer-based *digital audio editors*. These 2-channel and multichannel systems can be found in almost every imaginable production environment because they're often fast and cost-effective.

In essence, a digital audio editor makes use of a personal computer's existing processing, disk storage and data I/O hardware systems to perform a wide range of audio editing, processing and production tasks. By offering a graphic, on-screen production environment, these software programs are capable of handling audio production tasks that include mono, stereo and multi-track recording, signal processing, complex mix operations . . . all the way through to the final mastering process (a number of them can even burn the CD directly from the program).

As their name implies, *2-channel editors* (Figures 6.59 and 6.60) are capable of recording, processing and assembling sounds into a finished mono or stereo soundfile. These are capable of (but not limited to) handling such tasks as desktop/internet audio, processing files (EQ, effects, noise reduction, etc.), altering a files format (changing sample/bit rates and saving files in different formats, i.e., MP3) and mastering a soundfile for burning to CD.

Figure 6.59. Bias Peak
2-channel editor for the Mac.

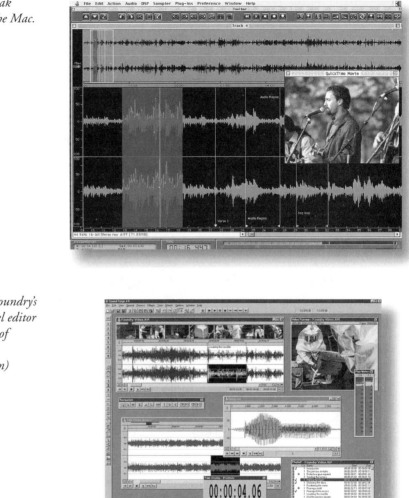

Figure 6.60. SonicFoundry's
Sound Forge 2 channel editor
for the PC. (Courtesy of
Sonic Foundry, Inc.,
www.sonicfoundry.com)

Multichannel hard-disk recording editors (Figures 6.61 through 6.64) work in much the same way as their 2 channel counterparts, except that any number of soundfiles or defined regions can be pasted into any of the program's on-screen "tracks." Each region or group of regions can then be assigned to one or more of the system's multichannel outputs.

Figure 6.61. *Cool Edit Pro 2 channel and multichannel editor for the PC. (Courtesy of Syntrillium Software Corp., www.syntrillium.com)*

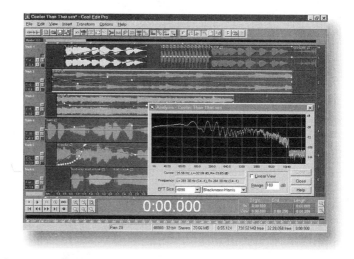

Figure 6.62. *Steinberg Wavelab 3 multichannel editor for the Mac/PC. (Courtesy of Steinberg, www.steinberg.de)*

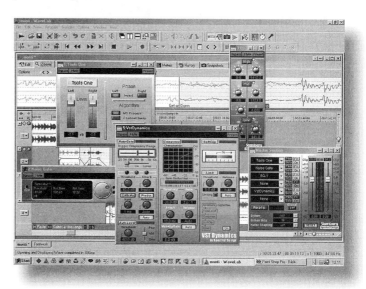

Such flexible, multichannel systems are often extremely useful in MIDI production because they allow vocals, acoustic instruments, effects and other continuous soundfiles to be integrally locked to a MIDI sequence (via MTC or by importing a MIDI file directly into the program). Each soundfile or region can be edited, processed, looped and slipped in time to easily match on-screen action or dialog. It can also be used in a broadcast setting where station IDs, spots and various effects can be built up. The system's individual on-screen track structure and multiple outputs often provide extensive signal processing and mixing control within an on-screen audio production environment.

Figure 6.63. *Bias Deck multichannel editor for the Mac.*

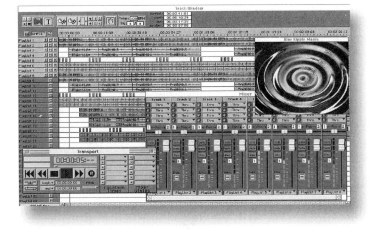

Figure 6.64. *MOTU Spark multichannel editor for the Mac.*

DSP Plug-Ins

In addition to offering a range of built-in signal processing effects, most digital audio editors allow you to effect a region or soundfile by using third-party DSP applications, known as *plug-ins* (Figures 6.65 through 6.68). These task-oriented programs have come about as the result of a set of programming standards that have been adopted for each operating system (OS) or program manufacturer. As of this writing, the most popular platforms include the following:

- *DirectX.* DSP platform for the PC, running under Windows 95, 98, NT and 2000.
- *AudioSuite.* DSP platform for the Mac.
- *VST.* DSP platform for the Mac or PC, running under programs that conform to the Steinberg VST plug-in platform.
- *TDM.* Real-time DSP platform for the Mac or PC, running under Digidesign soft/hardware.

Figure 6.65. Plug-in examples from the Waves Gold Native plug-in pack. (Courtesy of Waves, http://www.waves.com)

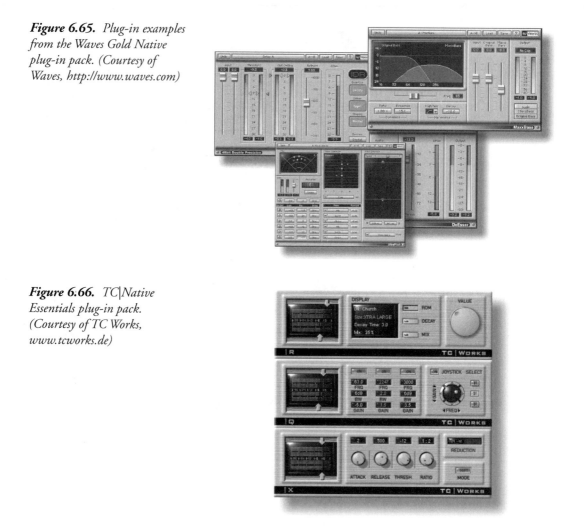

Figure 6.66. TC|Native Essentials plug-in pack. (Courtesy of TC Works, www.tcworks.de)

These popular software applications (which are being programmed by major manufacturers and third-party startups alike) have helped to shape the face of hard disk recording, by allowing us to pick and choose those plug-ins that best fit our personal production needs. As a result, new companies, ideas and task-oriented products are constantly popping up on the market . . . literally on a monthly basis.

Figure 6.67. *Ray Gun noise-reduction plug-in. (Courtesy of Arboretum Systems, Inc., www.arboretum.com)*

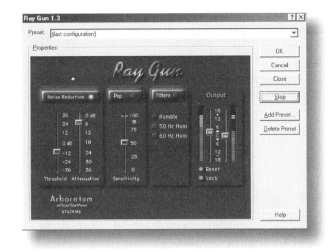

Figure 6.68. *Acoustic Mirror ambience modeler plug-in. (Courtesy of Sonic Foundry, www.sonicfoundry.com)*

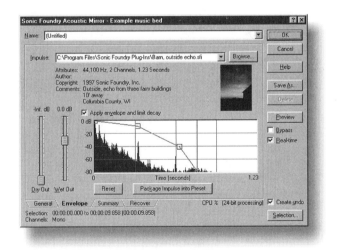

Digital Audio Workstations

In recent years, the term *digital audio workstation (DAW)* has increasingly come to signify a dedicated or computer-based hard disk recording system that offers advanced multitrack editing, signal processing and integrated peripheral features in one package (Figures 6.69 and 6.70).

Figure 6.69. *Soundcscape R.ed hard disk editing workstation. (Courtesy of Soundscape Digital Technology LTD., www.soundscape-digital.com)*

Figure 6.70. *Pro Tools hard disk editing workstation for the Mac. (Courtesy of Digidesign, a division of Avid, www.digidesign.com)*

Although a workstation can perform a wide range of audio-related functions, one of its greatest advantages is the ability to integrate a wide range of applications and devices into a single, connected audio production environment. In effect, these systems are ideally suited to integrate audio, video, MIDI and hardware peripherals together under a single, multifunctional umbrella that can freely communicate data and perform tasks related to sequencing, sample/playlist editing, sampling, hard disk recording, digital signal processing, synthesis/resynthesis, mastering, music printing, etc.

Throughout music and audio production history, we've become used to the idea that certain devices were only meant to perform a single task: a recorder records and plays back, a limiter limits and a mixer mixes. In response to this, I always liken digital audio editors and workstations to chameleons that can change their functional "colors" to match the task at hand. In effect, a digital audio workstation isn't so much a device as a systems concept that can perform a wide range of multichannel audio production tasks with ease and speed. Some of the characteristics that can (or should be) displayed by such systems include the following:

- *Integration.* One of the major functions of a dedicated or computer-based workstation is to provide centralized control over digital audio recording, editing, processing and signal routing, as well as to provide both transport and/or time-based control over MIDI/electronic music systems, external tape machines and videotape recorders.

- *Communication.* A workstation should be able to communicate and distribute digital audio data (such as AES/EBU, S/PDIF, SCSI, SMDI and/or the MIDI sample dump standard) throughout the system. Digital timing (wordclock) and synchronization (SMPTE time code and MTC) should also be supported.

- *Speed and flexibility.* These are probably a workstation's greatest assets. After you become familiar with a particular system, most production tasks can be tackled in far less time than would be required using similar analog equipment. Many of the extensive signal processing features would simply be next to impossible to accomplish in the analog domain.

- *Expandability.* The "ideal" workstation should be able to integrate new hardware and/or software components into the system with little difficulty.

- *User-friendly operation.* An important element of a digital audio workstation is its central interface unit . . . you! The operation of a workstation should be relatively intuitive and should forego any attempt at obstructing the creative process by speaking "computerese."

When choosing a system for yourself or your facility, be sure to take all these considerations into account. Each system has its own strengths and weaknesses. When in doubt, it's always a good idea to research the system as much as possible before you commit to it. Feel free to contact your local dealer for a test drive. As with a new car, purchasing a digital audio workstation can potentially be an expensive proposition that you'll probably have to live with for a while. Once you've made the right choice, you can get down to the business of making music.

Hands-on Control

Often, one of the more common complaints that some people have against the digital audio editor and workstation environment (particularly when relating to the use of on-screen mixers), is the lack of a hardware controller that gives the user access to hands-on controls. This has been addressed by major manufacturers and third-party companies in the form of a *hardware controller interface* (Figures 6.71 through 6.73).

These controllers generally mimic the design of an audio mixer, in that they offer slide or rotary gain faders, pan pots, solo/mute and channel select buttons, and they often include a full transport remote. The channel select buttons are often used to assign a series of grouped controls that relate to EQ, effects and dynamics to the controls on a mixer's virtual on-screen channel strip. Often, these controllers are designed as a group of eight strips that can be switched in banks to access to any number of grouped inputs on a virtual mixer (i.e., 1–8, 8–16, 17–24, etc.).

Commands are most commonly transmitted between the controller and audio editor via device-specific MIDI sys-ex messages. As such, in order to be able to integrate a controller into your system, the digital editing program must be specifically programmed to accept the control codes from the controller device that's being used. More recently, controller surfaces have sprung on the market that are able to communicate these messages to its host via the easy-to-use USB protocol.

Figure 6.71. Tascam US-428 hardware controller interface. (Courtesy of Tascam, www.tascam.com)

Figure 6.72. Motormix hardware controller interface. (Courtesy of CM Automation, www.cmautomation.com)

Figure 6.73. Hui hardware controller interface. (Courtesy of Mackie Designs, www.mackie.com)

Those of you who already have a digital mixer (such as Yamaha 01, 02r, etc.) may be able to remotely control your digital audio editor or workstation directly from the mixer's surface. That's to say, if the editor has been programmed to accept the mixer's control codes, these devices can be connected to an assigned MIDI in port . . . and away you go!

Going Native

With the advent of faster CPUs, many computer systems are now able to handle the workload of dealing with basic program operation, signal processing and passing audio to multiple I/O ports, without the need for special, dedicated hardware. As such, newer software systems (called *native digital audio editors*) have now begun to place the burden of processing and I/O signal routing entirely upon the personal computer and its operating system (Figures 6.74 and 6.75).

Figure 6.74. *Mx51native surround-sound digital audio workstation. (Courtesy of Minnetonka Audio Software, www.minnetonkaaudio.com)*

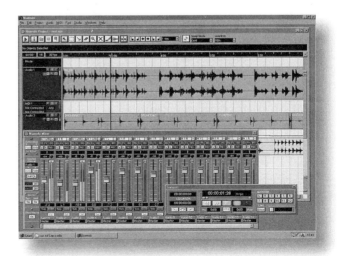

Figure 6.75. *Nuendo native surround-sound digital audio workstation. (Courtesy of Steinberg, www.nuendo.com)*

Native DAWs differ from their hardware-dependent counterparts in two ways:

- Dedicated hardware is often more expensive than off-the shelf components that are often used by native processors. The use of dedicated peripherals can drop thousands off the price of a high-end system.

- In most cases, program and plug-in software code must be specifically written for a dedicated hardware system. If you're using ProTools with the TDM bus, you're OK, because there are tons of third-party plug-in options. However, lesser-known systems can leave you with a limited number of hardware, plug-in and I/O options.

In defense of editors that are built around dedicated hardware, these processors will often take much of the processing off of the computer's CPU and can often handle more tracks and real-time processing functions without bogging down the system (although present-day hot-rod processor/hardware systems are often fast enough to handle most jobs without difficulty).

As of this writing, one of the biggest drawbacks to native technology (at least when slower or problematic systems are involved) has to do with the issue of latency. Quite literally, this relates to problems that are associated with delays (measured in milliseconds) that accumulate as signals pass from the digital audio interface and through the many processing and I/O routing chains. During a mix, these delays (at least, if all tracks are delayed equally) won't be a problem. However, when recording a synth track, you might actually hear the delayed monitor sound shortly after hitting the keys (not a happy prospect). With the advent of faster computers, improved device drivers and better programming, latency has now been reduced to levels that are so small as to not be noticeable . . . a trend that will assuredly improve with the tides of technological advancement.

Loop-Based Groove Editors

Groove editors are loop-based music production software systems (Figure 6.76) that are designed to let you drag and drop prerecorded or user-created loops and audio tracks into a graphic multitrack interface. With the help of custom, royalty-free loops (available from the manufacturer and from third-party companies), users can create an ambient, groove or any other type of track by simply dragging them onto the program's track view, and then arranging the loops into a multitrack project.

One of the most interesting aspects of these editors is their ability to adjust the musical key and tempo of a loop. This means that loops can be automatically adjusted in length and pitch, so as to fit in time with previously existing loops (a not-so-trivial feat that often takes a lot of patience to successfully pull off manually). At any point during the creation of a composition, tracks, vocals or live instruments can be recorded directly to a track to give it a live performance feel. Once done, the project can be mixed down (with volume, pan and effects for each track) to a finished soundfile (some will even burn the file to CD from within the program).

Figure 6.76. Acid Pro loop-based music composition editor for the PC. (Courtesy of Sonic Foundry, www.sonicfoundry.com)

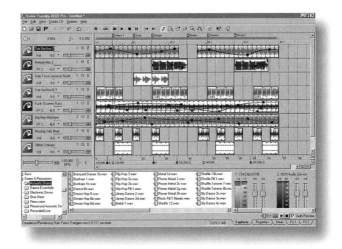

As with most editors, these software groove machines can be locked to sequencers and other devices via MIDI time code and support multiple sound cards and output ports.

Optical Recording Systems

The ability to record and/or play back audio from a removable, optical storage medium has several distinct advantages from the standpoints of portability, low medium cost and available hardware choices . . . especially when considering the dominant worldwide stature of the CD, CD-ROM, CD-R, CD-RW, DVD . . . and to a lesser extent, the minidisc.

CD Burning

One of the biggest revolutions to place audio, data and various media forms into the hands of just about everybody has occurred with the widespread acceptance of the *CD-R* (recordable). With the availability of dedicated CD burning hardware (Figure 6.77) and the wildly popular CD burning drives for the computer, musicians, engineers, DJs, etc., are now able to produce their own discs with relative ease.

Figure 6.77. ML-9600 high-resolution master disk recorder. (Courtesy of Alesis Studio Electronics, Inc., www.alesis.com)

Currently, two types of CD-recording media are commonly found: the CD-R and *CD-RW* (rewritable). These media use a dye whose reflectivity can be altered, so that data can be burned to disc using a number of available writing options:

- *Disc-at-once.* This mode continuously writes the data onto CD without any interruptions. All of the information is transferred from hard disk to the CD in a single pass, with the lead-in, program and lead-out areas being written to disc as an uninterrupted event.

- *Track-at-once.* This allows a session to be written as a number of discrete events (called tracks). With the help of special software, the disc can be read before the final session is fixated (a process that "closes" the disc into a final form that can be read by any CD or CD-ROM drive).

- *Multisession.* Discs written in this mode allow several sessions to be recorded onto a disc (each containing its own lead-in, program data and lead-out areas), thereby allowing data to be recorded onto the free space of a previously recorded CD. It should be noted that older drives might not be able to read this mode and will only read the first available session.

While the altering of the data pits on a CD-R is permanent, CD-RWs can be erased and rewritten any number of times (often figured in the thousands). When using a specially designed CD-RW drive (which can also burn standard CD-Rs), this medium type is excellent for creating data backups and media archiving. In addition, many of the newer CD and MP3 disc players are capable of reading rewritable CD-RW media.

Packet Writing

With the adoption of the Universal Disk Format (UDF, a CD burning technology more commonly known as *packet writing*), small packets of variable or fixed length data can be added to a CD-R disc "one file at a time." When used in conjunction with software that can convert data to UDF, a CD drive can actually be turned into a physical drive letter on your computer. This means that you can simply drag-and-drop a file, directory or selected files to a CD-R or CD-RW (just as you would to any other drive) or save a sound or program file directly to disc. Since the data "packets" are smaller than the CD-R's memory buffers, it's impossible to have a buffer underrun error (a data transfer condition that will render a disc useless). This means that you could even copy files from a floppy directly to your CD without any problem! Before packet writing came along, this simply wasn't possible.

Burning/Mastering Software

Currently, a number of software programs and digital audio editors are capable of burning audio directly to a CD recorder drive. While most of these programs (Figures 6.78 and 6.79) let you drag soundfiles into a cue list (up to 99 index tracks can be recorded on a single disc), other programs will let you to place index cue markers within a single soundfile (allowing you to place index markers anywhere within a song), vary the length of silence between tracks (which is usually defaulted to 2 seconds, but can be defined by some programs to be of any length) or even processed with EQ, noise reduction and other parameters.

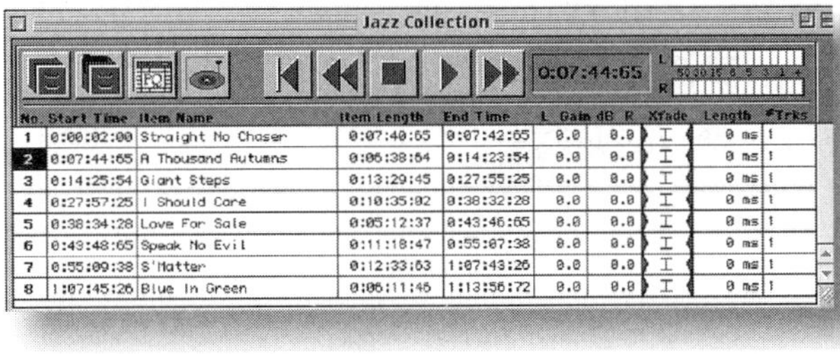

Figure 6.78. *MasterList CD burning software for Pro Tools. (Courtesy of Digidesign, a division of Avid, www.digidesign.com)*

Figure 6.79. *Hotburn Pro CD Burning software for the PC. (Courtesy of Asimware Innovations, www.asimware.com)*

On a final note, many burning programs allow you to enter artist, title and track names. Since a growing number of computer-based CD players can recognize these fields, adding this information can help an audience know more about the artist, copyright name/dates and the names of songs.

Minidisc

Meant as a consumer item for recording CDs onto a rewritable optical medium, the *minidisc* has actually begun to grow in popularity as a medium for recording and storing CD and MP3 audio. Since up to 74 minutes of compressed audio can be stored onto this optical medium (the compression codec is similar to that used with MP3 encoding—further info on MP3 can be found in Chapter 8), several portastudio systems have been designed to use the minidisc for audio storage (Figure 6.80).

Figure 6.80. *Tascam MD-301mkII minidisc recorder. (Courtesy of Tascam, www.tascam.com)*

Given the fact that the minidisc is a rewritable, random-access medium, it provides instant access to high-quality digital audio soundfiles, which can be recorded, bounced, processed and mixed down without the serious degradations in quality that occur when using cassette tape–based portastudios.

CHAPTER 7

◆

MIDI and Electronic Music Technology

Today, MIDI systems are being used by professional and nonprofessional musicians alike to perform an expanding range of production tasks, including music production, audio-for-video and film postproduction, stage production, etc.

This industry acceptance can, in large part, be attributed to the cost-effectiveness, power and general speed of MIDI production. Once a MIDI instrument or device comes into the production picture, there's often less need (if any at all) to hire outside musicians for a project. This is due to the fact that MIDI's multichannel production environment lets a musician compose, edit and arrange a piece with a high degree of flexibility, without the need for recording and overdubbing sounds onto multitrack tape. By this, I'm not saying that MIDI replaces the need for acoustic instruments, mics and traditional performance settings . . . I'm referring to the fact that it's a powerful tool that allows a large number of musicians in a surprisingly wide range of music styles to create music and audio production in ways that are innovative and highly personal.

This affordable potential for future expansion and increased control capabilities over an integrated production system has spawned the growth of an industry that, for the first time in music history, allows an individual to cost-effectively realize a full-scale sound production, not only in his or her own lifetime . . . but in a relatively short time. Since MIDI is both a real-time and non-real-time performance medium, it's also possible to listen to and edit a production at every stage of its development, all within the comfort of your own home or personal project studio.

MIDI systems can be designed to handle a wide range of production tasks with a maximum degree of flexibility and ease that best suits an artist's main instrument, playing style and even the user's personal working habits (choice of equipment, overall hardware/software interface and/or design layout). Each of these advantages is a tribute to the power and flexibility that's inherent within the modern MIDI production environment.

MIDI Production Environments

Currently, a vast number of electronic musical instruments, effects devices, computer systems and other MIDI-related devices are available on the new and used electronic music market. This diversity lets you select the type of device that best suits your own particular musical taste and production styles.

MIDI production systems exist in any number of shapes and sizes and can be designed to match a wide range of production and budget needs. For example, it's common for working and aspiring musicians to install MIDI-based systems in their own homes (Figure 7.1). These production environments can range from those that take up a corner of an artist's bedroom to larger systems that are integrated into dedicated project studios. Systems such as these can be specially designed to handle a multitude of applications and have the important advantage of allowing the artist to produce his or her music in a comfortable environment . . . whenever the creative mood hits. Such production luxuries (which would have literally cost an artist a fortune in the not-too-distant past) are now within the reach of almost every musician.

Figure 7.1. *A home project studio. (Courtesy of Mackie Designs, www.mackie.com)*

MIDI has also dramatically changed the sound, technology and production habits of the professional recording studio. Before MIDI and the concept of the home project studio, the recording studio was one of the only production environments that would allow an artist or composer to combine instruments and sound textures into a final recorded product. With the advent of MIDI, modular digital multitracks (MDMs) and hard disk recording, much of the music production process can be preplanned and rehearsed, or even totally produced and recorded before you step into the studio. This luxury has reduced the number of hours that are required for laying tracks onto multitrack tape to a cost-effective minimum.

Electronic music has long been an indispensable tool for the scoring and audio postproduction of TV commercials, industrial videos, and TV and full-feature motion picture sound tracks (Figure 7.2). For productions that are on a budget, an entire score can be created in the artist's project studio using MIDI, hard disk tracks and digital recorders . . . all at a mere fraction of what it might otherwise cost to hire the musicians, studio and mixdown rooms.

Figure 7.2. James Newton Howard's Studio, LA. (Courtesy of Bryston Ltd, www.bryston.ca)

Electronic music production and MIDI are also at home on the stage. In addition to using pre-programmed drum machines and synthesizers on the stage, most or all of a MIDI instrument's and effects device's parameters can be controlled from a presequenced or real-time controller source. This means that all the necessary settings for the next song (or section of a song) can be automatically called up before it's played, or one or more of a song's parameters can be changed during a live performance.

One of the "media" in multimedia is definitely MIDI. With the advent of *General MIDI* (a standardized spec that makes it possible for any sound card or "GM" compatible device to play back a score using the originally intended sounds and program settings), it's possible (and common) for MIDI scores to be integrated into multimedia games, text document, CD-ROMs and even Web sites.

Since MIDI is simply a series of performance commands (unlike digital audio, which actually encodes the audio information itself), the data processing overhead is extremely low. This means that almost no processing power is required to play MIDI, making it ideal for playing real-time music scores while browsing text, graphics or other media over the Web. . . . Truly, when it comes to weaving MIDI into the various media types, the sky (and your imagination) is the creative and technological limit.

What's a MIDI?

Simply stated, *Musical Instrument Digital Interface (MIDI)* is a digital communications language and compatible hardware specification that allows multiple electronic instruments, performance controllers, computers and other related devices to communicate with each other throughout a connected network (Figure 7.3). It's used to translate performance- or control-related events (such as playing a keyboard, selecting a patch number, or varying a modulation wheel) into equivalent digital messages and then transmits these messages to other MIDI devices where they can be used to control sound generators and performance parameters. The beauty of MIDI is that its data can be easily recorded into a hardware device or software program (known as a *sequencer)*, where it can be edited and transmitted to electronic instruments or other devices in order to create music.

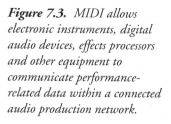

Figure 7.3. *MIDI allows electronic instruments, digital audio devices, effects processors and other equipment to communicate performance-related data within a connected audio production network.*

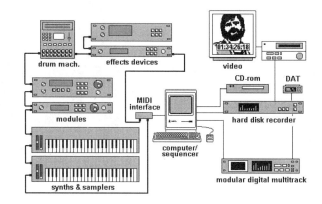

In artistic terms, this digital language is an important medium that lets artists create with a degree of expression and control that was, before its inception, often not possible on an individual level. Through the transmission of this performance language, an electronic musician can compose and develop a song in a practical, flexible, affordable and (one hopes) fun production environment.

In addition to composing and performing music, musicians can have complete control over a wide palette of sounds, their *timbre* (sound and tonal quality) and the overall *blend* (level, panning and other real-time controls). MIDI can also be used to vary the real-time performance and control parameters of electronic instruments, recording devices, control devices and signal processors during a performance.

The term *interface* refers to the actual data communications and hardware link that exists within a connected MIDI network. Through the use of MIDI, it's possible for data to be communicated to all the electronic instruments and devices within a connected network through the transmission of real-time performance and control-related messages. Furthermore, communication with instruments (or individual sound generators in an instrument) can occur using only a single MIDI data communications chain. This is possible as each data line can transmit performance and control messages over 16 discrete channels. This simple fact makes it possible for electronic musicians to record, overdub, mix and play back their performances in a working environment that loosely resembles the multitrack recording process. Once MIDI has been mastered, however, its repeatability and edit control offer production challenges and possibilities that can stretch beyond the capabilities and cost-effectiveness of the traditional multitrack recording environment.

System Interconnections

A MIDI cable (Figure 7.4) consists of a shielded, twisted pair of conductor wires that has a male 5-pin DIN plug located at each of its ends. The MIDI specification uses only 3 of the 5 pins, with pins 4 and 5 being used to conduct MIDI data, while pin 2 is used to connect the cable's shield to equipment ground. Pins 1 and 3 are currently not in use, but are reserved for possible changes in future MIDI applications. Twisted cable and metal shield groundings are used to reduce outside interference, such as radio frequency (RFI) or electrostatic interference, which can serve to distort or disrupt MIDI message transmissions.

Figure 7.4. *Wiring diagram and picture of a MIDI cable.*

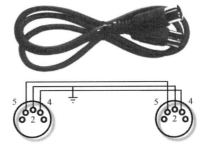

Prefabricated MIDI cables in lengths of 2, 6, 10, 20 and 50 feet often can be obtained from music stores that specialize in MIDI equipment. Fifty feet, however, is the maximum length that's stated by the MIDI specification in order to reduce the effects of signal degradation and external interference that might occur over extended cable runs.

MIDI In, Out and Thru Ports

Three types of MIDI ports are used to connect MIDI devices in a network: *MIDI in*, *MIDI out*, and *MIDI thru* (Figure 7.5):

- *MIDI in port.* Receives MIDI messages from an external source and communicates this performance, control, and timing data to the device.
- *MIDI out port.* Transmits MIDI messages from the device out to another MIDI instrument or device.

- *MIDI thru port.* Provides an exact copy of the incoming data at the MIDI in port and transmits this data out to another MIDI instrument or device that follows in the connected MIDI data chain.

Figure 7.5. *MIDI in, out and thru ports.*

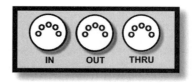

Certain MIDI devices don't include a MIDI thru port. These devices, however, may offer a software-based transmission function that can be selected as a MIDI out port or a *MIDI echo port.* As with the MIDI thru port, the selectable MIDI echo function provides an exact copy of any information received at the MIDI in port and then routes this data to the MIDI out/echo port.

As a general rule, there are only two valid methods of connecting one MIDI device to another (Figure 7.6): connecting the MIDI out port (or MIDI out/echo port) of one device to the MIDI in port of the next device, or connecting the MIDI thru port of one device to the MIDI in port of the next device.

Figure 7.6. *The two valid ways to connect one MIDI device to another.*

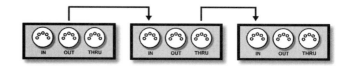

The Daisy Chain

One of the simplest and most common ways to distribute data throughout a MIDI system is the *daisy chain.* This method relays MIDI from one device to the next by retransmitting data that's received at a device's MIDI in port out to another device via its MIDI thru (or MIDI echo) port. In this way, MIDI data can be chained from one device to the next. For example, a typical MIDI daisy chain (Figure 7.7a) will flow from the MIDI out of a source device (such as a controller or sequencer) to the MIDI in port of the second device. The MIDI data being received by the second device is also routed out to its MIDI thru port, which is plugged into a third device's MIDI in port. The data received by the third device is then relayed out to its MIDI thru port, which is plugged into the fourth device's MIDI in port . . . and so on, until the final device in the chain is reached.

A computer can also be (and often is) designated as the master source in a daisy chain, so that a sequencing program can be used to control the playback, channelizing and signal processing functions of an entire system. In Figure 7.7b, the MIDI out of a master MIDI keyboard controller is routed to the computer. The computer's MIDI out port (or ports) can then be relayed throughout the system in a single or multiple line daisy chain fashion.

Figure 7.7. *Example of a connected MIDI system using a daisy chain.*
a. *Typical daisy chain hookup.*
b. *Example of how a computer can be connected into a daisy chain.*

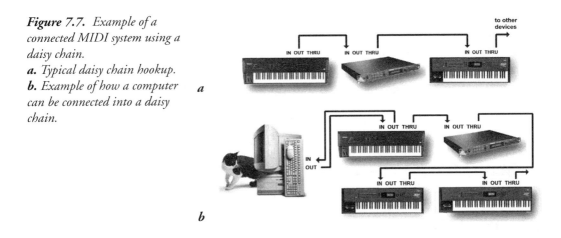

The MIDI Message

MIDI digitally communicates musical performance data between devices as a string of MIDI messages, which are transmitted through a single MIDI line at a speed of 31.25 kbaud (bits/sec). This data can only travel over a single MIDI line in one direction, from a single source to a destination (Figure 7.8a). In order to make two-way communication possible, a second MIDI data line must be used to communicate data back to the first device (Figure 7.8b).

Figure 7.8. *MIDI data can only travel in one direction through a single MIDI cable.*
a. *Data transmission from a single source to a destination.*
b. *Two-way communication using two MIDI cables.*

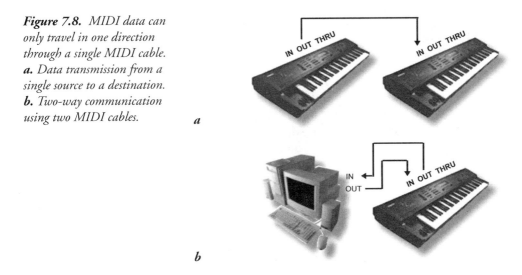

MIDI messages are made up of groups of 8-bit words (known as *bytes*), which are transmitted in a serial fashion to convey a series of instructions to one or all MIDI devices within a system.

There are only two types of bytes that are defined by the MIDI specification: the *status byte* and the *data byte*. Status bytes are used as an identifier to tell the receiving device which particular

MIDI function and channel is being addressed. The data byte information is used to encode the actual numeric values that accompany the status byte. Although a byte is made up of 8 bits, the most significant bit (MSB—the leftmost binary bit within a digital word) is used solely to identify the byte type. The MSB of a status byte is always 1, while the MSB of a data byte is always 0 (Figure 7.9). For example, a 3-byte MIDI note-on message (which is used to signal the beginning of a MIDI note) in binary form might read as shown in the following table:

	Status Byte	Data Byte 1	Data Byte 2
Description	Status/Channel #	Note #	Attack Velocity
Binary Data	(1001 0100)	(0100 0000)	(0101 1001)
Numeric Value	(Note On/Ch. #5)	(64)	(89)

Figure 7.9. *The most significant bit of a MIDI data byte is used to identify between a status byte 1 and a data byte 0.*

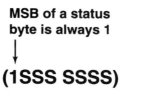

MSB of a status byte is always 1

(1SSS SSSS)

MSB of a data byte is always 0

(0DDD DDDD)

Thus, a 3-byte note-on message of (10010100) (01000000) (01011001) will transmit instructions that would be read as "transmitting a note-on message over MIDI channel #5, using keynote #64 at an attack velocity (the note's volume level) of 89."

MIDI Channels

Just as it's possible for a public speaker to single out and communicate a message to one individual in a crowd, MIDI messages can be directed to a specific device or range of devices in a MIDI system. This is done by embedding a nibble (four bits) within the status/channel number byte that makes it possible for performance or control information to be communicated to a specific device or one of the sound generators in a device over its own channel. Since the nibble is 4 bits wide, up to 16 discrete MIDI channels can be transmitted through a single MIDI cable (Figure 7.10).

Figure 7.10. *Up to 16 channels can be transmitted through a single MIDI cable.*

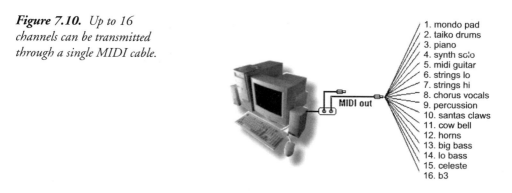

MIDI out

1. mondo pad
2. taiko drums
3. piano
4. synth solo
5. midi guitar
6. strings lo
7. strings hi
8. chorus vocals
9. percussion
10. santas claws
11. cow bell
12. horns
13. big bass
14. lo bass
15. celeste
16. b3

Whenever a MIDI device is instructed to respond to a specific channel number, it will only react to messages that are transmitted on that channel and will ignore channel messages that are transmitted on any other channel.

For example, let's assume that we have a MIDI keyboard controller, a synth and a sampler, which are linked together in a MIDI chain (Figure 7.11). In this instance, we've assigned the first synth to respond to data that's being transmitted over channel 3, while the sampler is set to respond to messages that are sent over channel 5. Setting the controller to transmit over channel 3 means that the first synth can be played when the controller's keys are pressed, while the sampler ignores the messages. Likewise, setting the controller to channel 5 will allow the sampler to be played, while the synth remains silent. Splitting the controller's keyboard (so that the lower octaves will transmit on channel 3, while the upper octaves transmit over channel 5) will allow us to play different musical parts on both instruments at once.

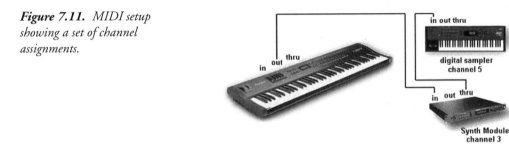

Figure 7.11. *MIDI setup showing a set of channel assignments.*

Using another example, we could create a short song using a keyboard synth that has a built-in sequencer (a device that's capable of recording, editing and playing back MIDI data), a drum machine and another "synth." We could start off by recording a percussion track into the sequencer that's set to transmit through the chain over MIDI channel 10 (the channel that the controlling synth's percussion is set to respond to). Once done, we can set the master synth to transmit notes on channel 3 and then begin recording a melody. We can then go back and add a third musical part on the sampler (which is set to channel 5). Once the song is complete, we can sit back and listen to and have control over each instrument (using a multitude of musical and mixing parameters), in what can be viewed as a virtual multitrack working environment . . . From here, the number of possible channel and/or instrument combinations could become almost infinite.

Channel Messages

Channel messages are used to transmit real-time performance data throughout a connected MIDI system. Channel messages are generated whenever the controller of a MIDI instrument is played, selected or varied by the performer. Such control changes could be generated by playing a keyboard, pressing program selection buttons or moving modulation/pitch wheels (in addition to a multitude of other controller surface types).

Each channel message contains a MIDI channel number in its status byte and therefore can be addressed by a device that's set to the assigned channel number. There are seven Channel Voice message types: Note On, Note Off, Polyphonic Key Pressure, Channel Pressure, Program Change, Control Change, and Pitch Bend Change. These message types are explained in the following list.

- *Note On.* Indicates the beginning of a MIDI note. This message is generated each time a note is triggered on a keyboard, drum machine or other MIDI instrument (by pressing a key, striking a drum pad, etc.). A Note On message consists of three bytes of information: a MIDI channel number, a MIDI pitch number and an attack velocity value (messages that are used to transmit the individual volume levels [0–127] of each note as it's played).

- *Note Off.* Indicates the release (end) of a MIDI note. Each note played through a Note On message is sustained until a corresponding Note Off message is received. A Note Off message doesn't cut off a sound; it merely stops playing it. If the patch being played has a release (or final decay) stage, it begins that stage on receiving this message.

- *Polyphonic Key Pressure.* These are transmitted by instruments that can respond to pressure changes that are applied to the individual keys of a keyboard. A Polyphonic Key Pressure message consists of three bytes of information: a MIDI channel number, a MIDI pitch number and a pressure value.

- *Channel Pressure* (or *Aftertouch*). This is transmitted and received by instruments that respond to a single, overall pressure that's applied to the keys. In this way, additional pressure on the keys can be assigned to control such variables as pitch bend, modulation and panning.

- *Program Change.* This message changes the active voice (generated sound) or preset program number in a MIDI instrument or device. Using this message format, up to 128 presets (a user- or factory-defined number that activates a specific sound-generating patch or system setup) can be selected. A Program Change message consists of two bytes of information: a MIDI channel number (1–16) and a program ID number (0–127).

- *Control Change.* Used to transmit information that relates to real-time control over a MIDI instrument's performance parameters (such as modulation, main volume, balance and panning). Three types of real-time controls can be communicated through control change messages: continuous controllers, shown in Figure 7.12 (which communicate a continuous range of control settings, generally in value ranging from 0–127); switches (controls having an ON or OFF state with no intermediate settings); and data controllers (which enter data either through numerical keypads or through stepped up/down entry buttons). A full listing of control-change parameters and their associated numbers can be found in Figure 7.13.

Figure 7.12. Continuous controller data value ranges.

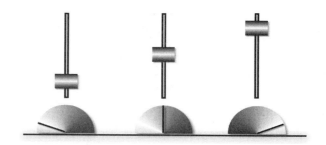

Figure 7.13. *Listing of controller ID numbers, outlining both the defined format and conventional controller assignments.*

14-BIT CONTROLLER MOST SIGNIFICANT BIT			7 BIT CONTROLLERS (continued)		
Controller Hex	Number Decimal	Description	Controller Hex	Number Decimal	Description
		.	.	.	.
00H	0	Undefined	4FH	79	Undefined
01H	1	Modulation Controller	50H	80	General Purpose Controller #5
02H	2	Breath Controller	51H	81	General Purpose Controller #6
03H	3	Undefined	52H	82	General Purpose Controller #7
04H	4	Foot Controller	53H	83	General Purpose Controller #8
05H	5	Portamento Time	54H	84	Undefined
06H	6	Data Entry MSB	.	.	.
07H	7	Main Volume	.	.	.
08H	8	Balance Controller	5AH	90	Undefined
09H	9	Undefined	5BH	91	External Effects Depth
0AH	10	Pan Controller	5CH	92	Tremolo Depth
0BH	11	Expression Controller	5DH	93	Chorus Depth
0CH	12	Undefined	5EH	94	Celeste (Detune) Depth
.	.	.	5FH	95	Phaser Depth
.	.	.	**PARAMETER VALUE**		
0FH	15	Undefined	Controller Hex	Number Decimal	Description
10H	16	General Purpose Controller #1	60H	96	Data Increment
11H	17	General Purpose Controller #2	61H	97	Data Decrement
12H	18	General Purpose Controller #3	**PARAMETER SELECTION**		
13H	19	General Purpose Controller #4	Controller Hex	Number Decimal	Description
14H	20	Undefined	62H	98	Non-Registered Parameter Number LSB
.	.	.	63H	99	Non-Registered Parameter Number MSB
.	.	.	64H	100	Registered Parameter Number LSB
1FH	31	Undefined	65H	101	Registered Parameter Number MSB
14-BIT CONTROLLER LEAST SIGNIFICANT BIT			**UNDEFINED CONTROLLERS**		
Controller Hex	Number Decimal	Description	Controller Hex	Number Decimal	Description
20H	32	LSB Value for Controller 0	66H	102	Undefined
21H	33	LSB Value for Controller 1	.	.	.
22H	34	LSB Value for Controller 2	.	.	.
.	.	.	78H	120	Undefined
3EH	62	LSB Value for Controller 30	**RESERVED FOR CHANNEL MODE MESSAGES**		
3FH	63	LSB Value for Controller 31	Controller Hex	Number Decimal	Description
7-BIT CONTROLLERS			79H	121	Reset All Controllers
Controller Hex	Number Decimal	Description	7AH	122	Local Control On/Off
40 H	64	Damper Pedal (sustain)	7BH	123	All Notes Off
41H	65	Portamento On/Off	7CH	124	Omni Mode Off
42H	66	Sostenuto On/Off	7DH	125	Omni Mode On
43H	67	Soft Pedal	7EH	126	Mono Mode On (Poly Mode Off)
44H	68	Undefined	7FH	127	Poly Mode On (Mono Mode Off)
45H	69	Hold 2 On/Off			
46H	70	Undefined			
.	.	. *continues*			

- *Pitch Bend.* Transmitted by an instrument whenever its pitch bend wheel is moved in either the positive (raise pitch) or negative (lower pitch) direction from its central (no pitch bend) position.

System Messages

As the name implies, *system messages* are globally transmitted to every device in a MIDI chain. This is possible because MIDI channel numbers aren't addressed in the byte structure of a system message. Consequently, any device will respond to these messages, regardless of what MIDI channel or channels the device is assigned to.

System Common messages are used to transmit MIDI Time Code, Song Position pointers, and Song Select, Tune Request and System-Exclusive data throughout the MIDI system or the 16 channels of a specified MIDI port. The following list describes most of the existing System Common messages:

- *MIDI Time Code (MTC).* Provides a cost-effective and easily implemented way of translating SMPTE time code into a format that conforms to the MIDI 1.0 specification. MTC Messages allow time-based code and commands to be distributed throughout the MIDI chain.

- *Song Position Pointer (SPP).* Allows a sequencer or drum machine to be synchronized to an external source (such as a tape machine) from any measure position within a song. Although SPP is used less often than MTC, it can synchronize a location in a MIDI sequence (in measures) to a matching position point on an external device (such as a drum machine or tape recorder) by providing a timing reference that increments once for every six MIDI clock messages, with respect to the beginning of a song.

- *Song Select Message.* Used to request a specific song from the internal sequence memory of a drum machine or sequencer (as identified by its song ID number). After being selected, the song responds to MIDI Start, Stop and Continue messages.

- *Tune Request.* Used to request that an equipped MIDI instrument initiate its internal tuning routine.

- *End of Exclusive (EOX) Message.* Indicates the end of a System Exclusive message.

System-Exclusive Messages

The *system-exclusive (sys-ex) message* allows MIDI manufacturers, programmers and designers to communicate customized MIDI messages between MIDI devices. It's the purpose of these messages to give manufacturers, programmers and designers the freedom to communicate any device-specific data of an unrestricted length, as they see fit. Sys-ex data is commonly used for the bulk transmission and reception of program/patch data, sample data and real-time control over a device's parameters.

The transmission format of a sys-ex message (Figure 7.14) as defined by the MIDI standard includes a sys-ex status header, manufacturer's ID number, any number of sys-ex data bytes and an End of Exclusive (EOX) byte. When a sys-ex message is received, the identification number is read by a MIDI device to determine whether or not the following messages are relevant. This is easily accomplished, as a unique 1- or 3-byte ID number is assigned to each registered MIDI manufacturer. If this number doesn't match the receiving MIDI device, the subsequent data bytes will be ignored. Once a valid stream of sys-ex data has been transmitted, a final EOX message is sent, after which the device will again begin to respond normally to incoming MIDI performance messages.

Figure 7.14. *System-exclusive data (one ID byte format).*

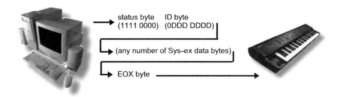

In actual practice, sys-ex makes it possible for MIDI manufacturers, programmers and designers to communicate customized MIDI messages between MIDI devices. The general idea behind sys-ex is that it uses MIDI messages to transmit and receive program, patch and sample data or real-time parameter information between devices. It's sort of like having an instrument or device that's a digital chameleon. One moment it's configured with a certain set of sound patches and setup data and then, after having received a new sys-ex *data dump*, you could easily end up with an instrument that's literally full of new and exciting (or not-so-exciting) sounds and settings. Here are a few examples of how sys-ex can be put to good use:

- *Transmitting patch data between synths.* Sys-ex can be used to transmit patch and overall setup data between synths of identical make and (most often) model. Let's say that we have a Brand X Model Z synthesizer, and as it turns out, you have a buddy across town that also has a Brand X Model Z. That's cool, except your buddy's synth has a completely different set of sound patches in her synth . . . and you want them! Sys-ex to the rescue! All you need to do is go over and transfer your buddy's patch data into your synth (to make life easier, make sure you take your instruction manual along).

- *Backing up your present patch data.* This can be done by transmitting a sys-ex "dump" of your synth's entire patch and setup data to disk, to sys-ex utility software (often shareware) or to your MIDI sequencer. This is so important that I'll say it again . . . Back up your present patch data before attempting a sys-ex dump!!! If you forget and download a sys-ex dump, your previous settings will be lost until you contact the manufacturer, download the dump from their Web site or take your synth back to your favorite music store to reload the data.

- *Getting patch data from the Web.* One of the biggest repositories of sys-ex data is the Internet. To surf the Web for sys-ex patch data, all you need to do is log on to your favorite search engine site, enter the name of your synth and hit "return." You'll be surprised how many hits will come across the screen, many of which are chock-full of sys-ex dumps that can be downloaded into your synth.

I should also point out that sys-ex data that's grabbed from the Web, disk, disc or any other medium will often be encoded using several sys-ex file format styles (unfortunately, none of which are standardized). A sys-ex dump that was encoded using sequencer Z might not be recognized by sequencer Y. For this reason, dumps are often encoded using easily available, standard sys-ex utility programs.

- *Varying sys-ex controller or patch data in real time.* Patch editors or hardware MIDI controllers can be used to vary system and sound generating parameters, in real time. Both of these controller types can ease the job of experimenting with parameter values or changing mix moves by giving you physical or on-screen controls that are often more intuitive and easier to deal with than programming electronic instruments that'll often leave you dangling in cursor and 3" LCD screen hell.

MIDI Machine Control

MIDI Machine Control (MMC) is a protocol that's been designed to allow MIDI sequencers, hard disk recorders, tape and video transports and other recording systems to be remotely controlled (from a hardware device or computer program) via MIDI. This is carried out by transmitting specially designated system-exclusive messages throughout the MIDI system to devices that can respond to MMC (Figure 7.15). An increasing number of devices have begun to support this protocol (including lights and other nonmusical equipment).

Figure 7.15. *Example of a MIDI Machine Control–equipped system.*

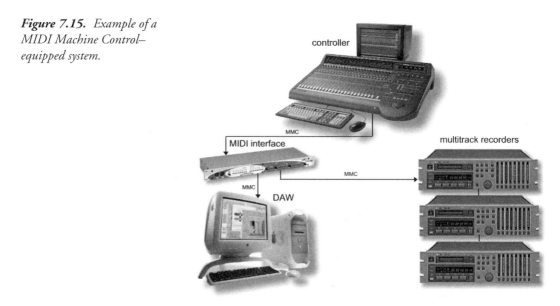

MMC control messages are able to communicate with individual devices within the connected network by assigning identification (ID) numbers to each relevant device. For example, a hard disk recorder might have an ID of 1, while a MIDI sequencer would have an ID of 2, etc. As with all synchronous systems, any number of slaves can be assigned to one master controlling device or program.

Electronic Musical Instruments

Since their inception in the early 1980s, MIDI-based electronic instruments have played a central and important role in the development of music technology and production. These devices (which fall into almost every instrument category), along with the advent of cost-effective analog and digital audio recording systems have helped to shape the music production industry into what it is today. In fact, it's the combination of these technologies that has made the personal project studio into an important driving force behind modern-day music.

Inside the Toys . . .

Although electronic instruments often differ from one another in looks, form and function; they almost always share a standard set of basic building-block components (Figure 7.16). These include the following:

- *Central processing units (CPU).* One or more dedicated computers (often in the form of a specially manufactured microprocessor chip) that contain all the necessary brains to control the hardware, voice data and sound-generating capabilities of an instrument or device.

- *Performance controllers.* These include keyboards, drum pads, wind controllers, etc., for entering performance data directly into the electronic instrument in real time, or for transforming a performance into MIDI messages. Not all instruments have a built-in controller. These devices (commonly known as *modules*) contain all the necessary processing and sound generating circuitry; however, they save space by eliminating redundant keyboards or other controller surfaces.

- *Control panel.* The all-important human interface of data-entry controls and display panels let you select and edit sounds, route and mix output signals and control the instrument's basic operating functions.

Figure 7.16. *The basic components of an electronic musical instrument.*

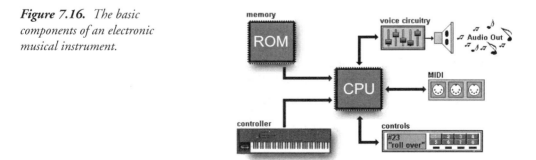

- *Memory.* Used for storing important internal data (such as patch information, setup configurations and/or digital waveform data). This digital data can be stored as *read-only memory* or *ROM* (data that can only be retrieved from a factory-encoded chip, cartridge or CD-ROM). Data can also be stored into *random access memory* or *RAM* (memory that can be stored onto or retrieved from a memory chip, cartridge, hard disk or recordable optical medium).

- *Voice circuitry.* Depending on the device type, this section can either generate analog sounds (voices), or it can be used to instruct digital samples that are permanently recorded into memory to be played back according to a set of specified parameters. In short, it's used to generate or reproduce a sound patch, which can then be amplified and heard via speakers or headphones.

- *Auxiliary controllers.* These are external controlling devices that can be used in conjunction with an instrument or controller. Examples of these include foot pedals (providing continuous-controller data), breath controllers, and pitch-bend and modulation wheels. Certain controllers may only have two or a limited number of switching states (such as a foot controller, sustain pedals or vibrato switches).

- *MIDI communications ports.* These are used to transmit and/or receive MIDI data.

Most (if not all) of an instrument's components building blocks can be accessed or communicated between several devices (in one form or another) via MIDI. In fact, there are many ways in which MIDI can be used to communicate performance, patch and system setup and even audio data between devices. For example, playing a synth's keyboard could transmit performance messages to a synth module. Or, all of an instrument's sound-patch data settings could be transmitted to or received from a sequencer or other instrument as MIDI system-exclusive messages (sys-ex). Or, sampled audio could be exchanged between samplers using the MIDI sample dump standard or SCSI (a computer protocol that communicates data at high speeds).

For the remainder of this section, we'll be discussing the various types of MIDI instruments and controller devices that are currently on the market. These instruments can be grouped into such categories as keyboards, percussion, MIDI guitars and strings, woodwind instruments and controlling devices.

Keyboards

By far the most commonly encountered instrument within almost any MIDI production facility belongs to the keyboard family. This is due, in part, to the fact that keyboards were the first electronic music devices to gain wide acceptance and that MIDI was initially developed to record and control many of their performance and control parameters.

The two basic keyboard-based instruments are the synthesizer and the digital sampler.

The Synth

A *synthesizer* (Figure 7.17) is an electronic instrument that uses multiple sound generators to create complex waveforms that can be combined (using various waveform synthesis techniques) into countless sonic variations. These synthesized sounds have become a basic staple of modern music and can vary from ones that sound "cheesy," to ones that closely mimic traditional instruments . . . all the way to those that generate other-world, ethereal sounds that literally defy classification.

Figure 7.17. Alesis A6 Andromeda Synth. (Courtesy of Alesis Studio Electronics, Inc., www.alesis.com)

Synthesizers (also known as *synths)* generate sounds and percussion sets using a number of different technologies or program algorithms. The earliest synthesizers were analog in nature and generated sounds using a technology known as *frequency modulation (FM) synthesis.*

The FM process usually involves the use of at least two signal generators (commonly referred to as "operators") to create and modify a voice. Often, this is done through the analog or digital generation of a signal that modulates or changes the tonal and amplitude characteristics of a base carrier signal. More sophisticated FM synths can use up to 4 or 6 operators per voice and also often use filters and variable amplifier types to alter the signal's characteristics into a sonic voice that either roughly imitates acoustic instruments or creates sounds that are totally unique.

Another technique that's used to create sounds is *wavetable synthesis.* This technique works by storing small segments of digitally sampled sound into a read-only memory chip. Various sample-based synthesis techniques use sample looping, mathematical interpolation, pitch shifting and digital filtering to create extended and richly textured sounds that use a very small amount of sample memory.

These sample-based systems are often called "wavetable" synthesizers because a large number of prerecorded samples are encoded within the instrument's memory and can be thought of as a "table" of sound waveforms that can be looked up and utilized when needed. Once selected, a range of parameters (such as wavetable mixing, envelope, pitch, volume, pan and modulation) can be modified to control an instrument's overall sound characteristics.

Synthesizers are also commonly designed into rack- or half-rack-mountable *modules* (Figure 7.18) that contain all of the features of a standard synthesizer, except that they don't have a keyboard controller. This space-saving feature means that more synths can be placed into your system and can be controlled from a master keyboard controller or sequencer, without cluttering up the studio with redundant keyboards.

Figure 7.18. *Orbit Dance Planet synth module. (Courtesy of E-MU/Ensoniq, www.emu.com)*

Software Synthesis and Sample Resynthesis

Since wavetable synthesizers derive their sounds from prerecorded samples that are stored in a digital memory media, it logically follows that these sounds can also be stored on hard disk (or any other medium) and loaded into the RAM memory of a personal computer. This process of downloading wavetable samples into a computer and then manipulating these samples is used to create what is known as a virtual or *software synthesizer* (Figure 7.19).

Figure 7.19. *Model E VST plug-in virtual analog synth. (Courtesy of Steinberg N. America, www.us.steinberg.net)*

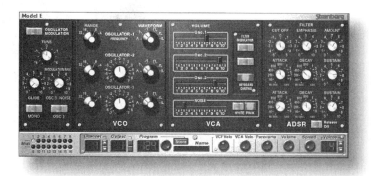

These systems are often capable of generating sounds, importing digital audio samples and linking together various signal processing modules to create sounds of almost any texture or type that you can possibly imagine.

These modules consist of such traditional synthesis building blocks as oscillators, voltage-controlled amplifiers, voltage-controlled filters, and mixers to modify, combine and attenuate the signal's overall harmonic content structure. Since the system exists in software, a newly created sound patch can be easily saved to disk for later recall.

Sound Card Synths

By far, the greatest number of installed synthesizers are those that have been designed into generic PC sound cards. These devices (which can be found in almost every home) are often designed into a single chipset and generate sounds using a simple form of digitally controlled FM synthesis. Although more expensive sound cards will often use wavetable synthesis to create richer and more realistic sounds, both card types will almost always conform to the General MIDI specification, which has universally defined the overall patch and drum-sound structure so that MIDI files will be uniformly played by all synths with the correct voicings and levels. Further information about General MIDI can be found in Chapter 8.

Samplers

A *sampler* (Figure 7.20) is a device that's capable of converting a segment of audio into a digitized form, loading this data into its internal RAM memory and editing it into a looped (or nonlooped) form that can be played and then stored to diskette, disk or disc for later recall.

Figure 7.20. *Akai S5000 professional digital stereo sampler. (Courtesy of Akai Professional, www.akaipro.com)*

Basically, a sampler can be thought of as a wavetable synthesizer that lets you record, edit and reload the samples into its internal memory. Once loaded, these sounds (whose length and complexity are often limited only by RAM size and your imagination) can be looped, modulated, filtered and amplified according to user or factory setup parameters, so as to modify their overall waveshape and envelope. Signal processing capabilities, such as basic editing, looping, gain changing, reverse, sample-rate conversion, pitch change and digital mixing capabilities, are also often designed into these devices.

A sampler's design will often include a keyboard or set of trigger pads that let you polyphonically play samples as musical chords, sustain pads, triggered percussion sounds or sound effect events. These samples can be played according to the standard Western musical scale (or any other scale, for that matter) by altering the playback sample rate. Pressing a low-pitched key on the keyboard will cause the sample to be played back at a low sample rate, while pressing a high-pitched one will cause the sample to be played back at rates that would put Mickey Mouse to shame. By choosing the proper sample-rate ratios, samples can be simultaneously played at various pitches that correspond to standard musical chords and intervals.

Once a set of samples have been recorded or recalled from disk, each sample in a multiple-voice system can be split (using a process known as mapping) across a performance keyboard (Figure 7.21). In this way, a sound can be "mapped" to a specific "zone" or range of notes and/or into velocity "layers," allowing the system to be programmed so that various key pressures will trigger two or more different samples. For example, a single key on a loaded piano setup might be layered to trigger two samples—pressing the key lightly would reproduce a softly recorded sample, while pressing it harder would produce a louder, more percussive sample.

A professional sampler often includes such features as integrated signal processors (which can be assigned to all or individual samples), multiple outputs (offering isolated channel outputs for added mixing and signal-processing power, or for recording individual voices to multitrack tape) and integrated MIDI sequencing capabilities.

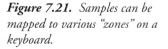

Figure 7.21. *Samples can be mapped to various "zones" on a keyboard.*

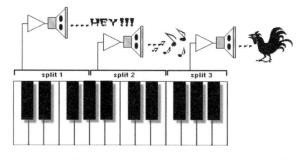

Distributing data to and from a sampler can be done in many ways. Almost every sampler has a floppy or fixed/removable hard drive for saving waveform or system data, and many are equipped with a port for adding a CD-ROM drive. Distribution between samplers or to/from a sample editing program can be handled through the use of the slower MIDI sample-dump standard, or via a high-speed SCSI (small computer system interface) port. More information on these transmission formats can be found in Chapter 6.

Percussion

One of the first applications of sample technology was to record drum and percussion sounds, which made it possible for electronic musicians (mostly keyboard players) to add percussion samples to their own compositions. Out of this sprang a major class of sample and synthesis technology that lets an artist create drum and percussion sounds directly from their synths, drum machines or samplers.

Over the years, MIDI has brought sampled percussion within the grasp of every electronic musician, whatever his/her performance skills—from the frustrated drummer, to professional percussionist/programmers who use their skills to perform live or to build up complex sequenced drum patterns.

The Drum Machine

The *drum machine* (Figures 7.22 and 7.23) is most commonly a sample-based digital audio device that can't record audio into its internal memory. Instead it uses ROM-based, prerecorded samples to reproduce high-quality drum sounds. These factory-loaded sounds often include a wide assortment of drum sets, percussion sets, rare and wacky percussion hits, and effected drum sounds (i.e., reverberated, gated, etc.). Who knows, you might even encounter "Hit me!" screams from the venerable King of Soul—James Brown.

These prerecorded samples can be assigned to a series of playable keypads that are often located on the machine's top face, providing a straightforward controller surface that usually has velocity and aftertouch dynamic capabilities. Drum voices can be assigned to each pad and can be edited using control parameters such as tuning, level, output assignment and panning position.

Figure 7.22. *Alesis SR16 16-bit stereo drum machine. (Courtesy of Alesis Studio Electronics, Inc., www.alesis.com)*

Figure 7.23. *Kits such as the V-Drums from Roland provides standard playing surfaces that lets you be the drum "machine." (Courtesy of Roland Corp. US, www.rolandus.com)*

The assigned percussion sounds can be played live or from a programmed sequence track. Alternately, certain drum machines have a built-in sequencer that's been specifically designed to arrange drum/percussion sounds into a rhythmic sequence (known as a *drum pattern).* These patterns often consist of basic variations on a rhythmic groove or can be built from patterns that are taken from an existing library of wide playing styles (such as rock, country, or jazz). Drum machines that have this feature will often let you combine patterns together into a continuous song. Once a song is assembled, it can be played back using an internal MIDI clock source or can be synchronously driven from another device (such as a sequencer) using an external MIDI clock source.

Although a number of drum machine designs include a built-in sequencer, it's more likely that these studio workhorses will be triggered from a MIDI sequencer. This lets us take full advantage of the real-time performance and editing capabilities that a sequencer has to offer. For example, sequenced patterns can easily be created in step time (where notes are entered and assembled into a rhythmic pattern one note at a time) and can then linked together into a song that's composed of several rhythmic variations. Alternately, drum tracks can be played into a sequencer on-the-fly, creating a live feel, or you can merge step- and real-time tracks together to create a more human-sounding composite rhythm track. In the final analysis, the style and approach to composition is entirely up to you.

Most drum machine designs include multiple outputs that allows individual or groups of voices to be routed to a specific output on a mixer or console. This feature allows these isolated voices to be individually processed (using equalization, effects, etc.) or to be recorded onto separate tracks of a multitrack tape recorder.

Alternative Percussion Voices

In addition to the sounds that can be found in a drum machine, a virtually unlimited number of percussion sounds can be obtained from other sources. For example, most synth wavetables will include several drum and/or percussion setups that are often mapped over the entire keyboard surface. Sampler libraries will almost always include a never-ending range of percussion instruments and drumsets. Soundfiles can be loaded into a hard disk editor to be built up as rhythm tracks, or you can "lift" percussion loops from the inexhaustible number of loop discs that are available on CD. Basically, the sky's only limited by your imagination. . . . Heck, you could even be bold and record your own samples!

Sequencers

Apart from electronic musical instruments, one of the most important tools that can be found in the modern-day project studio is the *MIDI sequencer*. Basically, a sequencer is a digital device that's used to record, edit and output MIDI messages in a sequential fashion. These sequential messages are generally arranged in track-based format that follows the modern production concept of locating separate instruments (and/or instrument voices) onto separate tracks. This traditional interface makes it easy for us humans to view MIDI data as though they were tracks on a tape recorder that moves along a linear timeline.

These "tracks" contain MIDI-related performance and control events that are made up of such channel and system messages as note on/off, velocity, modulation, aftertouch, and program/continuous-controller messages. Once a performance has been recorded into a sequencer's memory, these events can be graphically or audibly edited into a musical performance, played back and saved to any digital storage media for recall at any time.

Hardware Sequencers

Hardware sequencers are stand-alone devices that are designed for the sole purpose of sequencing MIDI data. These systems include a specially designed CPU and operating system, memory, MIDI ports and integrated controls for performing sequence-specific functions. Some of these systems provide a limited number of tracks that can be instantly accessed via a track arm/function select button, while others can access a greater number of tracks from a numeric keypad.

Ease of use and portability are often the advantages of a hardware sequencer, most of which are designed to emulate the basic functions of a tape transport (record, play, start/stop, fast forward and rewind). Generally, a moderate amount of editing features are available, including note editing, velocity and other controller messages, program change, cut and paste and track merging capabilities, tempo changes, etc.

Hardware sequencers commonly use an LCD to display programming, track and edit information. This display type is often limited by its size and resolution and is generally limited to information that relates to one parameter or track at a time.

Integrated Sequencers

Some of the newer and more expensive keyboard synth and sampler designs include a built-in sequencer. These portable workstations have the advantage of letting you take both the instrument and sequencer on the road without having to drag along a whole system.

Integrated sequencers often have the disadvantage of not offering a wide range of editing tools beyond the basic transport functions, punch-in/out commands and other basic edit functions. However, for basic sequencing, they're often more than adequate and can be used with other instruments that are connected in a MIDI chain.

Software Sequencers

By far, the most common sequencer type is the *software sequencer* (Figures 7.24 and 7.25). These programs run on all types of personal computers and take advantage of the hardware and software versatility that a computer can offer in the way of speed, hardware flexibility, digital signal processing, memory management and signal routing.

Figure 7.24. *Cakewalk Pro Audio 9™ piano roll screen. (Courtesy of Twelve Tone Systems, www.cakewalk.com)*

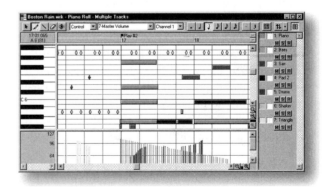

Figure 7.25. *Cubase VST/24 audio production system. (Courtesy of Steinberg N. America, www.us.steinberg.net)*

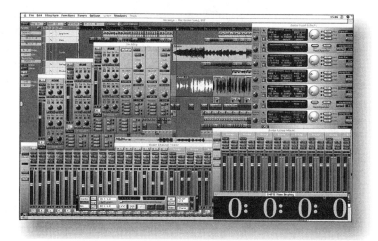

Often, a computer-based sequencer offers numerous functional advantages over its hardware counterpart. Among these are increased graphic capabilities (which gives direct control over track- and transport-related functions), standard computer cut and paste techniques, a windowed graphic environment (allowing easy manipulation of program and edit-related data), routing of MIDI to multiple ports in a system, and the graphic assignment of instrument voices via program change messages (not to mention the ability to save and recall files using standard computer memory media).

A Basic Intro to Sequencers

When dealing with any type of sequencer, one of the most important concepts to grasp is the fact that these devices don't store sound directly—instead, they encode MIDI messages that instruct an instrument to play a particular note, over a certain channel, at a specific velocity and with any optional controller values. In other words, a sequencer stores music-related commands that follow in a sequential order, which tell instruments and/or devices how their voices are to be played or controlled. This means that the amount of encoded data is a great deal less memory-intensive than its hard disk audio or video recording counterparts. Because of this, the data overhead that's required by MIDI is very small. A computer-based sequencer can work simultaneously with the playback of digital audio tracks, video images, Internet browsing, etc., all without unduly slowing down a computer's CPU. In short, MIDI and the MIDI sequencer are an ideal media form for the computer.

As you might expect, many hardware, integrated and software sequencers are currently on the market. Each type and model offers a unique set of advantages and disadvantages. It's also true that each sequencer has its own basic operating "feel," and thus, choosing one over another is totally up to personal preference.

Recording

Commonly, a sequencer is used as a digital workspace for creating personal compositions in environments that range from a bedroom to more elaborate project studios. Whether they're hardware- or software-based, most sequencers use a working interface that's designed to emulate the traditional multitrack recording environment. A tape-like transport lets you move from one location to the next using standard Play, Stop, FF, REW and Rec command buttons. Beyond using traditional record-enable button(s) to arm selected recording track(s), all you need to do is select the MIDI output port (if more than one exists), MIDI channel, instrument patch and other setup information . . . press the "record" button and then start playing.

Once you've finished laying down a track, you can jump back to the beginning of a sequence and listen to your original track while continuing to lay down additional tracks until the song begins to form.

Almost all sequencers are capable of *punching in* and *out* of record while playing a sequence. This common function lets us drop in and out of record on a selected track (or series of tracks) in real time. Although punch-in/out points can often be manually performed "on-the-fly," most sequencers can perform a "punch" automatically. This is generally done by graphically or numerically entering in the measure numbers that mark the punch-in/-out location points. Once done, the sequence can be rolled back to a point a few measures before the punch-in point and the artist can play along while the sequencer automatically performs the switching functions.

In addition to recording a performance one track at a time or all at once, most sequencers allow note values to be entered into sequence one note at a time. This feature (known as *step time)* allows you to give the sequencer a basic tempo and note length (i.e., quarter note, sixteenth note, etc.) and then manually enter the notes from a keyboard or other controller. This data entry style is often (but not always) used with fast, high-tech and dance styles, where a real-time performance just isn't possible or accurate enough for the song.

Whether you're recording a track in real-time or in step-time, you'll often want to select the proper song tempo before recording a sequence. I bring this up because most sequencers are able to output a *click track* that can be used as an accurate, audible guide for keeping in time with the song's selected tempo.

Editing

One of the more important features that a sequencer has to offer is its ability to edit sequenced tracks. Editing functions and capabilities often vary from one sequencer to another.

The main track-editing window is used to display such track information as the existence of track data, track names, MIDI port assignments for each track, program change assignments, volume controller values and other transport commands.

Depending on the sequencer, the existence of MIDI data on a particular track at a particular measure point (or over a range of measures) is indicated by the highlighting of a track range in a way that's highly visible. For example, in Figure 7.24, you'll notice that many of the measure boxes are highlighted. This means that these measures contain MIDI messages, while the other nonhighlighted measures don't. By navigating around the various data display and parameter boxes, it's possible to use cut and paste and/or direct edit techniques to vary almost every facet of the musical composition, including velocity changes, modulation and pitch bend, note and song transposition, quantization and humanizing (factors that eliminate or introduce human timing errors that are generally present in a live performance), as well as control over program or continuous controller messages . . . and the list goes on.

Playback

Once a sequence is composed and saved to disk (or disc), all of the tracks can be transmitted through the various MIDI ports and channels to the instruments or devices to make music, create sound effects for film tracks or even control device parameters in real-time.

Since MIDI data exists as encoded real-time control commands, you can listen to the sequence and make changes at any time. You could change the patch voices, alter the final mix, or change and experiment with such controllers as pitch bend, modulation or aftertouch, even change the tempo and key signature. In short, this medium is infinitely flexible in the number of versions that can be created, saved, folded, spindled and mutilated until you've arrived at the sound and feel that you want. Once done, you'll have the option of using the data for live performance and/or mixing the tracks down to a final recorded medium (such as DAT or hard disk and ultimately to CD), either in the studio or at home.

The MIDI Interface

Although computers and electronic instruments communicate using the digital language of 1s and 0s, computers simply don't understand the language of MIDI without the use of a device that translates the serial messages into a data structure that it can understand. Such a device is known as a *MIDI interface*.

A wide range of MIDI interfaces currently exist that can be used with most computer systems and OS platforms. Here are some of the interface types that you can expect to find:

- *Passive.* A passive interface is a simple external device that plugs into a computer's serial or parallel port. Its sole purpose in life is to give you a MIDI in and one or two MIDI out ports, period! Generally, no battery or power supply is needed, as this interface type often gets its power from the computer itself. The distinct advantage of this type of interface is that it's simple, reliable and cheap.

- *Sound card.* When we're talking about MIDI interfaces, who could pass up the one interface that surpasses all others in numbers (by the millions)? I'm referring, of course, to all of those SoundBlaster PC sound cards and their lookalikes that have a 16-channel MIDI interface built right into them. Almost every sound card, no matter how cheap, will have a 15-pin connector that can be used as a game joystick port or for plugging in a MIDI in, out and thru port adapter.

- *Synth interface.* Certain keyboards and synth modules have MIDI interface capabilities built right into them. This often comes in the form of an interface (or extra 16-channel port) that connects the computer's serial or USB port directly to the instrument.

- *Multiport interface.* The interface of choice for most professional electronic musicians is the external multiport MIDI interface (Figure 7.26). These rack-mountable devices often plug into the computer's serial or parallel port to provide 4, 8 or more independent MIDI in and out ports that can easily distribute MIDI data through separate lines over a connected network.

Figure 7.26. MIDIman USB 8x8 multiport MIDI interface. (Courtesy of Midiman, Inc., www.midiman.net)

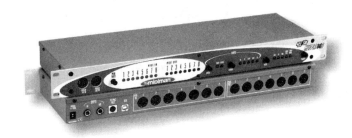

As an added note, multiport interfaces can often be used as a software-controlled patch bay for routing MIDI data from one instrument or device to others on another signal path. Special MIDI processing functions can also be found on most multiport interfaces. Examples of this include the ability to merge several MIDI inputs together into a single data stream, filter out specific MIDI message types (used to block out unwanted commands that might adversely change an instrument's sound or performance) or re-channelize data that's being transmitted on one MIDI channel to another channel that can be recognized by an instrument or device.

Another important function that's handled by most multiport interfaces is synchronization. Synchronization (sync for short) allows other, external devices (such as analog or digital tape recorders, videocassette recorders or other types of recorded media) to be played back using the same timing reference, allowing MIDI and other performance-related events to occur simultaneously.

Interfaces that includes sync features will often read and write SMPTE time code, convert SMPTE to MIDI Time Code (MTC) and allow recorded time-code signals to be cleaned up when copying code from one analog device to another (jam sync). Certain types might be able to read specialized time-code formats that are used by modular digital multitrack (MDM) machines. This option lets you synchronize ADAT or other MDM machine types to a MIDI sequencer and instruments without the need for buying additional hardware. Further reading on all the various types of synchronization can be found in Chapter 9.

Other Software Applications

In addition to sequencing packages that are designed to handle most of the production needs that a musician might encounter in the day-to-day world, other types of software applications exist that offer tools for carrying out specialized tasks. A few of these packages include drum pattern editors, algorithmic composition programs, patch editors and music printing programs.

Drum-Pattern Editor/Sequencers

At any one time, there are generally a handful (OK, maybe two handfuls) of companies that have software or hardware devices that are specifically designed to create and edit drum patterns. These systems often rely on user input and quantization to construct any number of percussion grooves. More often than not these editors rely on a grid pattern display that places MIDI notes or subpatterns representing drum sounds along the vertical axis, while time is represented in metric divisions along the horizontal axis (Figure 7.27). Typically, by clicking on each grid point with a mouse or other input system, individual drum or effect sounds can be built into patterns that are rhythmically diverse and interesting.

Figure 7.27. *The *Drums Wizard drum sequencer. (Courtesy of midibrainz software, www.midibrains.com)*

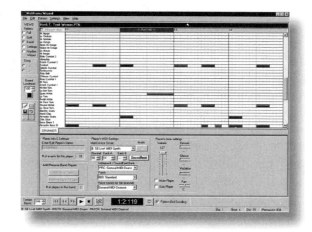

Once created, these and other patterns can be linked together to create a partial or complete song. These editors commonly offer such features as the ability to change MIDI note values (thereby changing drum voices), quantization and humanization, and adjustments to note and pattern velocities. Once completed, the sequenced drum track (or series of tracks) can often be saved in the standard MIDI file format for import into a standard music sequencer.

Algorithmic Composition Programs

Algorithmic composition programs (Figure 7.28) are interactive sequencers that directly interface with MIDI controllers or standard MIDI files to generate a performance in real time, according to programmed computer parameters. In short, once you give it a few basic musical guidelines, it can begin to generate performances or musical parts on its own, so as to help you gain new ideas for a song, create automatic accompaniment, make improvisational exercises, create special performances or just plain have fun.

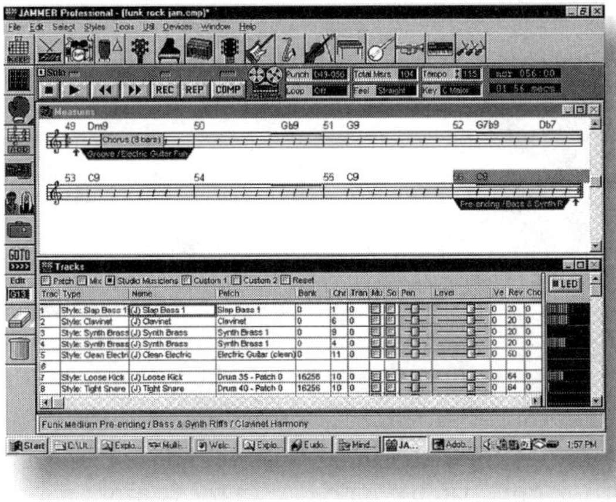

Figure 7.28. Jammer Professional Algorithmic composition program. (Courtesy of Soundtrek, www.soundtrek.com)

This sequencer type can be programmed to control the performance according to musical key, generated notes, basic order, chords, tempo, velocity, note density, rhythms, accents, etc. Alternatively, an existing standard MIDI file can be imported and further manipulated in real time, according to new parameters that can be varied from the computer keyboard, mouse or external MIDI data. Often such interactive sequencers will accept input from multiple players, allowing it to be performed as a collective jam. Once a composition has been satisfactorily generated, a standard MIDI file can be created and exported into any sequencer.

Patch Editors

The vast majority of MIDI instruments and devices store their internal patch data within RAM memory. Synths, samplers or other devices contain information on how the oscillators, amplifiers, filters, tuning and other presets are to be configured, in order to create a particular sound timbre or effect. In addition to controlling sound patch parameters, a unit's internal memory can also store such setup information as effects processor settings, keyboard splits, MIDI channel routing, controller assignments, etc. These settings, which can be manually edited from the device's front panel, can be easily accessed by recalling the patch number and/or name via the device's bank of preset buttons, alpha dial or keypad entry.

Another (and sometimes more straightforward) way to gain real-time control over the parameters of a specific instrument or wide range of MIDI devices is through the use of a patch editor (Figure 7.29). A *patch editor* is a software package that's used to provide on-screen controls and graphic windows for emulating hands-on controls that can often be varied in real-time.

Figure 7.29. *Midi Quest v 8.0*
universal editor/librarian.
(Courtesy of Sound Quest Inc.,
www.squest.com)

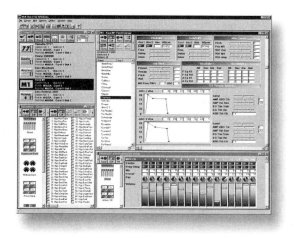

Direct communication between the software/computer system and the device's microprocessor commonly occurs via the real-time transmission and reception of MIDI sys-ex messages. Almost all popular voice and setup-editing packages include provisions for receiving and transmitting bulk patch data between the computer and MIDI device. This makes it possible to save and organize large numbers of patch-data files to floppy or hard disk. Many patch editors can also print out patch parameter settings, or make a print of all of the patches within a particular bank or overall instrument banks.

In addition to software editing packages, there are also hardware solutions for gaining quick and easy access to device parameters via sys-ex. Although they aren't common, certain synthesizer manufacturers will offer device-specific hardware control surfaces that plug directly into the synth. Another option (that's much less device-specific) are MIDI data fader controllers (Figure 7.30). These controllers are often equipped with data faders and "soft" buttons to mix volume levels and vary settings in the MIDI domain. In certain situations, these controllers can be used in a hands-on setting to directly control the volume and mix parameters of a sequencer's and/or hard disk recording system's mix screen.

Figure 7.30. *Phat.Boy MIDI*
Performance Controller,
designed by Keyfax. (Courtesy of
Keyfax Software,
www.keyfax.com)

Music-Printing Programs

In recent years, the field of transcribing musical scores onto paper has been strongly affected by both the computer and MIDI technology. This process has been enhanced through the use of newer generations of software that makes it possible for music notation data to be entered into a computer either manually (by placing the notes onto the screen via keyboard and/or by mouse movements) or via direct MIDI input. Once entered, these notes can be edited in an on-screen environment by using a *music-printing program* (also known as a *music notation program*) that lets the artist change and configure a musical score or lead sheet using standard cut-and-paste edit techniques. In addition, most printing programs can play the various instruments in a MIDI system directly from the score. A final and important program feature is their ability to print out hard copies of a score or lead sheets in a wide number of print formats and styles.

These programs (Figure 7.31) allow musical data to be entered into a computerized score in a number of manual and automated ways (often with varying degrees of complexity and ease). Although scores can be manually entered, most music-transcription programs will generally accept MIDI input, allowing a part to be played directly into a score. This can be done in real time (by playing a MIDI instrument or finished sequence into the program), in step time (entering the notes of a score one note at a time from a MIDI controller) or from an existing standard or program-specific MIDI file.

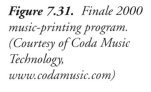

Figure 7.31. *Finale 2000 music-printing program. (Courtesy of Coda Music Technology, www.codamusic.com)*

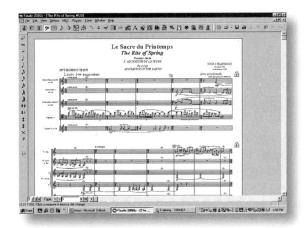

Another way to enter music into a score is through the use of an optical recognition program. These programs let you place sheet music or a printed score into a standard flatbed scanner, scan the music into a program and then save the notes and general layout as a NIFF notation file. This newly created format standard can be directly read by most professional printing programs for further editing, printing and playback.

One of the biggest drawbacks to automatically entering a score via MIDI (either as a real-time performance or from a MIDI file) is the fact that music notation is an interpretive art. "To err is human," and it's commonly this human feel that gives music its full range of expression. It's very difficult, however, for a program to properly interpret these minute, yet important imperfections and place the notes into the score exactly as you want them. (For example, it might interpret a held quarter-note as either a dotted quarter-note or one that's tied to a thirty-second note.) Even though these computer algorithms are getting better at interpreting musical data and quantization can be used to tell a computer to *round* a note value to a specified length, a score will still often need to be manually edited to correct for misinterpretations.

Mixing in the MIDI Environment

Although standard mixing practices are still the norm in most multitrack recording studios, MIDI itself can add to the power and flexibility of a mix by giving you total automation over many of the standard mixing functions, directly from your sequencer. That's right! There's no hidden cost and no additional hardware. . . . Truth is, most sequencing packages can give you access to an amazing amount of computerized automation in a simple, easy-to-use environment.

This automation is accomplished by transmitting MIDI channel messages directly to the receiving instruments or devices in such a way as to provide extensive, real-time control over such parameters as level, panning and modulation (Figure 7.32).

Figure 7.32. *System-wide mixing via MIDI channel messages.*

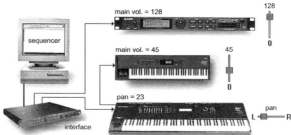

The vast majority of electronic instruments respond to such message parameters as individual note volume, overall track volume, pan position, timbre, effects, program changes, etc. This is done by controlling such MIDI voice parameters as velocity and continuous controller messages, which can be transmitted with values that generally range from 0 (minimum) to 127 (maximum).

- *Velocity messages.* "Velocity" is used to transmit the individual volume levels of each note as they're played. This parameter gives expression to a composition and most sequencers allow these levels to be edited by selecting the affected note or region.

- *Continuous controller messages.* These are often used to control dynamic mixing events such as main volume (an instrument's overall output volume), panning, pitch bend, modulation and aftertouch. (A full listing of these controllers can be found in Figure 7.13.)

As these messages are often transmitted over individual MIDI ports and channels, they can be used to control individual voices or a series of grouped voices, so as to form a virtual software mixer. Since the control data is directly embedded within the sequence, the automation can be saved with the sequence, allowing the mix to be automatically reconfigured whenever the sequenced is opened.

Mixing through MIDI

In modern-day production, it's common for sequencers to work in synchronous tandem with a digital mixer, digital audio workstation or hardware controller surface. Of course, since there are so many systems on the market, the ways in which a sequencer and hardware/software mix controller will interact often vary from system to system.

Console Automation

Although MIDI isn't generally implemented into larger, professional analog consoles, steps are being taken toward integrating MIDI into many project-based analog mixers. Their implementation can range from the simple MIDI control over mute functions, to the creation of static snapshots for reconfiguring user-defined mixer settings . . . all the way to complete dynamic control over all mix functions via MIDI.

Developments in digital technology have brought down the price of digital mixers and consoles to an affordable point. Since all audio, routing and processing functions are digital, full system automation and snapshot recall can be recorded directly into a mixer's internal memory without any difficulty. In addition, most or all of the mixer's controller surface will transmit and respond to MIDI controller messages. This often means that these control moves can be recorded into and reproduced from a MIDI sequence, or can be used as a hardware mixing control surface for a digital audio editor (whose automation controls respond to MIDI controller messages) and/or a MIDI sequencer.

Hardware Controllers

A number of hardware mixing surfaces (which can be seen in Figures 6.70 through 6.72) have also been designed for use with digital audio editors and MIDI sequencers. Although these devices often have the look and feel of a console (particularly a digital one) or a small mixing surface, they don't include any main audio paths; instead, they've been expressly designed only to communicate MIDI controller data to a software editor or sequencer. These remote controllers are usually fully automated, often being equipped with moving faders and indicator lights that alert you to channel and system status.

Effects Automation in MIDI Production

One of the most common ways to automate an effects device that responds to MIDI (from pre-recorded sequence or during a live performance) is through the use of program-change commands. In the same way that electronic instruments can store sound patches into their memory-location registers for later recall, most modern effects devices can store program patches into memory, where they can be automatically recalled using a MIDI program change command.

The use of program-change commands and continuous-controller messages make it possible for signal processing patches and parameters to be altered during the playback of a MIDI sequence. In most cases, a program change that relates to the desired effects patch can be inserted at the beginning of a sequenced track, so that the proper effect will be automatically recalled once the sequence is opened.

In closing, mixing in the MIDI environment isn't only fun, it's often a lesson in understanding the power and flexibility of automated mixing. To begin with, once you've assigned a MIDI channel and MIDI port for each instrument and/or effects device, you can then go about the business of assigning sounds, mixing via MIDI, editing a sequence, synchronizing the various media and devices in your system . . . and most of all, making music. Ideally (as we all know, technology isn't always transparent or bug-free), once you've created a composition, reopening a MIDI file will automatically reconfigure your system's settings so that all of the proper sounds and mix positions will be played back. . . . Ideally, all you need to do is press "play."

CHAPTER 8

Multimedia and the Web

It's no secret that modern-day computers have gotten faster, sleeker and sexier in their overall design. Their hardware and software systems have become more sophisticated (some would add overly complex and crash-prone to this list). However, one of the crowning achievements of the modern computer is the degree of software and media integration that has come to be universally known by the household buzzword *multimedia*.

The combination of working and/or playing with multimedia has found its way into modern computer culture in forms that use various types of hardware and software combinations that combine to bring you a unified experience that involves text, graphics, music, MIDI, video and other media types.

The obvious reason for having these various media types is the human desire to create content with the intention of sharing experiences with others. This has been done for centuries in the form of books, and for decades in the provider-to-client form of movies and television. In the here and now, the Web has created a vehicle that has allowed individuals (and corporate entities alike) to simultaneously communicate a multimedia experience to millions, and then allow each individual to manipulate that experience, learn from it and even respond in a bidirectional fashion. The Web has indeed unlocked the potential for experiencing multimedia events and information in a way that makes each of us a participant, not just a spectator. To me, this is truly the revolution at the dawning of the millennium!

The Multimedia Environment

When you get right down to it, multimedia is nothing more than a unified programming and operating system (OS) environment that allows multiple forms of program data and playback media to simultaneously coexist and to be routed to its appropriate hardware port(s) for output, playback and/or processing (Figure 8.1).

Figure 8.1. Example of a multimedia program environment.

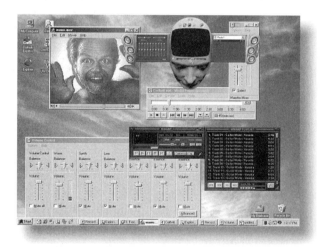

Although many complex design considerations must be in place for a multimedia computer to work properly, once you comprehend two basic concepts . . . the rest will be much easier to understand. These are the concepts of: *multitasking* and the *device driver*.

Multitasking Software

Basically, multitasking is a modern-day form of illusion. Just as a magic trick can be quickly pulled off with sleight of hand or a film that switches frames 24 times each second can create the illusion of continuous movement, the multimedia environment deceives us into thinking that all of the separate program and media types are working at the same time. In reality, the computer uses multitasking to alternately switch from one program to the next in a cyclic fashion. Similar to the film example, newer PCs have gotten so fast at cycling between programs and applications that it gives the illusion that they're all running at the same time.

Another important concept is that of the *device driver*. Briefly, a "driver" acts as device-specific software patch cord that routes media data of a specific type to its appropriate hardware output device (also known as a port), and vice versa (Figure 8.2). Thus, whenever any form of media playback is requested, it will be recognized as being a particular data type, is then routed to the appropriate device driver and is finally sent out to the selected output port (or ports).

Figure 8.2. *Basic interaction between a software application and hardware via a device driver.*

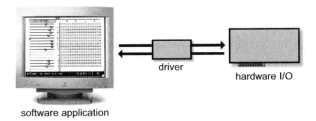

driver

hardware I/O

software application

From all of this, a multimedia computer can be seen as being a device that operates and switches between multiple programs and media players and then, on receiving or playing back any supported media type, routes this data to/or from the appropriate hardware port(s). In short, these devices can simultaneously deal with and direct various forms of media at speeds that are so fast as to be virtually seamless. Cool, huh?

Hardware

Just as there are lots of computer types and levels of complexity, there are an equally wide variety of options for configuring the hardware of a computer to meet the demands of multimedia production. As is often the case, these hardware choices will depend on the application. For the beginner or general home applications, entry-level sound cards are often more than sufficient. These cards are commonly designed to simultaneously handle digital audio, MIDI and basic synthesis sound generation (of either the FM or wavetable type).

For those who'll be producing music or high-quality media, the range and choice of higher-quality sound cards can be daunting . . . ranging from a simple, high-quality audio card that costs less than $100 to full-blown multichannel systems that can range well into the thousands.

A similar situation exists for MIDI. Most entry-level PC sound-card designs include a 16-channel MIDI interface (that doubles as a game port) and a rather chintzy-sounding FM synthesizer. Medium-priced cards might offer two MIDI ports (32 channels) and include a higher-quality wavetable synth. However, most professional systems intended for music production will have a dedicated 2- or multichannel sound-card system and a multiport MIDI interface that can access 8 or more ports, with literally hundreds of MIDI channels (not that you'll have to use them all).

In the final analysis, the choice of hardware is up to you, as are the ways in which you might want to apply them. It's always wise to take a long, hard look at what your production needs are and then research the equipment that will best fit those needs.

CD- and DVD-ROMs

One of the most important developments in the computer, information and entertainment fields has been the introduction of the *CD-ROM*. Unlike its older brother, the CD-audio disc, CD-ROMs can store up to 680 or 700 Mbytes (depending on the media) of computer data, including graphics, digital audio, MIDI, text or raw data. Thus, the CD-ROM isn't tied to any specific data format, which means that it's up to the manufacturer or programmer to specify what is contained on the disc. Consequently, a CD-ROM can contain an entire encyclopedia's worth of text, audio and graphics to such an interactive extent that this medium has become a driving force in all communications media and its impact will continue to be felt well into the millennium.

Similar to their former cousins, the DVD (which, after a great deal of industry deliberation, actually stands for . . . DVD) can contain any form of data. Unlike the CD, these discs are capable of storing a whopping 4.7 gigabytes for a single-sided, double-layered disc and 9.4 gigabytes for a double-sided, double-layered disc, which has made them the perfect medium for the DVD MPEG video encoding format, DVD-Audio and numerous DVD-ROM media discs.

The increased demands for multimedia games, educational products, etc., have spawned the computer-related industry of CD and DVD-ROM authoring. The term *authoring* refers to the creative, design and programming aspects of putting together a CD/DVD project. At its most basic level, a project can be authored, mastered and "burned" to disc from a single authoring program package. At higher levels, a number of related systems and company departments could be used to allow the various media to be assembled and prepared before the final CD or DVD is cut to disc.

The Web

One of the most powerful aspects of multimedia is the ability to communicate these experiences either to another individual or to the masses . . . for this, you need some kind of a network connection. The network that's most commonly found in the home, studio, classroom— you name it—is a connection to the *Internet*.

Basically, the Internet (Figure 8.3) can be thought of as a communications network that allows your computer to be connected to a server (a specialized computer that's designed to handle, pass and route large numbers of user connections, often using specially designed, high-speed modem connections). This user–server connection is established by way of a modem (modulate/demodulate) interface that can translate digital data into an encoded form that can be transmitted over phone or other transmission lines.

In a similar fashion, server computers are connected throughout the world using a high-speed network connection that allows site, connection and file requests to be transmitted from any site and/or server that's connected to the World Wide Web.

Figure 8.3. The Internet works by communicating requests and data from a user's PC to a single server that's connected to other servers around the world, which are likewise connected to other users' PCs.

Internet browsers transmit and receive information on the Web via a URL (Uniform Resource Locator) address. This address in then broken down into three parts, the protocol "http," the server name "www.modrec.com" and the requested page or file name "index.htm." The connected server is able to translate the server name into a specific IP (Internet provider) address that's used to connect your PC with the desired server, after which the requests to receive/send data are communicated and the information is passed.

E-mail works in a similar data transfer fashion, with the exception that the e-mail data isn't sent to or requested from a specific server (which might be simultaneously connected to a large group of user PCs)—rather, it's communicated from one specific e-mail address "yourname@yourprovider.com" to a destination address "myname@myprovider.com."

In my humble opinion, the CD-ROM was the first vehicle for bringing multimedia to the desktop. However the concept of pulling information, graphics and sound (most often, in that order) from the Internet's collective consciousness has finally brought multimedia into the mainstream.

Media Types

The basic media types that are used to bring the multimedia experience to our desktops include: text, graphics, video, MIDI and digital audio. The following sections are intended to give a basic overview of these types and how they're applied.

Graphics

Graphic imaging occurs on the computer screen in the form of *pixels*. These are basically tiny dots that blend together to create color images in much the same way that dots are combined to give color and form to your favorite comic strip. The only differences are that the resolution is much greater on your color monitor and the dot size doesn't change.

In much the same way that wordlength affects the overall amplitude range of a digital audio signal, the number of digits in a pixel's word will affect the range of colors that can be displayed in a graphic image. For example, a 4-bit word has 16 possible combinations. Thus, a 4-bit word will allow your screen to have a total of 16 possible colors; an 8-bit word will yield 256 colors; 16 bits will give you 65,536; and a 24-bit wordlength will yield a whopping total of 16.7 million colors!

The capabilities of your graphics display will depend on your system's hardware (Figure 8.4), while the ability to create graphics is dependent on software. Newer programming applications are coming into existence that add a degree of "smarts" to your system's graphic abilities, in that they can add shading, change lighting, alter color . . . all in real time, according to a set of programmed parameters. This type of interactivity applies mostly to newer CD-ROM games that can be played on either the Windows or Mac operating systems (OSs).

Figure 8.4. *The Windows 98 Display Properties Box is used to change screen sizing and pixel resolution.*

The field of graphics can be broken down into several categories:

- *Raster graphics.* In raster graphics, each graphic image is displayed as a series of pixels. This image type is what is used when we're viewing a single graphic image (often in an uncompressed bitmap or compressed JPEG, GIF or TIFF format). The sense of motion can only come from raster images by successively stepping through a number of changing images in a second (which is how a video image creates the sense of motion).

- *Vector graphics.* This process often creates a sense of motion by projecting a raster image and them overlaying one or more objects that can be animated according to a series of programmed vectors. In other words, one or more objects can be displayed over a background image. By instructing each object to move from point A to point B, then point C, etc., according to a defined script, a sense of animated motion can be created without the need to project separate images for each frame. This script form reduces a file's data size dramatically and is used with several popular image display formats (including MacroMedia's Flash, Shockwave and Director).

- *Wireframe animation.* This form of animation uses a computer to create a complex series of wireframe image vectors of a real or imaginary object. Once programmed, these stick-like wireframe objects can be filled in with any type of skin, color, shading, etc., and then programmed to move with a staggering degree of realism. Obviously, with the increased power of modern computers and supercomputers, this graphic art form has attained higher degrees of artistry and/or realism within film, video and desktop visual production.

Desktop Video

With the proliferation of analog and digital video cameras (as well as reductions in price of video interface hardware for desktop computers), desktop video has begun to play an increasingly important role in creating multimedia production and content (Figure 8.5).

Figure 8.5. Playback of a digital video clip using the Windows Media Player.

Video is encoded onto the screen as a continuous series of successive frames, which are refreshed at rates that vary from 12 or fewer frames/second (fr/sec) to the standard video rate of 30 fr/sec. As with graphic files, a single full-sized video frame can be made up of hundreds of video scan lines that are themselves composed of hundreds or thousands of pixels . . . which are themselves encoded as a digital word of n bits. Multiply these figures by nearly 30 frames and you'll come up with data file sizes that can quickly fill even the biggest of hard drives.

Obviously, it's more common to find such file sizes and data throughput rates on higher-end desktop systems and professional video editing workstations. However, there are a combination of options that can help bring video to multimedia and even the Internet:

- *Window size.* The basics of making the viewable picture smaller is simple enough . . . reducing the frame size will reduce the number of pixels in a video frame, thus correspondingly reducing the data requirements.
- *Frame rate.* Although standard video frame rates run at around 30 fr/sec (United States/Japan) and 25 fr/sec (Europe), these rates can be lowered to 12 fr/sec to reduce the amount of data that's needed to encode a file.
- *Compression.* In a similar manner to audio, codecs can be applied to a video frame to reduce the amount of data that's necessary to encode the file by filtering out and smoothing over areas that consume data. In situations where high levels of compression are needed, it's common to accept degradations in the video signal in order to reduce the file size and/or data throughput to levels that are acceptable to the medium (i.e., the Web).

From this, you can see that there can be many options for encoding a desktop videofile. When dealing with video clips, tutorials and the like, it's common for the viewing window to be medium in size and be encoded at medium-to-lower framerates. This middle ground is often chosen to accommodate the standard data throughput that can be streamed off of a CD-ROM. These files are commonly encoded in either AVI (Microsoft's "audio-video interleave" format, which uses little to no compression), QuickTime (a common codec that was developed by Apple Computer and can be played by either a Mac or PC) and MPEG I or II (codecs that vary in level from those for use with multimedia to high-level formats that are used to encode DVD movies).

Both the Microsoft and Macintosh OS platforms include built-in or easily obtained applications that allow all or most of these file types to be played without additional hardware or software. Recording analog or digital video into your system will often require special hardware to translate images into a signal that the computer can understand and process.

MIDI

One of the unique advantages of MIDI, as it applies to multimedia, is the rich diversity of musical instruments and program styles that can be played back in real time, while requiring almost no overhead processing from the computer's CPU. This makes MIDI a perfect candidate for playing back soundtracks from multimedia games or over the Internet . . . you could even embed a soundtrack within a written document.

It's interesting that MIDI has taken a back seat to digital audio as a serious music playback format for multimedia. Most likely, this is due to several factors, including a basic misunderstanding of the medium, the fact that producing MIDI content requires a basic knowledge of music, and the frequent difficulty of synchronizing digital audio to MIDI in a multimedia environment—not to mention that the market's still saturated with poorly designed FM synthesizers.

Fortunately, companies such as Microsoft have taken up the banner of embedding MIDI within their media projects and have helped push MIDI a bit more into the Web mainstream. As a result, it's becoming more common for your PC to begin playing back a MIDI score, on its own or perhaps in conjunction with a more data-intensive soundfile.

Standard MIDI Files

The accepted format for transmitting files or real-time MIDI information in multimedia (or between sequencers from different manufacturers) is the *standard MIDI file*. This filetype (which is stored using the .mid or .smf extension) is used to distribute MIDI data, song, track, time signature and tempo information to the general masses. Standard MIDI files can support both single and multichannel sequence data and can be loaded into, edited and then directly saved from almost any sequencer package.

When exporting a standard MIDI file, keep in mind that they come in two basic flavors: type 0 and type 1. Type 0 is used whenever you'd like for all of the tracks in a sequence to be compressed into a single MIDI track. All of the original channel messages still reside within that track—it's just that the data has no definitive track assignments. This data type might be the best choice when creating a MIDI sequence for the Internet (where the sequencer or MIDI player application might not know or care about dealing with multiple tracks). The type 1 format, on the other hand, will retain its original track structure and can be imported into another sequencer type with its basic track information and assignments left intact.

General MIDI

One of the most interesting aspects of MIDI production is the absolute uniqueness of each professional and even semipro project studio. In fact, no two studios will be alike (unless they've been specifically designed to be the same or there's some amazing coincidence). Each artist will be unique in having his or her own favorite equipment, supporting hardware, favorite way of routing channels and tracks and assigning patches.

The fact that each system is unique has placed MIDI at odds with the need for absolute compatibility between systems in the world of multimedia. For example, after importing a standard MIDI file over the Net and loading it into a sequencer, you might hear a song that's being played with a totally irrelevant set of sound patches (it might sound interesting, but it won't sound anything like it was originally intended). If the MIDI file is loaded into a new computer, the sequence might again sound completely different, with patches that are so irrelevant that the guitar track might sound like a bunch of machine-gun shots from the planet Glob.

In order to eliminate (or at best reduce) the basic differences that exist from system to system, a standard known as *General MIDI (GM)* was created. In short, General MIDI assigns a specific instrument patch to each of the 128 available program change numbers. Since all electronic instruments that conform to the GM format use these patch assignments, placing these standardized program change commands at the header of each track will automatically configure the sequence to play with its originally intended sound. With this system, the differences from one multimedia synth to the next can finally be minimized. As an example, a standard MIDI file that conforms to General MIDI might contain the following tracks:

Track	Prog. Change #	Instrument Name
1	33	Acoustic Bass
2	1	Acoustic Grand Piano
3	76	Pan Flute
4	47	Orchestral Harp
. . .	. . .	. . .
10		Percussion

No matter what sequencer is used to play the file back, as long as the receiving instrument conforms to the GM spec, the sequence will be heard using its intended instrumentation.

Tables 8.1 and 8.2 detail the program numbers and patch names that conform to the GM format (Table 8.1 for non-percussion and Table 8.2 for percussion instruments). These patches include sounds that imitate synthesizers, ethnic instruments or sound effects, which have been derived from early Roland synth patch maps. Although the GM spec states that a synth must respond to all 16 MIDI channels, the first nine channels are reserved for instruments, while GM restricts the percussion track to MIDI Channel 10.

Table 8.1. GM Instrument Patch Map.

1. Acoustic Grand Piano	33. Acoustic Bass	65. Soprano Sax	97. FX 1 (rain)
2. Bright Acoustic Piano	34. Electric Bass (finger)	66. Alto Sax	98. FX 2 (soundtrack)
3. Electric Grand Piano	35. Electric Bass (pick)	67. Tenor Sax	99. FX 3 (crystal)
4. Honky-tonk Piano	36. Fretless Bass	68. Baritone Sax	100. FX 4 (atmosphere)
5. Electric Piano 1	37. Slap Bass 1	69. Oboe	101. FX 5 (brightness)
6. Electric Piano 2	38. Slap Bass 2	70. English Horn	102. FX 6 (goblins)
7. Harpsichord	39. Synth Bass 1	71. Bassoon	103. FX 7 (echoes)
8. Clavi	40. Synth Bass 2	72. Clarinet	104. FX 8 (sci-fi)
9. Celesta	41. Violin	73. Piccolo	105. Sitar
10. Glockenspiel	42. Viola	74. Flute	106. Banjo
11. Music Box	43. Cello	75. Recorder	107. Shamisen
12. Vibraphone	44. Contrabass	76. Pan Flute	108. Koto
13. Marimba	45. Tremolo Strings	77. Blown Bottle	109. Kalimba
14. Xylophone	46. Pizzicato Strings	78. Shakuhachi	110. Bag pipe
15. Tubular Bells	47. Orchestral Harp	79. Whistle	111. Fiddle
16. Dulcimer	48. Timpani	80. Ocarina	112. Shanai
17. Drawbar Organ	49. String Ensemble 1	81. Lead 1 (square)	113. Tinkle Bell
18. Percussive Organ	50. String Ensemble 2	82. Lead 2 (sawtooth)	114. Agogo
19. Rock Organ	51. SynthStrings 1	83. Lead 3 (calliope)	115. Steel Drums
20. Church Organ	52. SynthStrings 2	84. Lead 4 (chiff)	116. Woodblock
21. Reed Organ	53. Choir Aahs	85. Lead 5 (charang)	117. Taiko Drum
22. Accordion	54. Voice Oohs	86. Lead 6 (voice)	118. Melodic Tom
23. Harmonica	55. Synth Voice	87. Lead 7 (fifths)	119. Synth Drum
24. Tango Accordion	56. Orchestra Hit	88. Lead 8 (bass + lead)	120. Reverse Cymbal
25. Acoustic Guitar (nylon)	57. Trumpet	89. Pad 1 (new age)	121. Guitar Fret Noise
26. Acoustic Guitar (steel)	58. Trombone	90. Pad 2 (warm)	122. Breath Noise
27. Electric Guitar (jazz)	59. Tuba	91. Pad 3 (polysynth)	123. Seashore
28. Electric Guitar (clean)	60. Muted Trumpet	92. Pad 4 (choir)	124. Bird Tweet
29. Electric Guitar (muted)	61. French Horn	93. Pad 5 (bowed)	125. Telephone Ring
30. Overdriven Guitar	62. Brass Section	94. Pad 6 (metallic)	126. Helicopter
31. Distortion Guitar	63. SynthBrass 1	95. Pad 7 (halo)	127. Applause
32. Guitar harmonics	64. SynthBrass 2	96. Pad 8 (sweep)	128. Gunshot

Table 8.2. GM Percussion Patch Map (Ch. 10).

35. Acoustic Bass Drum	51. Ride Cymbal 1	67. High Agogo
36. Bass Drum 1	52. Chinese Cymbal	68. Low Agogo
37. Side Stick	53. Ride Bell	69. Cabasa
38. Acoustic Snare	54. Tambourine	70. Maracas
39. Hand Clap	55. Splash Cymbal	71. Short Whistle
40. Electric Snare	56. Cowbell	72. Long Whistle
41. Low Floor Tom	57. Crash Cymbal 2	73. Short Guiro
42. Closed Hi Hat	58. Vibraslap	74. Long Guiro
43. High Floor Tom	59. Ride Cymbal 2	75. Claves
44. Pedal Hi-Hat	60. Hi Bongo	76. Hi Wood Block
45. Low Tom	61. Low Bongo	77. Low Wood Block
46. Open Hi-Hat	62. Mute Hi Conga	78. Mute Cuica
47. Low-Mid Tom	63. Open Hi Conga	79. Open Cuica
48. Hi Mid Tom	64. Low Conga	80. Mute Triangle
49. Crash Cymbal 1	65. High Timbale	81. Open Triangle
50. High Tom	66. Low Timbale	

Note: In contrast to Table 8.1, the numbers in Table 8.2 represent the percussion keynote numbers on a MIDI keyboard, not program change numbers.

Digital Audio

Digital audio is obviously a component that adds greatly to the multimedia experience. It can augment a presentation by adding a dramatic music soundtrack, help us to communicate through speech, or give realism to a soundtrack by adding sound effects.

Because of the large amount of data that's required to pass video, graphics and audio from a CD-ROM, the Internet or other media, the bit- and sample-rate structure of an audiofile is usually limited compared to that of a professional-quality sound file. The general accepted sound file standard for multimedia production is either 8-bit or 16-bit audio at a sample rate of 11.025 or 22.050 kHz. This standard has come about mostly because older single- and 2-speed CD-ROMs generally couldn't pass the professional 44.1-kHz. sample rate with full motion video or other graphics types without annoying and spasmodic interruptions. In addition, larger, pro-rate samplefiles could take minutes or even hours to download over the Internet. Fortunately, with improvements in MPEG compression techniques, hardware speed and design, the overall sonic and production quality of sound has become more accessible and has greatly improved in quality.

Table 8.3 details the differences between file sizes as they range from voice-quality 8-bit/11-kHz files all the way to the professional 24bit/96 kHz rates that are used to encode sounds with CD and DVD quality.

Table 8.3. Sampled Audio Bit Rate and File Size.

Sample Rate	Bit Structure	Bytes/minute
11 kHz	8-bit/mono	660 KB/minute
11 kHz	8-bit/stereo	1.3 MB/minute
11 kHz	16-bit/stereo	1.3 MB/minute
11 kHz	16-bit/stereo	2.6 MB/minute
22 kHz	8-bit/mono	1.3 MB/minute
22 kHz	8-bit/stereo	2.6 MB/minute
22 kHz	16-bit/mono	2.6 MB/minute
22 kHz	16-bit/stereo	5.3 MB/minute
44.1 kHz	8-bit/mono	2.6 MB/minute
44.1 kHz	8-bit/stereo	5.2 MB/minute
44.1 kHz	16-bit/mono	5.2 MB/minute
44.1 kHz	16-bit/stereo	10.3 MB/minute
48 kHz	16-bit/mono	5.6 MB/minute
48 kHz	16-bit/stereo	11.2 MB/minute
44.1 kHz	24-bit/mono	7.8 MB/minute
44.1 kHz	24-bit/stereo	15.6 MB/minute
48 kHz	24-bit/mono	8.4 MB/minute
48 kHz	24-bit/stereo	16.8 MB/minute
96 kHz	24-bit/mono	16.8 MB/minute
96 kHz	24-bit/stereo	33.6 MB/minute

Soundfile Formats

Although several of formats exist for saving samplefile and soundfile data onto computer-based storage media, only a few formats have been universally adopted by the industry. These standardized formats make it easier for files to be exchanged between compatible sample-editor programs.

By far, the most common filetype is the Wave (or .WAV) format that's used with Windows-based systems. This filetype can be used to save waveform data in any bit- and sample-rate that's supported by your sound card or system. Another file format that's most commonly used with Mac computers is the Audio Interchange File Format (or .AIF). As with Wave files, .AIF files can encode both monaural and stereo sounds at a variety of sample- and bitrates. Finally, a format that's commonly used in media production is the Sound Designer II or .snd format. This filetype was developed by Digidesign for their Pro-Tools and other digital editor systems, which are widely used in pro audio, broadcast and multimedia production.

Because of the wide range of sound file formats that a multimedia developer might encounter, most professional editors are capable of both reading and saving files in a number of cross-platform formats, as well as in a variety of bit- and sample-rate structures (Figure 8.6).

Figure 8.6. Digital audio editors can often save waveform data in any number of formats. (Courtesy of Syntrillium Software Corp., www.syntrillium.com)

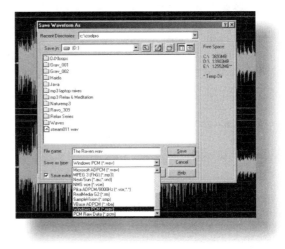

Multimedia and the Web

The household phrase "surfin' the Web" has become synonymous with jumping onto the Net, browsing the sites and grabbin' onto all of those hot digital graphics, waves and videos that might wash your way. Dude; while this analogy often holds true for text-based information, most folks who are surfin' at slow and even moderate modem speeds know that once you get out into deeper waters where the graphic, video and audio "waves" become much larger, pulling in the big ones often takes so much time that you're often left treading water, which leads us to . . .

Streaming in Cyberspace

Let's face it, folks, even though the Internet has profoundly changed many of our lives, the relatively slow modem transfer speeds of 28.8 and even 56 kbytes/sec have taken us back to the pre-multimedia days when graphics were sluggish and real-time video was but a dream.

Although data speeds have increased over the years, the "bottleneck" of the Web has forced hardware and programming designers to be innovative in coming up with ways to improve data transmission. No matter what the media type, these solutions often take any of three forms: reducing the bitrate, reducing the sample rate, or employing compression techniques.

It's generally accepted that the Web can't transmit .WAV or .AIF sound files in real time with any sense of stability. The only option (without special hardware and/or additional phone lines) is to download an entire file and then play it back off the hard drive. However, through the use of special compression techniques, the problem of "streaming" sound over the Internet in real time has been tackled by a small number of companies, which allows audio to finally become an integral part of the Internet's multimedia experience.

Much of the real-time audio glory goes to a Seattle-based company called RealNetworks (www.real.com), with the introduction of their server/player application—RealPlayer. In fact, with free distribution of millions of player applications that can stream the Real Audio sound file format to Internet sites worldwide, the RealNetworks family of streaming products has pretty much become the de facto standard for streaming real-time audio over the Web.

From a technical point of view, RealAudio data is transmitted using any of more than 12 proprietary encoding levels that range from transmission rates of 8 kbps (lo-fi mono voice quality over a 14.4K modem) all the way up to 1.5 Mbps (over a T1 or T3 high-speed line).

Although there are several compression levels to choose from, the most common types compress audio into a "voice mode" format that transmits data over 14.4K lines in a way that's optimized for human speech. The second, "music mode," compresses data in a way that's less harsh and introduces less artifacts over a greater dynamic range, thereby creating an algorithm that can more faithfully reproduce music with "near-FM" quality over 28.8K, 56K or faster lines. At the originating Internet site, the RealAudio server can automatically recognize which modem speed is currently in use and then transmit the data in the best possible audio format. This reduced data throughput ultimately means that the RealAudio player will take up very little of your computer's resources, allowing you to keep on working while audio is being played.

On a personal note, several years ago, I threw a big birthday bash on a Mexican theme . . . and thanks to RealGuide's radio tuner (http://realguide.real.com/tuner), music was piped in live from my favorite radio station out of Mexico City. Nosotros pasamos buen tiempo . . . ole!

Other forms of real-time and non-real-time streaming are also available on the Web. The most notable of the real-time streamers are Liquid Audio (www.Liquidaudio.com) and Microsoft's fast-growing Windows Media Player formats. By far, the best way to familiarize yourself with these is to load one or more of the players (generally, each of the player types offers a basic player that can be downloaded for free) . . . and start surfin' the Web for real-time music, radio and video.

Downloadable Audio

As was mentioned, high-quality uncompressed .WAV and .AIF files can't generally be streamed over standard Internet lines and are often so large that they take too much time to download. Although the way in which audio data is streamed has improved over the years, the level of quality for most real-time players at standard modem speeds is often below what we've come to think of as CD quality. This dilemma (at the time of writing) has lead to the popular acceptance (some would say tidal wave) of file formats, as well as supporting software and hardware players that allow high-quality audio to be downloaded, stored and played back on demand. These compressed audio formats currently include MP3, WMA, and AAC.

MP3

MPEG (which is pronounced "M-peg" and stands for the Moving Picture Experts Group, www.cselt.it/mpeg or www.mpeg.org) is a standardized encoding format that's used to encode digital audio and video into a compressed format for storage to various media and for transmission over the Web. As of this writing, the most popular format is the ISO-MPEG Audio Layer-3, commonly known as MP3 (Figure 8.7).

Figure 8.7. *A large number of soft-, share- and freeware players are available to play MP3 files from hard disk or any number of portable and/or desktop memory sources. (Courtesy of Nullsoft, Inc., www.winamp.com)*

Developed by the Fraunhofer Institute (www.iis.fhg.de) and Thomson Multimedia in Europe, MP3 has started the revolution of compressing digital audio by factors of 10:1 and greater, while still maintaining quality levels that approach those of a CD (depending on which compression levels are used). In fact, compression ratios of up to 24 times can be attained without seriously degrading the sound quality, and even higher levels of compression can be used for nonmusical tracks (such as voice).

Although faster Web connections are capable of streaming MP3 in real time, this format is most often downloaded to the end consumer off-line, where soundclips, songs and even complete albums can be downloaded, saved to disk or disc and even transferred into solid-state playback devices (such as portable MP3 players, cell phone players . . . you name it!).

WMA

Developed by Microsoft for use within its Windows Media Player, the Windows Media Audio (WMA) format allows for compression rates that can encode high-quality audio at low bitrate/filesize settings.

AAC

Jointly developed by Dolby Labs, Sony, ATT and the Fraunhofer Institute, the Advanced Audio Coding Scheme (as of this writing) is being touted as a "multichannel-friendly format for secure digital music distribution over the Internet." Stated as having the ability to encode CD-quality audio at lower bitrates than other coding formats, AAC is not only capable of encoding 1, 2 and 5.1 surround soundfiles, but can encode up to 48 channels within a single bitstream at bit/sample rates of up to 24/96. This format is also SDMI-compliant, allowing copyrighted material to be protected against unauthorized copy and distribution.

Codec Compression

Basically, the compression aspects of MP3, AAC, WMA and other schemes are based on the psychoacoustic principle that the human ear can't hear all of the frequencies that are present in a recording. In Chapter 2, we read that louder noises will often mask sounds that are both lower in level and relatively close to the louder signal's frequency. These perceptual coding schemes take advantage of this masking effect by filtering out noises and sounds that can't be detected. The process is said to be "lossy" or "destructive," as once the filtered data has been eliminated from the file it can't be replaced or introduced back into the file. For this reason, when encoding an important file, it's always a good idea to keep a copy of the original soundfile as a backup.

On an important personal note, I've found that not all of the MP3 codecs are created equal (even those that use the same bitrate settings). Although the original codec was created by the Fraunhofer Institute, several companies have gone about the task of creating their own "look-alike" coding structures that just don't stack up to the genuine article. For some time, I've been using a very high quality encoder that uses the original Fraunhofer codec . . . and (I'm gonna stick my neck out here) I've found the quality at 160 kbps to be CD in quality. Not near CD quality, but CD quality!

Along the same line, higher-quality MP3 files that have been created with a good codec can be copied several times without degrading. On the other hand, lower-fi codecs (and even hi-fi settings using a poorly designed codec) can degrade in quality when copied over several generations. All I'm saying here is caveat emptor . . . be aware of the quality difference between encoders, audition your MP3 encoder to listen for artifacts, make multigenerational copies of a file to see how it degrades and keep copies of your original soundfile, whenever possible.

In the final analysis, MP3 is capable of compressing sound files (or "*ripping*" as it's often referred to) to give CD-quality stereo sound. At 64, 128 or 160 kbps (kilobits per second) or higher, MP3 audio can arguably capture the quality of a CD. Of course, there are differences at lower rates, but the reduced file size, faster download time and portability of the medium has definitely thrust this format into the limelight. I wonder what all the industry buzz will be for those of you who are reading this book 5 or 10 years from now?

SDMI

As you can imagine, with the vast number of software (and hardware) systems that are able to "rip" CDs into MP3 and other compressed data formats, the powers that be in the recording industry have grown increasingly fearful of the rising prevalence of copyright infringement. Although many online music sites legally use these formats to allow potential buyers to preview music before buying or to simply put unreleased cuts onto the Web as a freebie . . . a number of sites have begun to amass large databases to music that have been illegally ripped off of CDs, without paying royalties to the record company or independent artist.

As a result, the RIAA (Recording Industry Association of America), major record labels and industry organizations have helped to form the SDMI (Secure Digital Music Initiative, www.sdmi.org). SDMI is an independent forum that brings together the worldwide recording, consumer electronics and information technology industries to develop open specifications for protecting digital music distribution. Ideally, this would be in the form of a digital "watermark" that identifies the copyright owner and provides for an electronic "key" that allows access to the music or information once the original copy has been legally purchased.

It's hoped that SDMI-compliant software and hardware will allow consumers to easily collect and play music that's been purchased from a variety of online and off-line sources.

Streaming Video

In addition to real-time audio, video can also be streamed over the Internet using various forms of codecs. Again, the most accepted provider for delivering video over the Web is RealVideo from Progressive Networks. RealVideo delivers newscast- and "event"-quality video over 28.8 kbps modems, full-motion video using 56 kbps and DSL cable modems, as well as near TV broadcast quality at higher broadband speeds.

Often, files that are streamed over the Web make use of file reduction techniques to downsize data throughput to levels that can play full-motion video in real time (or create the jerky illusion that at least some movement is going on). At this time of writing, these videos can range from the size of a postage stamp to a few diagonal inches in size. . . . This is bound to change as the overall bandwidth increases.

Thoughts on Being and Getting Heard in Cyberspace

Most of us grew up with or were born into the concept of the supermarket . . . where everything is wholesaled, processed, packaged and distributed to a single clearing house that's gotten so big that older folks can only shop with the aid of a motorized shopping cart. For more than five decades, the music industry has largely worked on a similar principle . . . find artists who'll fit into an existing marketing formula (or more rarely, create a new marketing image), produce and package them according to that formula and put tons of bucks behind them to get them heard and distributed. Not a bad thing in and of itself; however, for independent artists the struggle has been and continues to be one of getting themselves heard, seen and noticed—without the aid of the well-oiled machine.

With the creation of cyberspace, not only are established record industry forces able to work their way onto your desktop screen (and into your multimedia speakers), independent artist have a new medium for getting heard. Through the creation of a dedicated Web site, links from other sites, independent music dotcoms, as well as through creative gigging and marketing . . . new avenues have begun to open up to the independent artist.

Uploading to Stardom!

If you build it, they will come! . . . This early-1900s concept definitely doesn't apply to the Web. With an ever-increasing number of dot-whatevers going online every month, the idea of expecting people to come to your site, just because it's there, simply isn't realistic. Like anything that's worthwhile, it takes connections, persistence, a good product and good ol'-fashioned luck to be seen as well as heard!

If you're selling your music, T-shirts or whatever at gigs, on the streets, to family and friends . . . cyberspace can help to increase sales by making it possible (and even easy) to get your band, music or clients onto several independent music Web sites with links that go directly to your main Web site. This site can give background info, tour dates, music clips in downloadable or streaming formats of your choice . . . in short, it can help clue your audience in on what you and your music are all about.

Cyberproducts can be sold and shipped via the traditional mail or phone-in order channels; however, it's long been considered hip in the Web world to flash the silver, gold or platinum credit card to make your purchase. Due to the fact that attaining your own credit card processing and authorization system can be costly, a number of cyber companies have sprung up that offer secure credit card authorization, billing and artist payment plans for an overall sales percentage fee that ranges from 10–25 percent.

The preview and/or distribution format choice for releasing all or part of your music to the listening audience will ultimately depend on you and the format/layout style that's been adopted by the hosting site. For example, you could do any or all of the following:

- Place short, lo-fi segments onto a site that can be streamed and/or downloaded, to entice the listener to buy.
- Provide free access (at lo- or medium-fi) to a cut or entire project, while encouraging the listener to buy the hi-fi CD.
- Place several hi-fi cuts on your site for free as a teaser or as a gift to your fan base.
- Place the music on a secure site that's SDMI-compliant, using the pay per download system for either a song or the entire project.
- Sell the completed CD package on the site.

No matter what or how many cyber distribution methods you choose to get your music out, *always* take the time to read through the contractual fine print. In your excitement to get your stuff out there, you might not realize that you are signing the rights for free distribution of that particular project (or worse) away. This hints at the fact that you're dealing with the music "business" . . . and as with any business, you should always be aware of what you're signing.

Copyright Protection: Wanna Get Paid?

I hope that by the time you read this edition, many of the problems that currently exist with "ripping" off an artist from CD to a compressed file format and then posting that music onto the Web for everyone to download for free will be carefully considered and dealt with in a way that's fair to all those involved. Just as cyberspace is an amazing tool for getting unsigned and record label music out to the masses, ripping software (which is designed to copy audio material from disc and encode it in a compressed form) has proliferated to the point that near-CD-quality copies of copyrighted material can be downloaded for free.

Obviously, the established record company and royalty collection/distribution organizations aren't happy with the revenue losses. They have mounted numerous lawsuits and have established the already mentioned Secure Digital Music Initiative (SDMI) to help combat the act of ripping off another artist's livelihood and to provide a straightforward vehicle for promoting a viable e-commerce industry. The interesting thing to note is that since ripping has come into its own, CD sales have risen instead of declined . . . basically, the jury's out on the subject and it's future will be most likely dealt with, not in the courts, but in the general marketplace.

E-Radio

Because of the increased bandwidth of many Internet connections and improvements in audio streaming technology, many of the world's radio stations have begun to broadcast on the Web. In addition to offering a worldwide platform for traditional radio listening audiences, a large number of corporate and independent "Web" radio stations have begun to spring up, which can help to increase a musician's and record label's fan and listener base.

The Virtual Recording Gig

In addition to providing another vehicle for getting an artist's music out to the public-at-large, the Internet is making it easier for artists to e-collaborate over cyberspace.

Basically, there are numerous ways to collaborate over the Web. For starters, MIDI files can be e-mailed between musicians. However, even though the filesize is small, unless General MIDI is being used as the standard for setting up instrument wavetables or unless similar or like electronic instruments are being used, there can be a real problem with sound incompatibilities between the project studios. With a bit of care, you could ask that your buddy in Yugoslavia create a killer piano track for a song you're working on, by creating a General MIDI file of the song's parts for friend to use as a guide track . . . once the song has been e-mailed back, you could extract the track and place it back into your original working MIDIfile and assign a better sounding piano patch to it.

Digital audio has the disadvantage of having a much larger filesize; however, the positive side is that it's an actual recording of the original performance. There's no fudging around with patches or sound generators—what you hear is what you get! In order to get around the larger filesize problem, you might want to experiment with encoding copies of all the files in a session as MP3 files (if the software supports it) and then e-mail these (along with the session's data file) to the recipient. Alternatively (and more simply), you could mix down all the files in a session into a single compressed file. Your buddy could then overdub his/her track, which can then be compressed as a hi-fi, high-bitrate compressed file. Once the overdub track gets e-mailed back, you could simply import it back into the original session (you'll probably want to convert the file back into the session's native file format).

When encoding files that are to be imported back into a multitrack hard disk session, you should use an encoder that's as high quality as possible. Keep in mind that not all MP3 encoders (or those for other types of format, for that matter) are created equal. Try making multigenerational copies of a master file to see if there's significant (or any) signal degradation. . . . If there is, take time to search out a better encoder.

As of this writing, an interesting way to record, overdub or jam over the Internet in real and/or non-real time is to collaborate in an online session using Rocket Network. This system links standard audio recording tools (such as Digidesign's Pro Tools, Logic Audio, Cubase VST and Euphonics products) in a virtual network fashion, so that audio professionals and enthusiasts can meet online to collaborate and produce original audio from anywhere in the world.

When using a "RocketPowered" network, artists can upload/download parts of a collaborative project to the various computers that are connected to the same session. In addition, the sum of the session parts will be remotely stored at a central server location (known as the Internet Recording Studio), so that MIDI, audio tracks and session files can be up/downloaded by new players or retained as backups. In addition, a downloadable application lets you view all track download status info, as well as chat with and locate other session users or navigate between Internet recording studios . . . all in real time.

On A Final Note

One of the most amazing things about multimedia, cyberspace and their related technologies is the fact that they're ever-changing. By the time you read this book, many new developments will have happened. Old concepts fade away; new and possibly better ones will take over and then begin to take on a new life of their own.

Although I've always had a fascination with crystal balls and have often had a decent sense about new trends in technology . . . there's simply no way to foretell the many amazing things that lie before us in the fields of music, music technology, multimedia . . . and especially cyberspace. As with everything techno, I encourage you to read the trades, check out the Webzines and surf the Web to keep abreast of the latest and greatest tools that have recently arrived, or are about to rise on the horizon.

CHAPTER 9

Synchronization

In recent decades, the professional audio and visual markets have grown to incorporate audio into the visual media to help communicate ideas and emotion. With the dawning of music videos, computer games, surround sound movie production and a general maturing of the various media markets, the field of audio has moved from taking a back-seat role into one that plays an increasingly important role in the entertainment and educational media markets. As a result, it's become commonplace for the technologies of audio, video, film, multimedia and the electronic music arts to coexist within a single facility or production house. In video postproduction, for example, audio transports, video transports, digital audio workstations, automated console systems and electronic musical instruments routinely work together in tandem to help one or more operators to create and refine video soundtracks to perfection (Figure 9.1). The process that allows multiple audio and visual media to maintain a direct time relationship is known as *synchronization* or *sync*.

Synchronization is the occurrence of two or more events at precisely the same time. With respect to analog audio and video systems, sync is achieved by interlocking the transport speeds of two or more machines. In computer-related systems (such as digital audio and MIDI), internal or external sync between compatible devices is often maintained by using a clocking pulse that's directly embedded within the digital data line itself. Frequently, it's necessary for both analog- and digital-based devices to be synchronized together . . . as a result, some rather ingenious forms of systems communication and data translation have been developed.

Figure 9.1. *Example of an integrated audio production system.*

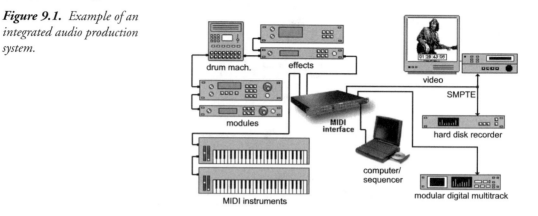

In this chapter, the various forms of synchronization for both analog and digital devices will be discussed, as well as current methods for maintaining sync between the two media.

Synchronization between Transports

Maintaining relative sync between analog tape transports doesn't require that all transport speeds involved in the process be constant; however, these devices must maintain the same *relative* speed at all points over the course of a program.

Because of differences in mechanical design, voltage fluctuations and tape slippage, it's a simple fact of life that analog tape devices aren't able to maintain a constant playback speed . . . not even over relatively short durations. For this reason, sync between two or more machines without some form of timing "lock" would be impossible over any reasonable program length. It therefore quickly becomes clear that if production is to utilize multiple forms of media and record/playback systems, a means of synchronously locking these devices in time is essential.

Time Code

The standard method of interlocking audio, video and film transports makes use of a code that was developed by *Society of Motion Picture and Television Engineers* (*SMPTE*). The use of this time code (or *SMPTE time code*) allows for the identification of an exact position on a tape or within a media program by assigning a digital *address* to each specified length. This address code can't slip, always retains its original location and allows for the continuous monitoring of tape position to an accuracy of between 1/24th and 1/30th of a second (depending on the media type and frame standards being used). The specified tape segments are called *frames*, a term taken from film production. Each audio or video frame is tagged with a unique identifying number, known as a *time code address*. This eight-digit address is displayed in the form 00:00:00:00, in which the successive pairs of digits represent hours:minutes:seconds:frames (Figure 9.2).

Figure 9.2. *Readout of a SMPTE time code address. (www.smpte.org)*

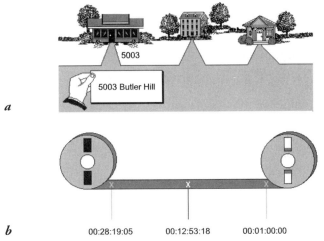

0:01:20:05

The recorded time code address is used to locate a position on magnetic tape (or any other recorded media) in much the same way as a letter carrier uses an address to deliver the mail to a specific residence (Figure 9.3a). Suppose that a time-encoded videotape begins at time 00:01:00:00 and ends at 00:28:19:05 and contains a specific cue point (such as a glass shattering) at 00:12:53:18 (Figure 9.3b). By monitoring the time code (in a fast shuttle mode), it's a simple matter to locate the precise position that corresponds to the cue point on the tape and then perform whatever function is necessary, such as inserting an effect into the sound track at that specific point . . . CRASH.

Figure 9.3. *Location of relative addresses.*
a. *A postal address.*
b. *Time code addresses and a cue point on longitudinal tape.*

a

5003

5003 Butler Hill

b   00:28:19:05   00:12:53:18   00:01:00:00

The standard method for encoding time code in audio production is to record (*stripe*) an open audio channel (usually the highest available track) with SMPTE time code that can then be read directly from the track in either direction and at a wide range of tape speeds.

The Time Code Word

The total of all time-encoded information that's encoded into each audio or video frame is known as a *time code word*. Each word is divided into 80 equal segments called bits, which are numbered consecutively from 0 to 79. One word covers an entire audio or video frame, such that for every frame there is a unique and corresponding time code address. Address information is contained in the digital word as a series of bits that are made up of binary ones and zeros, which are electronically encoded within the signal in the form of a modulated square wave. This method of encoding information is known as *biphase modulation*. With biphase, a voltage transition in the middle of a half-cycle of a square wave is equal to a bit value of 1, while no transition within this same period signifies a bit value of 0 (Figure 9.4). The most important feature about this system is that detection relies on shifts within the pulse and not on the pulse's polarity. Consequently, time code can be read in either the forward or reverse direction, as well as at fast or slow shuttle speeds.

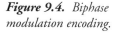

Figure 9.4. *Biphase modulation encoding.*

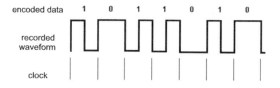

The 80-bit time code word is subdivided into groups of 4 bits (Figure 9.5), whereby each grouping represents a specific coded piece of information. Each 4-bit segment represents a binary-coded decimal (BCD) number that ranges from 0 to 9. When the full frame is scanned, all eight of these 4-bit groupings are read out as a single SMPTE frame number (in hours, minutes, seconds and frames).

User-Information Data

An additional 32 bits (called *user bits*) join the 26 digital bits that make up the time code address. This extra information has been set aside for the time code users to enter their own ID information. The SMPTE standards committee has placed no restrictions on the use of this "slate code," which can contain such information as date of shooting, shot or take ID, reel number, and so on.

Sync-Information Data

Another form of information that's encoded into the time code word is sync data. This exists as 16 bits at the end of the time code word, which are used to define the end of each frame. Because time code can be read in either direction, sync data is also used to tell the device which direction the tape or digital device is moving.

Time Code Frame Standards

In productions using time code, it's important that the readout display be directly related to the actual elapsed time of a program, particularly when dealing with the exacting time requirements of broadcasting. In the case of a black and white (monochrome) video signal, a rate of exactly *30 frames per second* (fr/sec) is used. If this rate (often referred to as *non-drop code*) is used on a B&W program, the time code display, program length and actual clock-on-the-wall time would all be in agreement.

This simplicity was broken, however, when the National Television Standards Committee set the frame rate for the color video signal in the United States and Japan at *29.97 fr/sec*. Thus, if a

Figure 9.5. *Biphase representation of the SMPTE time code word.*

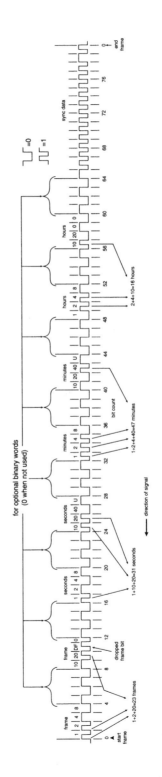

time code reader that's set up to read the monochrome rate of 30 fr/sec were used to read a color program, the time code readout would pick up an extra 0.03 frame for every second that passes. Over the duration of an hour, the time code readout would differ from the actual elapsed time by a total of 108 frames (or 3.6 seconds).

In order to correct for this difference and bring the discrepancy between the time code readout and the actual elapsed time back into agreement, a series of frame adjustments were introduced into the code. Because the object is to drop 108 seconds over the course of an hour, the code used for color has come to be known as *drop frame code*. In this system, two frame counts for every minute of operation are omitted, with the exception of minutes 00, 10, 20, 30, 40, and 50. This has the effect of adjusting the frame count so that it agrees with the actual elapsed program duration.

In addition to the color 29.97 drop-frame code, a 29.97 non-drop-frame color standard can also be found in video production. When using *non-drop time code*, the frame count will always advance one count per frame, without any drops. As you might expect, this mode will result in a disagreement between the frame count and the actual clock-on-the-wall time over the course of the program. Non-drop, however, has the distinct advantage of easing the time calculations that are often required in the video editing process (because no frame compensations need to be taken into account).

Another frame rate format that's used throughout Europe is the *European Broadcast Union (EBU) time code*. EBU utilizes SMPTE's 80-bit code word, but differs in that it uses a *25-fr/sec* frame rate. Because both monochrome and color video EBU signals run at exactly 25 fr/sec, an EBU drop-frame code isn't necessary.

The medium of film differs from all of these as it makes use of an SMPTE time code format that runs at *24-fr/sec*.

From these rates, it's easy to understand why confusion often exists around what frame rate is to be used on a project. Basically, the best reference when working on an "in-house" project that never uses time code referenced material that comes from the outside world is to choose the rate that makes sense for you and has a likelihood of being compatible with the outside world. For example, American electronic musicians who are working in-house will often choose to work at 30 fr/sec. Those in Europe have it easy, as 25 fr/sec is the logical choice for all video productions. Those who work with projects that come through the door from other production houses, on the other hand, will need to take special care to reference their time code to the same rate that was used by the originating media house. This can't be stressed enough: If care isn't taken to keep your time code references at the proper rates and with as little degradation as possible from one generation to the next . . . the various media might have trouble syncing up when it comes time to put the final master together . . . and that often spells *big* trouble.

LTC and VITC Time Code

Currently, two major systems exist for encoding time code onto magnetic tape: LTC and VITC.

Time code recorded onto an analog audio or video cue track is known as *longitudinal time code* (*LTC*). LTC encodes a biphase time code signal onto the analog audio or cue track in the form of a modulated square wave at a bit rate of 2400 bits/second.

The recording of a perfect square wave onto a magnetic audio track is, under the best of conditions, difficult. For this reason the SMPTE standard has set forth an allowable risetime of 25 ±5 microseconds for the recording and reproduction of code. This tolerance requires a signal bandwidth of 15 kHz, which is well within the range of most professional audio recording devices.

Variable-speed time-code readers often are able to decode time code information at shuttle rates ranging from 1/10th to 100 times normal playspeed. This is effective for most audio applications; however, in video postproduction, it's often necessary to monitor a videotape at slow or still speeds. As LTC can't be read at speeds slower than 1/10th to 1/20th normal playspeed, two methods can be used for reading time code. In the first of these, a character generator is used to "*burn*" time code addresses into the video image of a copy worktape. This superimposed readout allows the time code to be easily identified, even at very slow or still picture shuttle speeds (Figure 9.6).

Figure 9.6. *Video image showing "burned-in" time code window.*

In most situations, LTC code is preferred for audio, electronic music and midlevel video production, as it's a more accessible and cost-effective protocol.

The second method, which is used by major video production houses, is the *vertical interval time code* (*VITC*). VITC makes use of the same SMPTE address and user code structure as LTC, but it is encoded onto videotape in an entirely different signal manner. VITC actually encodes the time code information into the video signal itself . . . inside a field (known as the vertical blanking interval) that's located outside the visible picture scan area. Because the time code information is encoded into the video signal itself, it's possible for professional helical scan video recorders to read time code at very slow and even still-frame speeds. Because time code can be accurately read at all speeds, this added convenience opens up an additional track on a video recorder for audio or cue information and eliminates the need for a burned window dub.

Jam Sync/Restriping Time Code

Longitudinal time code operates by recording a series of square-wave pulses onto magnetic tape. As you now know, it's somewhat difficult to record a square waveform onto analog magnetic tape without suffering moderate to severe waveform distortion (Figure 9.7). Although time code readers are designed to be relatively tolerant of waveform amplitude fluctuations, such distortions are severely compounded when code is directly copied by one or more generations. For this reason, a feature known as *jam sync* has been incorporated into most time code synchronizers and MIDI sequencers with sync capabilities. Jam sync basically works by reading degraded time code information from a recorded LTC track and regenerates it into fresh code that can then be re-recorded onto a new track or sent to a new media form. In the jam sync mode, the output of the generator is slaved to an external time code source. After reading this incoming signal, the generator outputs an undistorted signal that's identical to the original time code address.

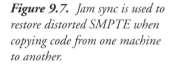

Figure 9.7. *Jam sync is used to restore distorted SMPTE when copying code from one machine to another.*

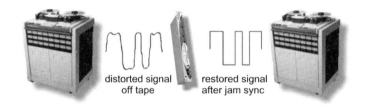

distorted signal off tape

restored signal after jam sync

Two forms of jam sync are currently in use: one-time and continuous. In *one-time jam sync*, the receipt of time code causes the generator's output to be initialized to the first valid address number. The generator then begins to count in an ascending order on its own, in a freewheeling fashion that ignores any deterioration or discontinuity in code and produces fresh, uninterrupted address numbers. *Continuous jam sync* is used in cases where the original address numbers need to remain intact and shouldn't be regenerated into a contiguous address count. After the reader has been activated, the generator updates the address count for each frame in accordance with incoming address numbers.

Synchronization Using Time Code

In order to achieve a frame-by-frame time code lock between multiple audio, video and film transports, it's necessary to use a device or integrated system that's known as a *synchronizer* (Figure 9.8). The basic function of a synchronizer is to control one or more tape, computer-based or film transports (designated as *slave* machines) so that their speeds and relative positions are made to accurately follow one specific transport (designated as the *master*).

Although the lines of distinction often break down, synchronization as a whole can be divided into two basic system types: those that are used in larger audio and video production and post-production facilities, and those that can be found in project- to medium-sized production facilities. The greatest reason for this division is not so much a system's performance as it is its price.

Systems that are used in video production and high-level systems make use of either a control synchronizer or an edit decision list (EDL) controller.

Figure 9.8. *The synchronizer in time code production.*

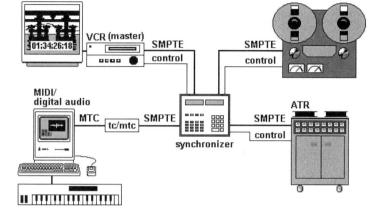

Control synchronizers are used to gain both manual and automated control over all of the transports that are connected in a system. By using a central computer and control panel (which can either exist as a stand-alone unit or be integrated into a recording console's design), a control synchronizer provides options such as the following:

- *Machine selection.* Lets us choose which machines are to be connected in the control slave chain, as well as allowing for the selection of a designated master.

- *Transport control.* Provides conventional remote control functions over any or all machines in the system.

- *Locate.* A transport command that causes all selected machines to automatically locate to a specific time code address.

- *Loop.* Enters the system into a continuous repeat cycle (play, rewind, play again) between any two address cue points that are stored in memory.

- *Offset.* Lets you enter a time into the system that can be used to correct for any time code discrepancy that might exist between two or more transports (lets you adjust relative time by ±X frames to achieve or improve sync).

- *Event points.* A series of cues can be entered into memory for triggering event(s) at specific time code addresses (i.e., triggering a sound effect or changing console snapshots from one mix scene to another).

- *Record punch in/out.* Allows the synchronizer to take control over transport record/edit functions, enabling tight record in/out points that can be repeated with frame accuracy.

An *edit decision list (EDL) controller/synchronizer* uses a process that has evolved from the online video editing process and is most commonly found in a video and audio-for-video postproduction suite. This synchronizer type goes one step further, in that it uses a user-built list of edit commands to electronically control, synchronize and switch all video and audio transports, and then exert control over their edit in/out points, tape positions, time code offsets, and so on . . . to assemble all of the various takes and scenes of a project into a finished, assembled master . . . all under automated, computer control.

Time Code Setup

In a basic audio production system, the only connection that's usually required between the master machine and a synchronizer is the LTC time code track (Figure 9.9). Generally, when connecting slaves to a synchronous system, two connections will need to be made between each slave transport and the synchronizer. These include provisions for the reproduced time code track and a control interface (which often uses the Sony 9-pin remote protocol for giving the synchronizer full logic transport and speed-related feedback information).

Figure 9.9. *System interconnections for synchronous audio production. (Courtesy of Timeline Vista, Inc., www.timelinevista.com)*

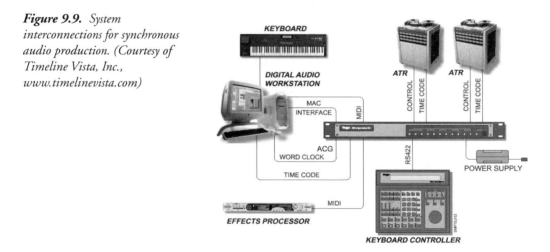

LTC signal lines can be distributed throughout the production system in the same way that any other audio signal is distributed. They can be routed directly from machine to machine or patched through audio switching systems via balanced, shielded cables. Because the time code signal is biphase or symmetrical, it's immune to problems of cables or a device that might be out-of-phase.

Time Code Levels

One problem that can plague systems using time code is crosstalk. Such problems arise from having a high-level time code signal leak into adjacent audio signals or analog tape tracks. Currently, no industry standard levels exist for the recording of time code onto magnetic tape. However, the levels shown in Table 9.1 can help you get a good signal level, while keeping analog crosstalk to a minimum.

Table 9.1. Optimum time code recording levels.

Tape Format	Track Format	Optimum Recording Level
ATR	Edge track (highest number)	–5 VU to –10 VU
¾-inch VTR	Audio 1 (L) track or time	–5 VU to 0 VU code

Tape Format	Track Format	Optimum Recording Level
1-inch VTR	Cue track or audio 3	–5 VU to –10 VU
Modular digital multitrack	Highest number track	–20 dB

Note: If the VTR to be used is equipped with Automatic Gain Compensation (AGC), override the AGC and adjust the signal gain controls manually.

MIDI-Based Synchronization

Just as synchronization is routinely used in audio and video production, the wide acceptance of MIDI and digital audio within the various media has created the need for synchronization in project studio and mid-sized production environments. Devices such as MIDI sequencers, digital audio editors, effects devices and digital mixing consoles make extensive use of synchronization and time code. However, advances in design have fashioned this technology into one that's much more cost-effective and easy-to-use—all through the use of MIDI.

The following sections will discuss several forms of synchronization that are often encountered in a MIDI-based sync and production environment.

MIDI's Internal Sync Clock

Although not related to SMPTE time code or any external reference, it's important to know that MIDI has a built-in (and often transparent) protocol for synchronizing all of the timing elements of each MIDI device within a connected system to a master timing clock. This protocol operates by transmitting real-time MIDI messages over standard MIDI cables. As with all forms of synchronization, one MIDI device must be designated to be the master device in order to provide the timing information to which all other slaved devices are locked.

MIDI Real-Time Messages

MIDI real-time messages consist of four basic types that are each 1 byte in length: timing clock, start, stop and continue messages.

- *Timing clock.* A clock timing that's transmitted to all devices in the MIDI system at a rate of 24 pulses per quarter note (24 ppq). This method is used to improve the system's timing resolution and simplify timing when working in nonstandard meters (i.e., 3/8, 5/16, 5/32, and so on).

- *Start.* Upon receipt of a timing clock message, the start command instructs all connected devices to begin playing from the beginning of their internal sequences. Should a program be in mid-sequence, the start command repositions the sequence back to its beginning, at which point it begins to play.

- *Stop.* Upon the transmission of a MIDI stop command, all devices in the system stop at their current positions and wait for a message to follow.

- *Continue.* Following the receipt of a MIDI stop command, a MIDI continue message instructs all instruments and devices to resume playing from the precise point at which the sequence was stopped. Certain older MIDI devices (most notably drum machines) aren't capable of sending or responding to continue commands. In such a case, the user must either restart the sequence from its beginning or manually position the device to the correct measure.

MIDI Time Code

MIDI time code (MTC) was developed in order for electronic music studios, project studios and virtually all other production environments to provide a cost-effective and easily implemented way to translate time code into time-stamped MIDI messages (and back). Created by Chris Meyer and Evan Brooks, MIDI time code allows SMPTE-based time code to be distributed throughout the MIDI chain to devices or instruments that are capable of understanding and executing MTC commands.

MIDI time code is an extension of MIDI 1.0, which makes use of existing message types that were either previously undefined or were being used for other, nonconflicting purposes. Since most modern recording devices include MIDI within their design, there's often no need for external hardware when making direct connections. Simply chain the MIDI cables from the master to the appropriate slaves within the system (either via physical cables or by virtually routing the MIDI data stream within a computer). Although MTC uses a reasonably small percentage of MIDI's available bandwidth (about 7.68 percent at 30-fr/sec), it's generally customary to separate these lines from MIDI cables that are communicating performance data. As with conventional SMPTE, only one master can exist within an MTC system, while any number of slaves can be assigned to follow, locate and chase to the master's speed and position.

Because MTC is easy to use and is often included free in many operating systems, this technology has grown to become the most common and most straightforward system for locking together such devices as digital audio workstations, modular digital multitracks and MIDI sequencers, as well as analog and videotape machines (through the use of a SMPTE-to-MTC converter).

MIDI Time Code Control Structure

The MIDI time code format can be broken into two parts: time code and MIDI cueing. The time code capabilities of MTC are relatively straightforward and make it possible for devices to be synchronously locked or triggered via SMPTE time code. MIDI cueing is a format that informs MIDI devices of events that are to be performed at a specific time (such as load, play, stop, punch in/out, reset). This protocol envisions the use of intelligent MIDI devices that can prepare for a specific event in advance and then execute the command on cue.

MIDI Time Code Messages

MIDI time code uses three message types: quarter frame messages, full messages and MIDI cueing messages.

- *Quarter frame messages.* These are transmitted only while the system is running in real or varispeed time and in either the forward or reverse directions. In addition to providing the system with its basic timing pulse, four frames are generated for every SMPTE time code field. This means that should you decide to use drop-frame code (30 frames per second), the system would transmit 120 quarter frame messages per second. Quarter frame messages exist as groups of eight messages that encode the SMPTE time in hours, minutes, seconds and frames (00:00:00:00). Because eight quarter frames are required for a complete time code message, the complete SMPTE time is updated every two frames. Each quarter frame message contains two bytes. The first byte is "F1," the Quarter Frame Common header, while the second byte contains a nibble (four hits) that represents the message number (0 through 7) and a nibble for each of the digits within a time field.

- *Full messages.* Quarter frame messages are not sent in the fast-forward, rewind, or locate modes because this would unnecessarily clog or outrun the MIDI data lines. When the system is in any of these shuttle modes, a fall message is used that encodes the complete time code address in a single message. After a fast shuttle mode is entered, the system generates a full message and then places itself in a pause mode until the time-encoded device has autolocated to its destination. After the device has resumed playing, MTC again begins sending quarter frame messages.

- *MIDI cueing messages.* MIDI cueing messages are designed to address individual devices or programs within a system. These 13-bit messages can be used to compile a cue or edit decision list, which in turn instructs one or more devices to play, punch in, load, stop and so on, at a specific time. Each instruction within a cueing message contains a unique number, time, name, type and space for additional information. At the present time, only a small percentage of the possible 128 cueing event types have been defined.

SMPTE/MTC Conversion

A SMPTE-to-MIDI converter is used to read SMPTE time code and convert it into MIDI time code (and vice versa). These conversion systems are available as a stand-alone device (Figure 9.10) or as an integrated part of a multiport MIDI interface/patch bay/synchronizer system. Certain analog and digital multitrack systems include a built-in MTC port within their design, meaning that the machine can be synchronized to a MIDI interface/sequencing system without the need for any additional hardware.

Figure 9.10. *PPS-2 SMPTE-to-MIDI synchronizer/converter. (Courtesy of JLCooper Electronics, www.jlcooper.com)*

Proprietary Synchronization Systems for Modular Digital Multitrack Recorders

Modular digital multitrack recorders, such as the Tascam DA-98 and Alesis ADAT, encode a proprietary form of time-encoded sync data onto tape, along with the audio information. This sync coding can be used to lock several MDMs together in order to give us more tracks. It can also be used to lock these devices to an external SMPTE source via an interface that can translate MTC or SMPTE into the MDM's sync code (and vice versa). In this way, one or more digital multitracks can be easily rigged up to act as a master or slave within a hard-wired or computer-based system.

Digital Audio's Need for a Stable Timing Reference

The process of maintaining a synchronous lock between digital audio devices or between digital and analog systems differs fundamentally from the process of maintaining relative speed between analog transports. This is due to the fact that a digital system generally achieves synchronous lock by adjusting its playback sample rate (and thus its speed and pitch ratio) so as to precisely match the relative playback speed of the master transport.

Whenever a digital system is synchronized to a time-encoded master source, a stable timing source is extremely important. An accurate timing reference may be required to keep jitter (in this case, an increased distortion due to rapid pitch shifts) to a minimum. In other words, the source's program speed should vary as little as possible to prevent any degradation in the digital signal's quality. For example, all analog tape machines exhibit speed variations, which are caused by tape slippage and transport irregularities (a basic fact of analog life known as wow and flutter). If we were to synchronize a disk-based recorder to an analog source that contains excessive wow and flutter, the digital system would be required to constantly speed up and slow down to precisely match the transport's speed fluctuations. One way to avoid such a problem would be to use a source that's more stable, such as a digital audio recorder.

Video's Need for a Stable Timing Reference

Whenever a video signal is copied from one machine to another, it's essential that the scanned data (containing timing, video, and user information) be copied in perfect sync from one frame to the next. Failure to do so will result in severe picture breakup or, at best, the vertical rolling of a black line over the visible picture area.

Copying video from one machine to another generally isn't a problem, because the VCR or VTR that's doing the copying obtains its sync source from the playback machine. Video post-production houses, however, often simultaneously use any number of video decks, switchers and edit controllers during the production of a single program. Mixing and switching between these sources can definitely result in nonsynchronous chaos, with the end result being a very unhappy client.

This sync nightmare is generally resolved by generating a single timing source. This generator produces an extremely stable timing reference (called *black burst*, or *house sync*) that has a clock frequency of exactly 15,734.2657 Hz. The purpose of this signal is to synchronize the video frames and time code addresses that are received or transmitted by every video-related device in a production facility, so that their video frame's leading edge occurs at exactly the same instant in time (Figure 9.11).

Figure 9.11. *Example of a system whose overall timing elements are locked to a black burst reference signal.*

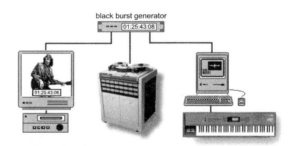

By resolving all video and audio devices to a single black burst reference, you're assured that relative frame transitions and speeds throughout the system will be consistent and stable. This holds true even for analog machines because any inherent wow and flutter can be smoothed out, due to the fact that they're being locked to an extremely stable timing reference.

Real-World Sync Applications for Using Time Code and MIDI Time Code

Before we delve into the many possible ways in which a system can be set up to work in a time code environment, it needs to be understood that each system has its own particular "personality," and that the connections, software and operation of one system might differ greatly from that of another. This is often due to factors such as system complexity and the basic hardware types that are involved, as well as the type of software systems that are installed in the computer.

Larger, more expensive setups that are used to create TV and film soundtracks will often involve extensive time code and system interconnections that can get fairly complicated. Fortunately, the use of MIDI time code has greatly reduced the cost and complexity of connecting and controlling a synchronous production system down to levels that can be easily managed by both experienced and novice electronic musicians. Having said these things, I'd still like to stress the fact that solving synchronization problems might require as much intuition, diligence, insight and art as it will technical skill.

For the remainder of this chapter, we'll be looking into some of the basic concepts and connections that can be used to get your system up and running. Beyond these, the best way to get your particular system working smoothly is to consult your manual, seek help from an experienced friend or call the tech department for the particular hardware or software that's giving you and your system the willies.

Master/Slave Relationship

Since synchronization is based upon the timing relationship between two or more devices, it follows that the easiest way to achieve sync is to have one or more devices (known as *slaves*) follow the relative movements of a single transport or device (known as a *master*). The basic rule to keep in mind is that there can be only one master in a connected system. However, any number of slaves can be set to follow the relative movements of a master transport or device (Figure 9.12).

Figure 9.12. *There can be only one master in a synchronized system; however, there can be any number of slaves.*

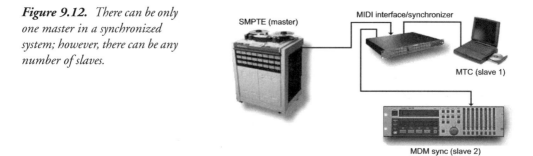

Generally, the rule for deciding which device will be the master in a production system can best be determined by asking which device will "want" to be the master. That's to say, which one will be the most stable master, from a timing standpoint?

Audio Recorders

In many audio production situations, whenever an analog tape recorder is connected in a time code environment, this machine will want to act as the master, as a fair amount of costly hardware is required to "lock" an analog machine to an external time source. This is due to the fact that the machine's speed regulator (generally a DC capstan servo) must be connected into a feedback loop that compares its present location with that of the actual SMPTE location. As a result, it would be better that other, nonvideo devices (such as MIDI or digital audio editors) be slaved to this source.

The course of action is to "*stripe*" the highest track on a clean tape with SMPTE time code and then route this signal to the SMPTE IN on your MIDI interface or synchronizer. If you don't have a multiport interface or if your interface doesn't have a SMPTE input, you'll need to get hold of a box that converts SMPTE to MTC and then plug that into the appropriate MIDI IN port.

VCRs

Video is an extremely stable timing source. As a result, a video machine should almost invariably act as a system master. In fact, without expensive hardware, a VCR can't easily be set to act as a slave, since the various sync references within the machine would be thrown off and the picture would break up or begin to roll.

From a practical standpoint, locking other devices to a VCR is done in much the same way as with an analog tape machine. Professional video decks generally include a separate track that's dedicated to time code (in addition to other tracks that are dedicated to audio). As with the earlier analog scenario, the master time code track must be striped with SMPTE before beginning the project. This process shouldn't be taken lightly, as the time code must conform to the time code addresses on the original video master or working copy (see the discussion of jam sync).

Basically, the rule of thumb is: If you're working on a project that was created "out-of-house," it's best that the videotape be striped by the original production team.

Striping your own code or erasing over the original code with your own would render the latter useless, as it wouldn't relate to the original addresses or include any variations that might be a part of the original master source. In short, make sure that your working copy includes SMPTE that's a regenerated copy of the original code. Should you overlook this, you might run into timing and sync troubles later in the postproduction phase, while assembling the music or dialog back together with the final video master.

MDMs

As a digital device, a modular digital multitrack machine is also an extremely stable timing reference and often works well as a master in an audio setting. Because of extensive sync and pitch shifting technology, these devices can also be slaved within a system without too much difficulty.

As with an analog machine, it's possible to record SMPTE onto the highest available track and then route this track to a valid SMPTE sync input. However, if you don't feel like loosing a physical track to SMPTE, you might want to pick up a sync interface that can translate the MDM's proprietary sync code into SMPTE or MTC.

Software Applications

In general a MIDI sequencer will be programmed to act as a slave device. This is due to its ability to "chase" a master source by easily changing its timing position.

Digital Audio Editors

Computer-based editors can often be set to act as a master or as a slave. This will ultimately depend on the software, as most professional editors can be set to chase (or be triggered to) a master source.

Routing Time Code to/from Your Computer

From a connections standpoint, most sequencer, digital audio and audio application software packages are flexible enough to let you choose from any number of available sync sources (whether connected to a hardware port, MIDI interface port or virtual sync driver). All you have to do is assign all slaves within the system to the device that's generating the system's master code (Figure 9.13).

Figure 9.13. *Sync Preferences dialog box for Acid. (Courtesy of Sonic Foundry, www.sonicfoundry.com)*

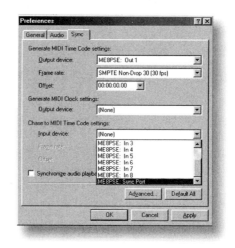

In most cases where the digital audio and MIDI sequencing applications are operating within the same computer, it's generally best to have your digital audio editor generate the master code for the system. From time to time, you might run into an editor that's unable to generate time code in any form. When faced with such an all-slave software environment, you'll actually need a physical time code master that can be routed to both your editor and MIDI sequencer. In practice, this source could be an MDM or analog recorder, but if you simply want to sync the two pieces of software together without a tape machine, the easiest solution is to use a multi-port MIDI interface to generate time code. In such a situation, all you need to do is to select the interface's sync driver as your sync source for both the editor and sequencer. Pressing the "Generate SMPTE" button from the interface's software application window, or from the interface's front panel, will lock both applications to the generated code, beginning at 00:00:00:00 or at any specified offset address.

One of the cooler ways to lock a MIDI sequencer to a digital audio editor is through the use of a virtual MIDI router. A VMR can be used to synchronize your system totally in the software domain, without the need for any external hardware or connections.

As of this writing, several generic VMRs exist for both the PC and Mac. The most notable of these for the PC are Hubi's Loop back Device (Web Search: Hubi's Loop Back Device) and the Virtual MIDI Router that comes with most Sonic Foundry products (www.sonic-foundary.com). For the latest info on what's available, I'd suggest that you surf the Web or contact your favorite digital audio hardware and software manufacturers.

Frame Rates

When synching devices together, it's important that you make sure that all of the slaves are set to the same frame setting as the master source. In the United States, these settings will most likely either be 30 frames a second (fr/sec) or 29.97 fr/sec. Thirty fr/sec (also known as non-drop code) is often used in music production that doesn't involve video production. On the other hand, 29.97 fr/sec is often used in production work where color video is involved. I think that it can be safely said that if you work in-house on projects that don't involve video, you're safe with 30 fr/sec. If you're working with a project that isn't in-house, *always* make sure that you know the original media's frame rate and stick with it (when in doubt, always confer with the original production house, as mixed time code rates can cause timing problems down the road that can be very difficult to fix).

SMPTE Offset Times

In the real world of audio production, songs don't always begin at 00:00:00:00. Let's say that you were handed an ADAT tape that needed a synth track laid down onto track 7 of a song that goes from 00:11:24:03 to 00:16:09:21. Instead of inserting more than 11 minutes of empty bars into your sequencer, you could simply open up your sync dialog box and insert an *offset* start time of 00:11:24:03.

In effect, this offset "slips" the relative times between the master and the slave, so that bar 1 on the sequence will actually begin at 00:11:24:03. The same can also apply when syncing any slaved device to an external time code source that doesn't begin at 00:00:00:00. In fact, offsets are particularly important when synchronizing devices to an analog or videotape source. As you probably know, it takes a bit of time for a tape transport to settle down and begin playing (this time often quadruples whenever videotape is involved). If the time code were to begin at the head of the tape, it's extremely unlikely that you would want to start a program at 00:00:00:00 (since playback would be delayed and extremely unstable at this time). More likely, you will want to begin the production at an offset time (often beginning a minute into the tape—00:01:00:00).

In closing, I'd like to point out that synchronization can be a simple procedure or it can be a fairly complex one, depending on your experience and the type of equipment that's involved. A number of books and articles have been written on the subject. If you're serious about production, I'd suggest that you do your best to keep up on the subject. Although the fundamentals often stay the same, new technologies and techniques are constantly emerging. As always, the best way to learn is simply by reading, jumping in and then doing it.

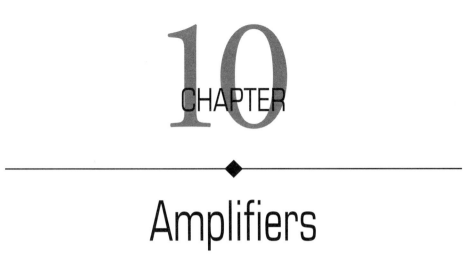

CHAPTER 10

Amplifiers

In the world of audio, amplifiers have many applications. They can be designed to amplify, equalize, combine, distribute or isolate a signal. They can even be used to match signal impedance between two devices. At the heart of any amplifier (amp) system is either a vacuum tube or semiconductor-transistor design. Just about everyone knows about these regulating devices, but not everyone grasps how they operate. Here are a few insights into these electronic wonders.

Amplification

In order to best understand how the theoretical process of amplification works, let's draw an analogy using a term that was originally used for the vacuum tube: valve (a word that's still in use in England and other Commonwealth countries). Let's begin by assuming that we have a high-pressure water pipe. Connected to this pipe is a valve that, when turned, can control large amounts of water pressure with very little effort (Figure 10.1). Using a small amount of energy, we can use our little finger to turn a trickle of water into a forceful gusher and back again.

In practice, both the vacuum tube and the transistor work much like this valve. For example, a *vacuum tube* (Figure 10.2) operates by placing a DC current between its plate and a heated cathode element. A wire mesh grid separates these two elements, which acts like a control valve for allowing electrons to pass from the plate to the cathode. The introduction of small current changes at the tube's grid (the control valve) will allow much larger, corresponding changes to be passed between the plate and the cathode (Figure 10.3).

Figure 10.1. *Current through a vacuum tube or transistor is controlled in a manner that's similar to the way water pressure is controlled by the valve tap on a water pipe.*

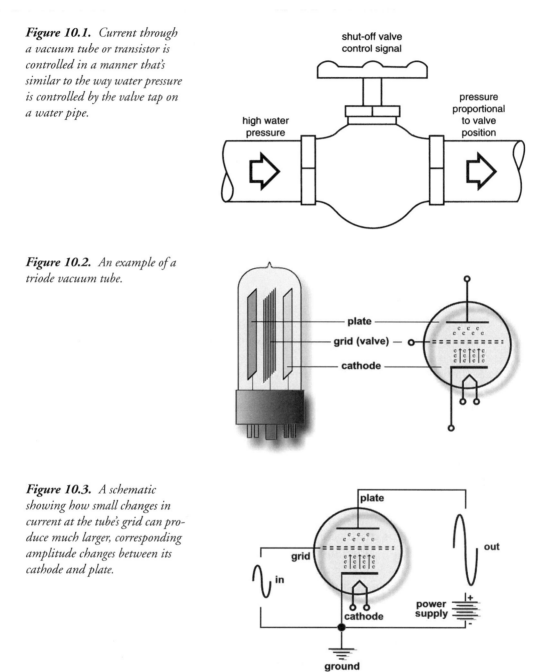

Figure 10.2. *An example of a triode vacuum tube.*

Figure 10.3. *A schematic showing how small changes in current at the tube's grid can produce much larger, corresponding amplitude changes between its cathode and plate.*

Although the *transistor* (a term derived from trans-resistor, referring to a device that can change resistance) operates using a different electrical principle, the valve analogy is still relevant. Figure 10.4 shows a basic amplifier schematic with a DC power source that's been set up across the transistor's collector and emitter points. Now, let's revisit the valve analogy. By presenting a small control voltage at the transistor's base, it's possible for a corresponding change in the resistance to be introduced between the collector and emitter, allowing a much larger and corresponding voltage/current source to be passed through to the device's output.

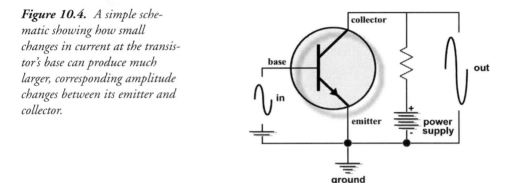

Figure 10.4. *A simple schematic showing how small changes in current at the transistor's base can produce much larger, corresponding amplitude changes between its emitter and collector.*

As a device, the transistor isn't inherently linear. That is to say, applying a signal to the base won't always produce a corresponding output change. The linear operating region of a transistor lies between the device's lower-end cutoff region and an upper saturation point (Figure 10.5). Within this operating region, changes in the base current will produce a corresponding change in the collector current and voltage. When operating close to either the cutoff or saturation points, the base current lines won't be linear and the output will be distorted. In order to keep the signal within this linear operating range, a DC bias signal is applied to the base of the transistor (for much the same reason a high-frequency bias signal is applied to a recording head). After a corrective bias voltage has been applied and sufficient amplifier design characteristics have been met, the amp's dynamic range will be limited by only two factors: noise (which results from thermal electron movement within the transistor and other circuitry) and saturation.

Amplifier *saturation* results from the input signal being at such a large level that its DC supply output voltage isn't large enough to produce the output current that's required without encountering severe waveform distortion. The result of overdriving an amp will produce a mild-to-ugly distortion effect known as *clipping* (Figure 10.6). For example, if an amp having a supply voltage of +24 volts (V) is operating at a gain ratio of 30:1, an input signal of 0.5 V will produce an output of 15 V. Should the input be raised to 1 V, the required output level would be increased to 30 V. However, since the maximum output voltage is limited to 24 V, wave excursions of greater levels will be chopped off or "clipped" at the upper and lower ends of the waveform, until the signal level falls below the maximum 24-V supply level. Whenever a transistor clips, severe odd-order harmonics are created that are immediately audible with transistors and most integrated circuit designs. Although clipping is a sought-after part of many an instrument's sound (tubes tend to have a more musical, even-order harmonic aspect to their clipping distortion), it's not a desirable effect to have in quality studio and monitoring gear.

Figure 10.5. *Operating region of a transistor.*

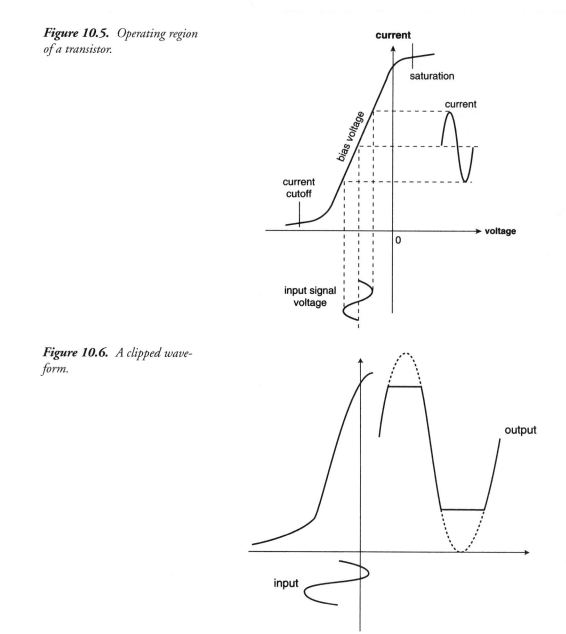

Figure 10.6. *A clipped waveform.*

The Operational Amplifier

The *operational amplifier* or *op-amp* is a stable, high-gain, high-bandwidth amp that has a high input impedance and a low output impedance. These qualities allow op-amps to be used as a basic building block for a wide variety of audio and video applications, simply by tagging additional components onto the basic circuit to fit the required design needs.

Figure 10.7 shows a typical op-amp design that's used for basic amplification. In order to reduce an op-amp's output gain to more stable, workable levels, a *negative feedback loop* is often required. Negative feedback is a technique that applies a portion of the output signal through a limiting resistor (which determines the gain) back into the negative or phase-inverting input terminal. As a result of feeding the output back to its input out of phase, the device's output signal level will be reduced. Controlling the gain of an amp using negative feedback has the additional effect of stabilizing the amp and further reducing distortion.

Figure 10.7. *Basic op-amp configuration.*

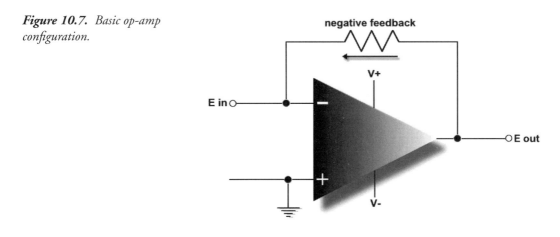

The Preamplifier

One of the mainstay amplifier types found at the input section of most professional mixer, console and outboard devices is the *preamplifier* or *preamp*. This amp type is often used in a wide range of applications, such as boosting a mic's signal to line level, providing variable gain for line level signals, isolating signals from extraneous input interference or improper grounding or signal voltage conditions and equalization . . . just to name a few.

Preamps are an important component in audio engineering as they can often set the "tone" of how a device or system will sound. Just as a microphone has its own sonic character, preamp designs will have their own "sound." Questions such as "Are the op-amps' blocks designed from quality components? Do they use tubes or transistors? Are they quiet or noisy?" are all important considerations in the overall sound of a device. This holds especially true when choosing a mic preamp (a device that boosts a mic's low output signal up to line level), because dynamic ranges in excess of 120 dB are often required for high-quality pickup conditions. Stated another way, a preamp with a 120-dB dynamic range will exhibit a 100-dB overall range when used in conjunction with a microphone that has a self-noise of 20 dB.

Equalizers

An equalizer is often nothing more than a frequency-discriminating amplifier. In most analog designs, equalization (EQ) is achieved through the use of resistive/capacitive networks that are located in an op-amp's negative feedback loop (Figures 10.8 and 10.9) in order to boost (amplify) or cut (attenuate) certain frequencies in the audible spectrum. By changing the circuit design, any number of EQ curves can be achieved.

Figure 10.8. *Low-frequency equalizer circuit.*

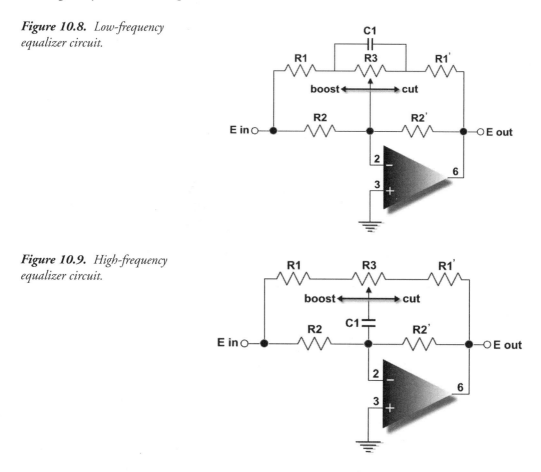

Summing Amplifiers

A *summing amp* (also known as an active combining amplifier) is designed to combine any number of discrete inputs while providing a high degree of isolation between these inputs (Figure 10.10). The summing amplifier is an important component in console design because of the large amounts of internal signal routing that require total isolation in order to separate each input from being inadvertently leaked into other input/output signal chains.

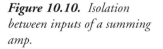

Figure 10.10. *Isolation between inputs of a summing amp.*

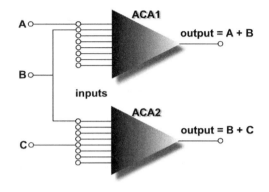

Distribution Amplifiers

Often, it's necessary for audio signals to be distributed from one device to several other devices or signal paths within a recording console or music studio. Whenever increased power is needed, a *distribution amp* may be required. Under such circumstances, a distribution amp might not provide gain (such a device is known as a unity gain amplifier), but instead will amplify the current that's delivered to one or more signal loads (Figure 10.11). One application for such an amp is in the distribution of headphone monitor feeds to large numbers of musicians in a studio.

Figure 10.11. *Distribution amp.*

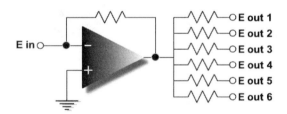

Isolation Amplifiers

An amplifier that's used to isolate signals that are combined at its input (as in a summing amp) might also be used to prevent unwanted electrical and ground potentials at its input from reaching its output. Such a device is called an *isolation amplifier*. An example of such an *iso amp* is an active direct box, which, in addition to reducing impedance, can prevent spurious ground and voltage potentials of an electric instrument or amplifier from being injected into the console's input preamps.

Impedance Amplifiers

Amps can also be used to change the impedance of a signal. An example of this application is the preamp of a condenser mic, which is used to reduce the mic capsule's impedance (which often has an impedance rating in excess of a billion ohms) to a workable impedance of around 200 ohms.

Power Amplifiers

As you might expect, *power amplifiers* (Figures 10.12 through 10.14) are used to boost the current of a signal to a level that can drive one or more loudspeakers at their rated volume levels. Although these are often very reliable devices, power amp designs have their own special set of inherent problems. These include the fact that transistors don't like to work at the high temperatures that are often generated during continuous operation at high levels. Such temperatures can also result in changes in the unit's response and distortion performance figures, requiring that protective measures (such as fuse and thermal protection) be taken to ensure that they keep working. Many of the newer amplifier models can protection the amp under a wide variety of circuit conditions (such as load shorts, mismatched loads, and even open [no-load] circuits) and are designed to work with speaker impedance loads that range from 4 to 16 Ω (with most speaker models being designed to present an 8-Ω nominal load).

Figure 10.12. QSC Audio's RMX 850 professional power amplifier. (Courtesy of QSC Audio Products, Inc., www.qscaudio.com)

Figure 10.13. Bryston 2blp amp, shown without lid cover. (Courtesy of Bryston LTD., www.bryston.ca)

An important precaution that should be taken when matching amplifier and speaker combinations is to make sure that the amp is capable of delivering sufficient power to properly drive the speaker system. If the speaker's sensitivity rating is too low or the power rating too high for

Figure 10.14. *Hafler P4000 professional power amplifier. (Courtesy of Hafler, www.hafler.com)*

what the amp can deliver, there could be a tendency to "overdrive" the amp at levels that could cause the signal to be clipped. In addition to sounding distorted, clipped signals will often contain a high-level DC component that could potentially damage the speaker's voice coil drivers.

Voltage- and Digitally-Controlled Amplifiers

Up to this point, the discussion has centered upon amps whose output levels are directly proportional to the signal level that's present at the input. Several exceptions to this principle are the *voltage-controlled amplifier* (VCA) and the *digitally-controlled amplifier* (DCA).

In the case of the VCA, the program output level is a function of an external DC voltage (which generally ranges from 0 to 5 V) that's applied to the device's control input (Figure 10.15). As the control voltage is increased, the analog signal will be proportionately attenuated. Thus, an external voltage can be used to change an audio path's signal level. Certain older console automation, automated analog signal processors and even newer digital console designs often make use of VCA technology.

Figure 10.15. *Simplified example of a voltage-controlled amplifier.*

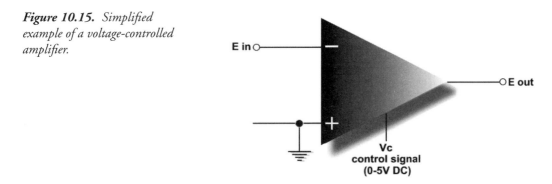

With the wide acceptance of digital technology in the production studio, it's now more common to find digital devices that use digitally-controlled amplifiers (also known as a digitally-controlled attenuators or DCAs) to control the gain of an analog signal. Although most digital devices change the gain of a signal directly in the digital domain (see the section on digital signal processing in Chapter 12), it's possible to change the gain of an analog signal from an external digital source (such as a computer, digital signal processor or controller device). The basic operation of a DCA is simple in principle—the overall gain of an analog amp can be altered by placing a series of digitally controlled step resistors into its negative feedback loop and digitally switching the amount of inserted resistance to achieve the desired levels.

An example of an 8-channel mixer that makes use of DCAs is the Analog Devices SSM2163 (Figure 10.16). This chip allows up to eight audio inputs to be mixed under digital control to a stereo output bus. Each input channel can be attenuated up to 63 dB in 1-dB intervals. In addition, any input can be assigned to or panned between the stereo outputs.

Figure 10.16. *The Analog Devices SSM2163 digitally controlled mixer makes use of DCAs to control up to 8 analog input signals.*
a. Simplified block diagram.
b. Analog signal path, showing a negative feedback step resistor network (located at the left).

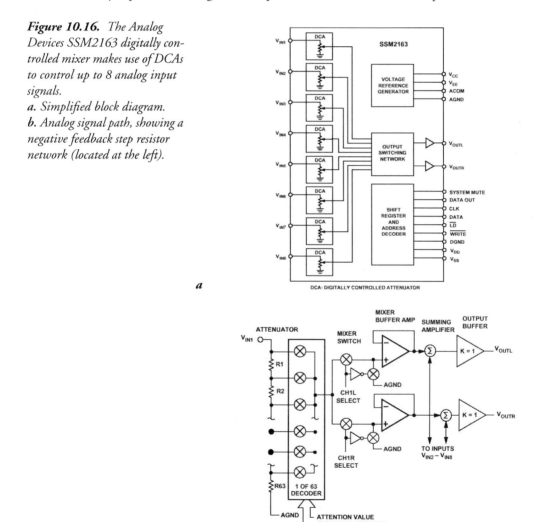

CHAPTER 11

The Audio Production Console

The basic purpose of an *audio production console* (Figure 11.1) is to give us full control over volume, tone, blending and spatial positioning for any or all signals that are applied to its inputs from microphones, electronic instruments, effects devices, recording systems and other audio devices. An audio production console (which also goes by the name of *board, desk* or *mixer*) should also provide a straightforward way to quickly and reliably route these signals to any appropriate device in the studio or control room, so they can be recorded, monitored and/or mixed into a final product. By analogy, a console and engineer can be likened to a palette and an artist, in that the console provides a creative control surface that allows an engineer to experiment and blend all the possible variables of sound.

Before the introduction of multitrack recording, all the sounds and effects that were to be part of a recording were mixed together at one time during a live performance. If the recorded blend (or *mix*, as it is called) wasn't satisfactory, or if one musician made a mistake, the selection had to be performed over until the desired balance and performance was obtained. However, with the introduction of multitrack recording, the production phase of a modern recording has radically changed into one that generally involves three stages: recording, overdubbing, and mixdown.

Figure 11.1. Simon Andrews *and Frank Fillepetti at NYC's Right Track Recording's Solid State Logic SL 9000J console in Studio C. (Courtesy of Solid State Logic, www.solid-state-logic.com and Right Track Recording, www.righttrackrecording.com)*

Recording

The recording phase involves the physical process of capturing live or sequenced instruments onto tape. Logistically, this process can be carried out in a number of ways:

- All the instruments to be used in a song can be recorded onto tape in one live pass.
- Live musicians can be used to "lay down" the basic foundation tracks (usually rhythm) of a song. At a later time, other instruments, vocals, etc., can be added.
- Electronic instruments, which were previously arranged and sequenced to form the basic foundational tracks of a song, can be recorded onto the various tracks of a multitrack recorder in such a way that other live instruments, vocal tracks, and so on, can be added at a later time.

The last two of these procedures are most commonly encountered in the recording of popular music. The resulting foundation tracks to which other tracks can be laid down at another time are called *basic, rhythm,* or *bed tracks.* These important tracks consist of instruments that provide the rhythmic foundations of a song and often consist of drums, bass, rhythm guitar and keyboards (or any combination thereof). An optional guide vocal track can also be recorded to help the musicians and vocalist capture the proper tempo and that all-important "feel" of a song. The number and type of instruments, the track layouts and other important decisions are generally determined by the producer and/or the artist. In order to achieve the best possible performance and final product, the producer and/or artist should also consider any technical advice that the engineer might have.

When recording popular music, each instrument is generally recorded onto separate tracks of a tape or hard disk recorder (Figure 11.2). This is accomplished by plugging each mic into an input strip on the console (either directly into the mixer itself or into an appropriate input on a mic panel that's located in the studio), setting the gain throughout the input strip's signal path to its optimum level and then assigning each signal to an appropriate console output . . . which is finally routed to a desired track on the multitrack recorder. Although monitoring is important during this phase, the beauty behind this process is that the final volume, tonal and placement changes can be made at a later time—during the mixdown stage.

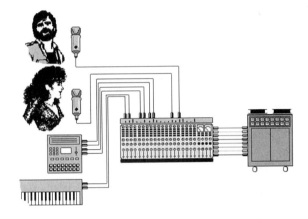

Figure 11.2. When recording popular music, each instrument is generally recorded onto a separate track (or stereo tracks) of a multitrack recorder.

Should the need arise, a group of instruments or instrument that requires multiple mics can be recorded onto a single track (or a stereo pair of tracks) by assigning each signal's input strip to the same console output bus (a process known as *grouping*). These combined signals can then be balanced in level, panned, equalized and processed, either by monitoring the console's main output for the selected tracks or by monitoring the signal returns from the tape machine itself (Figure 11.3).

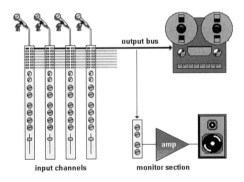

Figure 11.3. Several instruments can be "grouped" onto a single track (or stereo pair of tracks) by assigning each signal's input strip to the same console output bus.

input channels   monitor section

Unlike in the first example where each instrument is recorded onto its own track, a greater degree of care should be taken in setting the volume, tonality and placement whenever a number of signals are grouped to a single track or stereo track pair. As a general rule, it's much more difficult to make changes to previously recorded tracks that are a combined mix of several sound sources, because changes to one instrument almost always directly affect the group mix.

Each signal that's recorded onto analog multitrack tape should be recorded at the highest level possible without overloading the tape. Recording at the highest level helps achieve the best signal-to-noise ratio possible for each track on the tape so that the final product isn't impaired by tape hiss or preamp noise. Digital tracks are more forgiving (because of the increased headroom); however, it's always a good idea to record signals to tape or disk at recommended levels (often at least 12 dB below the maximum overload point).

The recording stage is vitally important to the outcome of the overall project. Since the goal is to capture the highest-quality sound to each track at an optimum level, the signals will often be recorded without regard to the musical balance levels of each instrument on other tracks. In other words, a chanted whisper from a vocalist will often be boosted to recorded levels that are equal to those of the electric guitar (which ideally has been placed in another, separate room). From this, you might guess that the process of monitoring recorded signals is extremely important.

When you ask producers about the importance of preparation, they will most likely place it near the top of the list for capturing a project's sound, feel and performance. The rhythm tracks (i.e., drums, guitar, bass and possibly piano) are often the driving backbone of a song, and recording them improperly can definitely get the project off to a bad start. Beyond making sure that the musicians are properly prepared and that the instruments are tuned and in top form (both of these being the producer and/or band's job), it's the engineer's job to help capture a project's sound to tape, disk or disc as best as he or she can during the recording phase. Recording the best possible sound (both musically and technically), without having to excessively rely on the "fix it in the mix" approach, will put everyone involved on the path towards obtaining the best possible product.

Monitoring

Because the artists, producer and engineer must be able to hear the instruments as they're recorded and played back, and because the levels have been recorded irrespective of their overall musical balance, a separate mix must be made in order to *monitor* what is being laid down to tape or disk. As you'll learn later in this chapter, a multitrack performance can be monitored in several ways. No particular method is right or wrong; rather, it's best to choose a method that matches your own personal production style. No matter which monitoring style is chosen, the overall result will generally be as follows:

- During the recording process, each signal being fed to a track on the multitrack recorder will be fed to the monitor mix section (Figure 11.4), so that the various instrument groups can be mixed (with regard to level, panning, effects, etc.) and then fed to the control room's main monitor speakers.

- A monitor mix can be created that can be heard over headphones by the musicians in the studio. In fact, two or more separate cue mixes might be available (depending on the musicians' listening needs).

Figure 11.4. *During recording, each signal can be fed to the monitor mix section, where the various instruments can be mixed and then fed to the control room's main speakers and/or the performer's headphones.*

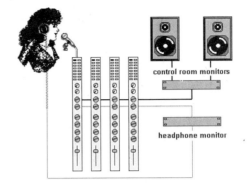

control room monitors

headphone monitor

Overdubbing

Instruments that aren't present during the original performance can be added to the existing multitrack project during a process known as *overdubbing*. At this stage, musicians listen to the previously recorded tracks over headphones and then play along while recording their new tracks. A new "take" can be laid down to tape or disk if one or more of the musicians have made minor mistakes during an otherwise good performance or if other instruments need to be added to the basic tracks in order to finish a project. These new performances are recorded in sync with the original performances and are recorded onto unrecorded tracks or onto open tracks that contain information that's no longer needed.

When overdubbing tracks onto an analog multitrack recorder, it's important to remember to place the tracks that are to be played back into the *sync mode* (a process whereby the record head is used as a playback head for the necessary channels in order to maintain a proper time relationship during playback). Most modern analog multitracks can be placed into a master sync mode, which automatically switches the machine between the input and sync monitor modes (thereby eliminating the above need for manual switching). For more information on sync playback, refer to Chapter 5, "The Analog Audio Tape Recorder."

Mixdown

After all the musical parts have been performed and recorded to everyone's satisfaction, the *mixdown* or *mix* stage can begin. At this point, the multitrack's playback outputs are fed to the console inputs. Often, this is done by switching the console into the mixdown mode or by changing the appropriate input switches to the line or "tape" position. The master tape is then repeatedly played while adjustments in level, panning, EQ, effects, etc., are made for each track. Throughout this artistic process, the individually recorded signals are blended into a composite surround, stereo or mono signal that's fed from the console outputs to the master mixdown recorder. When a number of mixes have been made and a single version has been approved, this recording (called the *final mix*) can be assembled (along with other songs or productions in the project) into the final product.

The Professional Analog Console

Most analog audio production consoles used in professional recording studios (Figures 11.5 through 11.7) are designed with similar controls and capabilities. Their control surfaces and functions mostly differ in appearance, location of controls, on-board dynamic processing and signal-routing capabilities, as well as how they incorporate automation and control-setting recall functions (if any exist).

Figure 11.5. *Mackie 8-bus.*
(Courtesy of Mackie Designs,
www.mackie.com)

Figure 11.6. *Amek 5.1.*
(Courtesy of Amek,
www.amek.com)

Figure 11.7. *Solid State Logic SL 9000J. (Courtesy of Solid State Logic, www.solid-state-logic.com)*

Before we delve into the details of how a console works, it's important to take a look at one of the most important concepts in all of audio technology: the *signal chain* (also known as the *signal path*). As is true with literally any audio system, the recording console can be broken down into functional components that are chained together into a large (and, one hopes, manageable) number of signal paths. By identifying and examining the individual components that work together to form this chain, it becomes easier to understand any mixer or console, no matter how large or complex.

The trick to understanding a console or mixer system is to realize that each component has an input and an output, which has an associated audio source and a destination. In such a signal flow chain, the output of each source device must be connected to the input of the following device . . . until the end of the audio path is reached. Whenever a link in this source-to-destination path is broken, no signal can pass.

Although this might seem like a simple concept, keeping it in mind can save your sanity when paths, devices and tangled cables get out of hand. It's as basic as knit one, purl two: An audio signal flows from the output of one component or device to the input of another . . . whose output then flows to the input of another . . . and so on, until the final destination in the chain is reached.

The signal for each input of a recording console or mixer will almost always flow vertically down a series of controls that are associated with a channel, in what is known as an *input strip* or *I/O module* (Figure 11.8). *I/O* stands for input/output and is so named because all the associated electronics for a single track/channel combination are often laid out onto a single circuit board that's physically attached to the input strip controls. As the electronics of an I/O module are self-contained, they can be fitted into a modular mainframe in a number of channel and layout configurations, so as to better match the present and future production needs of a particular studio. This plug-in nature also makes them easily interchangeable and removable for service.

Figure 11.8. *Console I/O module input strips.*
a. Mackie 8-bus (Courtesy of Mackie Designs, www.mackie.com)
b. Amek 5.1. (Courtesy of Amek, www.amek.com)

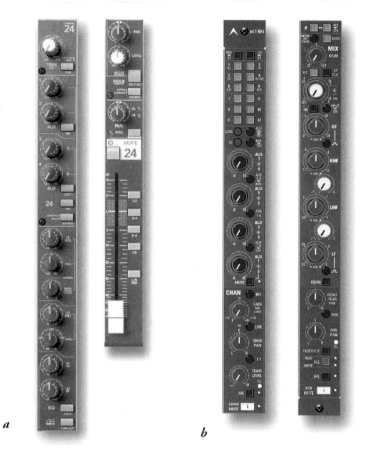

a   *b*

The following sections describe the various I/O stages of a professional audio production console. Although consoles tend to vary in layout, this introduction should offer some good insights into basic I/O design.

Channel Input

The *channel input section* (Figure 11.9) serves to optimize the signal gain levels at the input of an I/O module before being further processed and routed. Either mic or line inputs can be selected, which are varied in gain by one or two continuous level controls (often called *gain*

c. Solid State Logic SL 9000J.
(Courtesy of Solid State Logic,
www.solid-state-logic.com)

c

trims). Mic gain "trims" are typically capable of boosting a signal over a +20 to +70 dB range, while a line trim can be varied in gain over a –15 (15 dB pad) to +45 dB range.

Gain trims are a necessary component, as the output level of a microphone is typically very low (–45 to –55 dB) and requires that a high-quality, low-noise amplifier be used to raise the various mic output levels in order for the signal to be passed throughout the console at an optimum level (as determined by the console's design and standard operating levels). Whenever a mic or line signal is boosted to levels that cause the preamp's output to be driven above +28 dBm, severe clipping distortion will almost certainly occur. In order to avoid such overloads, the input gain must be reduced (either by simply turning the gain trim down or by inserting an

Figure 11.9. *Channel input section.*
a. Mackie 8-bus. (Courtesy of Mackie Designs, www.mackie.com)

a

b. Amek 5.1. (Courtesy of Amek, www.amek.com)

b

c. Solid State Logic SL 9000J.
(Courtesy of Solid State Logic,
www.solid-state-logic.com)

c

attenuation pad into the circuit). On the other hand, signals that are too low in level will unnecessarily add noise into signal path. Finding the right levels is often a matter of knowing your equipment, watching the meters and/or overload lights and using your experience.

Input attenuation pads (which are used to reduce a signal by a specific amount . . . i.e., –10 or –20 dB) will often precede the preamp's input. On larger consoles, the output is fed to a phase-reversal "Ø" switch, which is used to change the signal's phase by 180° in order to compensate for an out-of-phase microphone or cable. High- and low-pass filters may also follow the preamp, allowing such extraneous signals as tape hiss or subsonic floor rumble to be filtered out.

Auxiliary Send Section

The *auxiliary* (*aux*) *sends* are used to route and mix signals from an input strip to the various effect sends (and possibly monitor/headphone cues) of a console. These sections are used to create a submix of any (or all) of the various console input signals to a mono or stereo send, which can be routed to any destination (Figures 11.10 and 11.11). Commonly, up to eight individual aux sends can be found on an input strip.

Figure 11.10. *Although a console's input path generally flows vertically from the top to the main fader at the bottom, "aux" sends operate in a horizontal fashion . . . in that the various channel signals are mixed together to feed a mono or stereo output bus.*

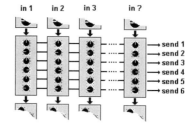

Figure 11.11. *Auxiliary send section.*
a. Mackie 8-bus. (Courtesy of Mackie Designs, www.mackie.com)

a

b. Amek 5.1. (Courtesy of Amek,
www.amek.com)

b

c. Solid State Logic SL 9000J.
(Courtesy of Solid State Logic,
www.solid-state-logic.com)

c

An auxiliary send can serve many purposes. For example, several mono sends can be used to individually drive reverb devices, signal processors or other devices, while another mono send could be used to drive a speaker that's placed in that great-sounding bathroom down the hall. A stereo pair of sends could be used to provide a headphone mix for a couple of musicians who decided they had to overdub another part. In fact, these sends can be used to accomplish almost any set of mixing tasks. For example, during a live studio performance, sends 3 and 4 could be used to feed a stereo DAT recorder and send 5 could be used to send a live satellite feed to a Moscow TV station, while send 1 feeds a miked 3" speaker that's been placed in the producer's motorcycle saddlebag. How you deal with "aux" sends is literally up to you, your needs and your creativity.

Equalization

The equalization (EQ) section (Figure 11.12), like the auxiliary sends, derives its feed directly from the channel input section. An input strip's equalizer is used to compensate for variations or discrepancies in frequencies that are present in the audio signal. Although the specifics vary from console to console, larger EQ section designs often include up to four continuously variable overlapping frequency-control bands, each having a variable bandwidth (Q) and a boost or cut control (often having values that can be boost or cut over a ±18 dB range). Most designs include an EQ in/out button that lets the engineer silently switch the equalizer in or out (bypass) of circuit.

Figure 11.12. *Equalization section.*
a. Mackie 8-bus. (Courtesy of Mackie Designs, www.mackie.com)

a

b. Amek 5.1. (Courtesy of Amek,
www.amek.com)

b

c. Solid State Logic SL 9000J.
(Courtesy of Solid State Logic,
www.solid-state-logic.com)

c

Insert Point

Many mixer and console designs provide a break in the signal chain that often follows after the equalizer. At this point, a *direct send/return* or insert access point (often abbreviated as *direct* or *insert*) can be used to send the line level audio signal out to an external processing device or recorder. The external signal can then be directly inserted back into the input strip's signal path, in a number of balanced or unbalanced configurations.

It's important to note that plugging a dynamics, EQ or effects processor into an insert point will only affect the signal that passing through the selected I/O channel. Should you wish to effect a number of channels, the auxiliary send or group outputs can be used to process a combined mix of several inputs.

Physically, the send and return jacks of a console or mixer can be accessed as two separate jacks on the top or back a mixer/console (Figure 11.13a), as a single, stereo jack (in the form of a tip-ring-sleeve jack that respectively carries the send, return and common signals (as shown in Figure 11.13b), or as access points on a console's patch bay.

Figure 11.13. Direct send/ return signal paths.
a. Two jacks can be used to send signals to and return signals from an external device.
b. A single TRS (stereo) jack can be used to insert an external device into an input strip's path.
c. Cable wiring diagram for a TRS send/return signal path. (Courtesy of Mackie Designs, www.mackie.com)

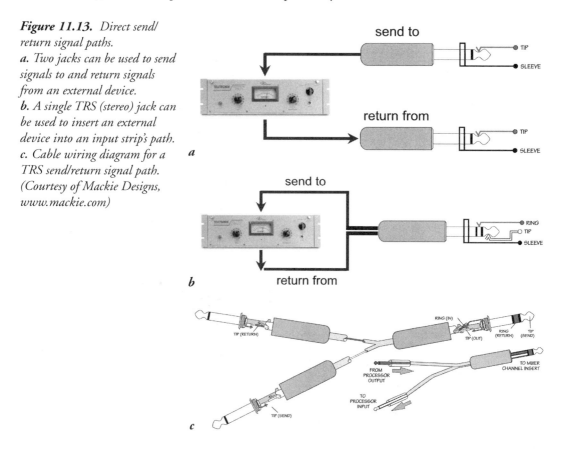

A number of mixers and console manufacturers (such as Mackie and Tascam) use unbalanced, stereo TRS jacks to directly insert a signal that can be accessed in several interesting ways (Figure 11.14; please consult the manual for your system's wiring layout):

- Inserting a mono send plug to the first "click" will connect the cable to the direct out signal, without interrupting the return signal path.

- Inserting a mono send plug all the way in will connect the cable to the direct out signal, while interrupting the return signal path.

- Inserting a stereo TRS plug all the way in allows the cable to be used as a send/return loop.

Figure 11.14. *The various insert positions for an unbalanced TRS send/return loop. (Courtesy of Mackie Designs, www.mackie.com)* *a. Direct out with no signal interruption to a mixer's main output path. Insert only to first "click." b. Direct out with an interruption to the mixer's main output path. Insert all the way to the second "click." c. For use as an insert loop (tip = send to external device, ring = return from external device, sleeve = common circuit ground).*

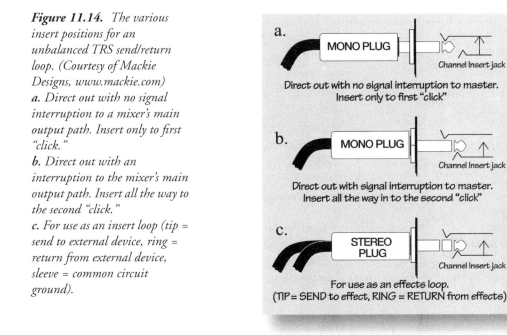

Direct out with no signal interruption to master.
Insert only to first "click"

Direct out with signal interruption to master.
Insert all the way in to the second "click"

For use as an effects loop.
(TIP = SEND to effect, RING = RETURN from effects)

Dynamics Section

Many top-of-the-line consoles have designed a dynamics section into each of their I/O modules (Figure 11.15). This allows each signal to be dynamically processed more easily, without the need to scrounge up tons of outboard devices. Often the full complement of compression, limiting and expansion (including gating) is provided. A complete explanation of dynamic control can be found in Chapter 12, "Signal Processors."

Figure 11.15. *Dynamics section of the Solid State Logic SL 9000J. (Courtesy of Solid State Logic, www.solid-state-logic.com)*

Monitor Section

Since all the signals have commonly been recorded onto tape at their highest possible levels (without regard to the relative musical balance on other tracks), a means for creating a separate monitor mix in the control room is necessary in order to hear a musically balanced version of the production. As a result, a separate *monitor section* is usually designed into the console to provide control over each input's level, pan and effects, etc., and then route this mix to the control room's mono, stereo or surround speakers.

The approach and techniques of monitoring tracks during a recording will often vary from console to console (as well as between individuals). Again, no method is right or wrong compared to another. It simply depends on what type of equipment you're working with, as well as your personal working style. The following sections briefly describe a few of the most common approaches to monitoring.

In-Line Monitoring

Many newer console designs incorporate an I/O *small-fader section* (Figure 11.16), which can be used to directly feed the recorded signal that's being fed to either the multitrack recorder or the monitor mixer (depending on its selected operating mode).

Figure 11.16. *Small-fader section.*
a. Mackie 8-bus. (Courtesy of Mackie Designs, www.mackie.com)
b. Amek 5.1. (Courtesy of Amek, www.amek.com)

a

b

Figure 11.16. *(continued)*
c. Solid State Logic SL 9000J.
(Courtesy of Solid State Logic,
www.solid-state-logic.com)

c

In the standard monitor mix mode (Figure 11.17a), the small fader is used to adjust the monitor level for the associated tape track. In the "flipped" mode (Figure 11.17b), the small fader is used to control the signal level that's being sent to tape, while the larger, main fader is used to control the monitor mix levels. This function allows multitrack levels (which aren't often changed during a session) to be located out of the way, while the more frequently used monitor levels are assigned to the more accessible master fader position.

Figure 11.17. *Small-fader*
monitor modes.
a. Standard monitor mode.

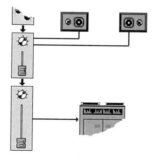

a

b. *"Flipped" monitor mode.*

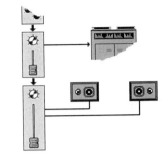

b

Separate Monitor Section

Certain English consoles (particularly those of older design) incorporate an entirely separate monitor section, which is generally placed at the console's right side (Figure 11.18). In essence, this section offers level, pan, effects and "foldback" (the older English word for a headphone monitor control), with the inputs to this section being generally driven by the console's multi-track output and tape return buses.

Figure 11.18. Older-style English consoles may have a separate monitor section, which is driven by the console's multitrack output and tape return buses.

input section   monitor section

During mixdown, this type of design has the advantage of offering a large number of extra inputs that can be assigned to the main output buses for use to mix effects returns, electronic instrument inputs and so on. During a complex recording session, this monitoring method might require an extra amount of concentration to avoid confusing the inputs that are being sent to tape with the corresponding tape return strips that are being used for monitoring (which is probably why this design style has fallen out of favor).

Direct Insertion Monitoring

If a console doesn't have any of the preceding monitor facilities (or even if it does), a simple and effective third option is still available to the user. This approach uses the direct send/returns of each input strip to insert the recorder directly into the input strip's signal path. Using this approach, the direct send for each of the associated tape tracks (which can be either before or after the EQ section) is routed to the associated track input on a multitrack recorder (Figure 11.19). The return signal is then routed from the recorder's output to the same input strips return path (where it injects the signal back into the strip's effects send, pan and main-fader sections).

Figure 11.19. By directly inserting the tape send and return into the signal path, a recorded track can be easily monitored.

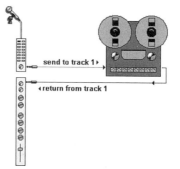

send to track 1 ▸

◂ return from track 1

Using this approach, the input signal directly following the mic/line preamp will be fed to tape (with levels being adjusted by the mic/line gain trim). The return tape path is then fed through the input strip's effects and output path so that the signal can be mixed (along with full effects) *without regard* for what's being recorded to tape, disk or other medium. Playing the tape back or placing any track into the sync mode won't affect the overall monitor mix at all because the tape outputs are already being used to drive the console's main outputs (which are being used to create a monitor mix). This approach can be used effectively on mixers and larger consoles alike . . . with the only potential drawback being the large number of patch cords that might be required to patch the inputs and outputs to and from the multitrack recorder.

Channel Assignment

Following the main "to-tape" fader, the signal is then usually routed to the strip's *track assignment matrix* (Figure 11.20), which is capable of distributing the signal to any or all tracks of a multitrack recorder. Although this section electrically follows either the main or small fader section (depending on the channel's monitor mode), the track assign buttons will often be located either at the top of the input strip or designed into the main output fader (often being placed at the fader's right-hand side).

Functionally, pressing any or all assignment buttons will route the input strip's main signal to the corresponding track output buses. For example, if a vocal mic is plugged into channel 14, the engineer might assign the signal to track 14 by pressing (you guessed it) the "14" button on the matrix. If a quick overdub on track 15 is also needed, all the engineer has to do is unpress "14" and assign the signal to track 15.

Figure 11.20. *Channel assignment section.*
a. Mackie 8-bus. (Courtesy of Mackie Designs, www.mackie.com)
b. Amek 5.1. (Courtesy of Amek, www.amek.com)

a

b

Figure 11.20. *(continued)*
c. Solid State Logic SL 9000J.
(Courtesy of Solid State Logic,
www.solid-state-logic.com)

c

Many newer consoles offer only one button for even and odd paired tracks, which can then be individually assigned by using the strip's main output pan pot. For example, pressing the button marked 5/6 and panning to the left routes the signal only to output bus 5 . . . to the right routes it to bus 6. This simple approach accomplishes two things:

- Fewer buttons need to be designed into the input strip (lowering cost and reducing moving parts).

- Stereo submixes can easily be built up by panning instruments over a stereo bus pair and then assigning the outputs to a pair of tracks on the multitrack recorder.

Output Fader

Each input strip contains an associated *main output fader* (which determines the level) and *pan pot* (which is often designed into or near this section . . . and determines the signal's left/right placement in the stereo and/or surround field). Generally, this section (Figures 11.21 and 11.22) includes a *Solo/Mute* feature, which offers the following functions:

- *Solo.* When this is pressed, the monitor outputs for all other channels are muted, allowing the listener to monitor only the selected channel (or soloed channels) without affecting the multitrack or main stereo outputs during the recording and mixdown process. During mixdown, this button can be programmed to mute the main bus outputs for all unselected channels (which might contain extraneous noises or unwanted notes).

- *Mute*. This function basically works opposite to the solo button, in that when it is pressed, the selected channel is muted or cut from the monitor outputs. Likewise, this function may be programmed to mute the signal in the main bus outputs, or in some designs, will noiselessly cut power to the selected input strip.

Figure 11.21. *Output fader section of the Solid State Logic SL 9000J. (Courtesy of Solid State Logic, www.solid-state-logic.com)*

Figure 11.22. *Pan pot configurations.*
a. Stereo pan left/right control.
b. Surround pan control/joystick.

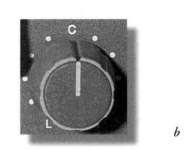 

a   *b*

Output Bus

In addition to the concept of the signal chain, one other important signal path concept should be understood: the *output bus*.

Conceptually, we can think of a *bus* as a single electrical conduit that runs the horizontal length of a console or mixer (Figure 11.23). This single signal path often is a heavy copper wire or a single wire on a ribbon connector cable that runs the entire width of the console. It can be thought of as an electrical junction point that allows any number of signals to be injected into the bus line (where it can be mixed in with other signals that are present). Signals can also be routed off the bus to one or more output destinations (such as a console output, track output or auxiliary effects send). Much like a city transit bus, this signal path follows a specific route and allows audio signals to get on or off at any point along its path.

Figure 11.23. *Example of an effects send bus, whereby multiple inputs are mixed and routed to a master send output.*

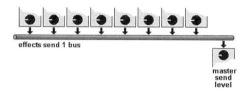

effects send 1 bus

master
send
level

From the discussion of the input strip, we've seen that a channel's audio signal, by and large, follows down from the top to the bottom of the strip. As we take the time to follow this path, it's easy to spot where audio is routed off the strip and onto a horizontal output bus path. Aux sends, monitor sends, channel assignments and main outputs are all examples of signals that are injected into buses, which are then sent to an output destination or destinations. For example, it's easy to see that the aux send #1 controls are horizontally duplicated across a console's surface. In the same way, these input gain controls are physically tied to an auxiliary send bus (aux #1) that routes the mixed level signals off to its output destination: aux send out 1. An example of a stereo bus is the console's main stereo output. After the relative signal volumes are determined by each strip's fader and pan pot position, the output is injected into the console's main left/right output buses. The output bus signals will then be a combination of all the different input and effects return signals, which are then routed to the master recorder.

Master Output/Group Faders

A number of mixer and console designs include a *master output fader* section that allows the overall levels of an output bus to be trimmed. For example, since the Mackie 8.bus console has eight buses, the output fader section (Figure 11.24) has eight output faders. Certain consoles having 16, 24 or more buses will actually have faders for each output channel bus, however, most console designs incorporate bus out trims pots into their input or monitor output strips.

Basically, a master output fader or rotary pot serves as a convenient point for controlling overall group output levels that are being sent to tape, a device or other recording media. For example, it would be much easier (and would cause less disruption to the mix levels) to turn down the 8 bus faders of the 8.bus, during an 8-track session, than it would to try to proportionately turn down all of the gain trims or channel faders on each strip. In addition, with the help of an oscillator these levels can be used to help easily match the console's output levels to the input levels of a multitrack recording device.

Figure 11.24. *Mackie 8-bus master output fader section. (Courtesy of Mackie Designs, www.mackie.com)*

During mixdown, another important use of such a master output section would be the ability to assign a series of *grouped* instruments or sources to a single output bus or stereo output buses . . . which can then be mixed into the main mono, stereo or surround outputs. By assigning any number of input channels to a single set of grouped outputs, a lot of time and frustration can be saved when changing the level of a master group fader during a mix. For example, let's say that we're mixing down a tape that has 10 drum tracks. Instead of assigning each track to the main L/R bus, let's assign them to output buses 3/4. All of the standard panning can be done at the input strip; however, the overall drum levels can now be controlled by two faders at the master fader section. Instead of having to turn down all the inputs (and risk losing the balance), we can simply turn down bus faders 3/4 . . . which are, of course, being fed to the main mix bus outputs.

Monitor Level Section

Most console and mixing systems include a central monitor section that controls levels for the various monitoring functions (such as control room monitor level, studio monitor level, headphone levels and talkback). This section (Figure 11.25) often makes it possible to easily switch between the various monitor speakers. It also can provide switching between the various sources and recording devices that are found in the studio (i.e., surround/stereo/mono output buses, tape returns, aux send monitoring, solo monitoring, etc.).

Figure 11.25. *Monitor level section.*
a. Mackie 8-bus. (Courtesy of Mackie Designs, www.mackie.com)
b. Amek Media 5.1. (Courtesy of Amek, www.amek.com)

a

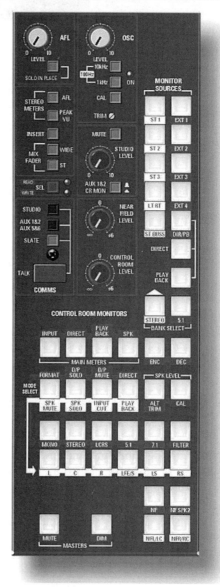

b

Patch Bay

A *patch bay* (Figure 11.26) is a panel that (under the best of conditions) contains accessible jacks that correspond to the various inputs and outputs of every access point within a mixer or recording console. Most professional patch bays (also known as patch panels) offer centralized I/O access to most of the recording, effects and monitoring devices or system bocks within the production facility (as well as access points that can be used to connect between different production rooms).

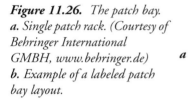

Figure 11.26. The patch bay.
a. Single patch rack. (Courtesy of Behringer International GMBH, www.behringer.de)
b. Example of a labeled patch bay layout.

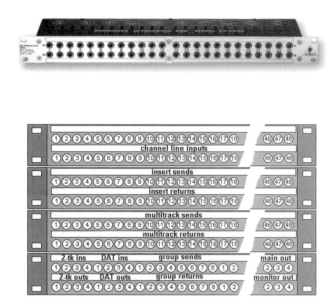

Patch bay systems come in a number of plug/jack types as well as wiring layouts. For example, prefabricated patch bays are available in tip-ring-sleeve (balanced) and tip-sleeve (unbalanced) ¼" phone configurations, as well as RCA (phono) connections. These models will often place jacks at the panel's front and rear, so that project studio users can reconfigure the panel by simply rearranging the plugs at the rear access points. Other professional systems (often those that use the professional telephone-type plugs) will require you to hand-wire the connections in order to wire or configure or reconfigure a bay (often an amazing feat of patience, concentration and stamina).

Patch jacks can be configured in a number of ways, so as to allow for several signal connection options, when connecting between inputs, outputs and external devices (Figure 11.27):

- *Open.* When no plugs are inserted, each I/O connection entering/leaving the panel is independent of the other and has no electrical connection.

- *Half-normalled.* When no plugs are inserted, each I/O connection entering the panel is electrically connected (with the input being routed to the output). When a jack is inserted into the top jack, the in/out connection is still intact, allowing you to "tap" into the signal path. When a jack is inserted into the bottom jack, the in/out connection is broken, allowing only the inserted signal to pass to the input.

- *Normalled.* When no plugs are inserted, each I/O connection entering the panel is electrically connected (with the input routing to the output). When a jack is inserted into the top jack, the in/out connection is broken, allowing the output signal to pass to the cable. When a jack is inserted into the bottom jack, the in/out connection is broken, allowing the input signal to pass through the inserted cable connection.

- *Parallel.* In this mode, each I/O connection entering the panel is electrically connected (with the input routing to the output). When a jack is inserted into the either the top or bottom jack, the in/out connection will still be intact, allowing you to "tap" into both the signal path's inputs and outputs.

Figure 11.27. Typical patch bay signal routing schemes. (Courtesy of Behringer International GMBH, www.behringer.de)

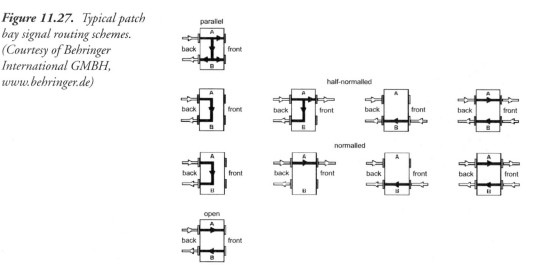

The purpose of breaking normalled connections is so that an engineer can use patch cords to connect different or additional pieces of equipment between two components that are normally connected. For example, a limiter might be temporarily patched between a mic preamp output and an equalizer input. The same preamp/EQ patch point could be used to insert an effect or any other type of device. These useful access points can also be used to bypass a defective component or to change a signal path in order to achieve a certain effect. . . . Versatility is definitely the name of the game, here!

Metering

Each input, line output and other level point (such as an effects send) is often measured by a meter that displays the signal level's strength (Figure 11.28). Meter and indicator types will often vary from system to system. For example, banks of readouts that indicate console bus output and tape return levels might use VU metering, PPMs (peak program meters, which can be found in European designs) or LED or liquid crystal displays. It's also not uncommon to find LED overload indicators on any number of devices. These lights give a quick and easy peak indication as to whether you've approached or have reached the component's headroom limits (a sure sign to back off your levels).

Figure 11.28. A set of LED, light-bar and VU meter displays.

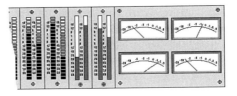

The basic rule regarding levels isn't nearly as rigid as you might think and will often vary depending on whether the device and/or recording medium is analog or digital. In short, if the signal level is too low, tape, system and even digital noise could be a problem, as the levels throughout the signal chain will probably not be optimized. If the level is too high, the preamps, saturated tape or clipped digital converters will often result in a distorted signal. Here are the basic rules of thumb:

In analog recording, proper recording level is achieved when the highest reading on the meter is near the zero level, although levels slightly above or below this might not cause difficulties (Figure 11.29). In fact, (slightly) overdriving some analog devices and tape machines will often result in a sound that's "rough" and "gutsy."

Figure 11.29. VU meter readings for analog devices.
a. Too low.
b. Too high.
c. Just right.

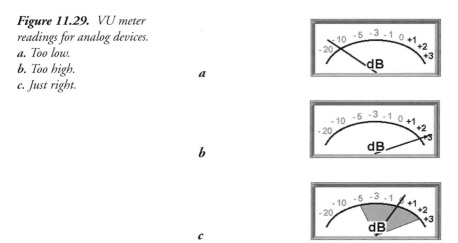

In digital recording, noise is often less of a practical concern. It's often a good idea to keep levels as high as possible, while keeping a respectful distance from the dreaded overload or "over" indicator Unlike analog, digital is usually very unforgiving of clipped signals and will generate grunge that's guaranteed to make you cringe! As there is no real standard for digital metering levels beyond these guidelines, it's best to get specifics about level from the device's manual (if there are any specifics).

Digital Console Technology

Console design, signal processing and signal routing technology have undergone a tremendous design revolution with the advent of digital audio. As the new millennium dawns, digital consoles and mixer designs are finding their ways into professional, project and other forms of audio production facilities at an amazing pace. These devices use large-scale integrated circuits and central processors to convert, process, route and interface to external audio and computer-related devices with relative ease. In addition, this technology makes it possible for many of the costly and potentially faulty discrete switches and level controls that are required for such functions as track selection, gain, and EQ to be replaced by assignable digital switches and control networks. The big bonus, however, is that since routing and other processing control functions are digitally encoded, it becomes a simple matter for routing and dynamic level settings to be saved in computer memory for complete and virtually instantaneous recall at any time. One of the other surprising aspects of this technology is the fact that digital consoles often cost the same or are more cost-effective than their analog counterparts.

Digital console designs are currently available in two forms: the fully digital console and the digitally controlled analog console. The former has begun to take center stage in many production facilities, while digitally controlled analog consoles have begun to wane in popularity. These system types are available in many configurations and have price tags that run the full gamut of affordability (including a growing number that can be thought of as real bargains). Although digital systems interface with the signal path in ways that differ from their analog counterpart, it's important to keep in mind that the signal chain is still conceptually the same: the audio signal follows from one control or processing block to the next, until the desired effect or destination is reached.

Digital Consoles

In a digital console design (Figures 11.30 through 11.33), the analog input signals are converted directly into digital audio data (or are directly inserted into the console's chain as digital data) and are thereafter distributed, routed and processed entirely in the digital domain. At the console's various sends, main and/or bus outputs, the signal may either be decoded back into analog form or remain in the digital domain for distribution to recorders or other audio devices. Because all audio, routing and processing data is digitally controlled, all of the system's dynamic and snapshot automation moves can be recorded directly into the console's memory or external computer memory, or (in the case of more economical systems) it can be stored as MIDI messages within a sequencer.

Centralized Control Panels

From a functional standpoint, the basic flow of a digital console or mixer's input strip is similar to that of an analog console . . . in that the input will first need to be boosted in level by a mic/line preamp and then will be converted to digital and passed through functional processing and

Figure 11.30. *Tascam TM-D1000 Digital Mixer. (Courtesy of Tascam, www.tascam.com)*

Figure 11.31. *Mackie Digital 8 Bus Console. (Courtesy of Mackie Designs, www.mackie.com)*

Figure 11.32. *Sony DMX-R100 Digital Console. (Courtesy of Sony Electronics, Inc., www.sony.com/proaudio)*

Figure 11.33. *SSL Axiom-MT Digital Multitrack Recording System. (Courtesy of Solid State Logic, www.solid-state-logic.com)*

routing "blocks" until the main channel output or assigned output grouping is reached. From a control standpoint, the human interface of routing and controlling a send level, changing an EQ setting or assigning a strip to a multitrack channel bus might be quite different in the way that it's carried out.

More often than not, a digital console operates by allowing the user to assign a channel's function (such as aux send, dynamics, EQ, or track assignment settings) to a *centralized control panel*, which can be used to vary the particular channel setting parameters. Such a panel can be multipurpose in its operation (Figure 11.34), allowing itself to be reconfigured (via software "soft" buttons, pan and level controls) in a chameleon-like fashion to fit the task at hand. Alternatively, console parameters can be controlled by physically placing a readout display at each control point on the input strip, at locations that might resemble their physical counterparts on an analog console. Using such a control/display system, you could simply grab the desired parameter and move it to its new position (as though it were an analog console). Generally, this system is more expensive than those that use a central control panel, as more readout indicators, control knobs and digital control interface systems are required. The obvious advantage, however, is having instant access (both physically and visually) to all these parameters at once.

Panel readouts will often vary between devices, with some having dedicated, configurable LCD or video readout displays that show a number of assigned settings (which can be varied via soft keys and parameter dials). Others are designed around touch-screen displays that can be reconfigured and altered by simply touching the desired parameter.

Each input strip on a console can gain access to the central control panel if the *channel select* is pressed on the desired channel. Once this is done, the current setting for that channel will be automatically recalled and displayed in the control panel readouts . . . and will be immediately available to the user for any and all parameter changes. In a number of cases, the select buttons can be function specific (meaning that you might need to press the EQ select button on a strip in order to call up that particular function).

Figure 11.34. *Centralized control panel for the Sony DMX-R100 Digital Console. (Courtesy of Sony Electronics, Inc., www.sony.com/proaudio)*

Do It Yourself Tutorial: Channel Select

1. Find yourself a digital mixer or console. (If one's not nearby, you could visit a friendly studio or music store.)
2. Feed several live or tape inputs into it and create a simple mix.
3. Press the channel select for the first input and experiment with the EQ and/or other controls.
4. Move on to the next channel and have fun building your mix.

A number of console designs are laid out with controls that mimic analog consoles, with the dedicated function controls flowing down the strip in a traditional fashion . . . except that the controls are digital and are used to vary one or more assigned parameters in the system's CPU. The truth of the matter, however, is that very few digital console designs are alike, and it may take a bit of time to get used to a new console's basic layout and operation.

From the preceding section, you can get a sense that the parameter settings of a digital mixer or console (including dynamic fader, switching and control functions) can be saved to the device's computer or internal processor as a snapshot or automated mix file for instantaneous recall at any later time. Literally, even the most complicated of mix moves can be recalled into the system in seconds!

Beyond the human interface aspect of digital mixing, the facets of interfacing these devices to the outside world is often eased by the fact that they're already fluent in digital—the primary language of digital recorders, time code synchronizers and external transport controls. Often these devices can provide systemwide word clock timing (see Chapter 6) and connections to and from external audio devices (in either the analog and/or digital domain, and often in several consumer, pro and multichannel formats), as well as location/transport control over tape, computer, video and other production devices within the facility.

Digitally Controlled Analog Consoles

Unlike the fully digital console, the signal path of a *digitally controlled analog console* (Figure 11.35) is distributed and processed in analog form; however, control over all console parameters is carried out in the digital domain. In most cases, this means that the console's control surface (containing all the knobs, faders, assignment buttons, and so on) will output its control parameters in digital form. This information is then transmitted to a central interface that has direct control over the system's analog I/O routing, switching and dynamic level functions (which is often located away from the console in a central processing cabinet).

Figure 11.35. *Euphonix CS2000 digitally controlled analog console. (Courtesy of Euphonix, www.euphonix.com)*

Because control over signal parameters is carried out in the digital domain (and the system is much less constrained by the analog signal path's layout requirements), the control interface and its layout can vary from the analog console's physical traditional form. Since the control surface is digital, it's possible for parameters to be assigned to a central control panel.

Console Automation

In multitrack recording, the final musical product generally isn't realized until the mixdown stage. Each console channel that's being fed from the multitrack tape (as well as from other sources) will often have its own particular volume, reverb, left-to-right pan, EQ and other settings. Multiply each of these settings by 24 or more tracks, then add the auxiliary effects send and return volumes, compression/limiting and other signal-processing functions . . . and it quickly becomes obvious that the mixdown process can easily turn into a beast that hard for an engineer to efficiently handle. As a result, mixdowns often need to be repeatedly rehearsed so the engineer can learn which controls must be operated, how much they should be varied and at what point in the program any changes should come. It's not uncommon for the engineer, producer and artists to spend 12 hours mixing a complicated piece of music before an acceptable mix can be obtained. Often mixes must be rejected because the engineer simply forgot to make one simple, but important, control change. Although the producer and engineer know how and when the control settings should be changed, the memory and physical dexterity required to execute them can sometimes simply exceed human abilities.

One solution to this problem is to mix with the aid of console automation. With such help, all of the settings and changes can be entered during the mixdown phase and then be executed (under computer control) during a final mix pass. This way of working has the obvious advantages of allowing those involved to continually improve the mix until the desired effect is achieved. In addition to providing an extra set of "hands," an automation system can be a virtual lifesaver should you need to go back into the studio at a later date to make changes to the mix . . . or if a new mix is needed for a different medium (such as radio, TV, music video or film).

Analog Console Automation in Action

In practice, an automation system can range from being able to sense only the position of a volume fader and level-related switching functions (often called snapshot automation, because this type represents a mixer or console setting at one point in time) to being a fully automated system that can store and recall all the dynamic functions on a production console. This amazing feat is accomplished by a system that scans the various controls of a mixing surface over the course of a mix. Once a control has been altered, the automation system will detect the associated moves and convert them into a series of corresponding digital words that can be stored directly into the system's computer automation memory.

When dealing with an analog-based mixing system, automation is often carried out by converting the positions and moves of the various dynamic controls on a mixing surface (such as volume, pan, EQ and sends) into a DC voltage that can then be converted into digital data for easy storage and processing. Stated more simply, an automated control might not pass audio, but instead will pass a DC voltage that can be converted into digital data. Once in digital form, the automation data can be used to control gain and other *analog* audio parameters as it passes through the various links in the console chain using any of several device types: the *VCA* (*voltage-controlled amplifier*), DCA (*digitally-controlled amplifier*) or *moving fader*.

- *VCA.* This amplifier type is used to control the gain of a signal as a function of an external DC control voltage. A level control or other parameter can be designed to pass a scaled DC voltage (often 0–5 VDC). This voltage (which is equal to the parameter's setting) is used to vary the gain of an audio signal. The automation system must then translate the voltage into digital data and then, on playback, convert the data back into an equivalent voltage level (which directly controls the parameter level).

- *DCA.* This amplifier type is used to control the gain of a signal as a function of a digital word or series of words. Using this controller type, a DC control voltage is converted into equivalent digital data, which is then used to directly alter the analog audio signal that is being passed. From an automation standpoint, this type is easier in that the automation data can largely remain in the digital domain, without having to be reconverted back into analog control voltages.

- *Moving fader.* In this system, a fader will have two separate control paths: one to pass audio and one to pass a DC control voltage. Changing the fader position will vary a control voltage (which is converted to digital data for the automation system). On playback, the control voltage is fed to a DC servo motor in the fader . . . which in turn causes the fader to automatically move to the proper fader position. There's little doubt that this type of automation brings out the kid in almost everybody. It's lots of fun to watch a mixer or console literally operate itself!

Unlike dynamic gain functions (which can change over time), switching automation is often much easier to encode/decode, as buttons tend to have an off/on (0/1) state that can be easily converted to digital, with the two states being handled by a simple, reliable and silent transistor switching network.

Digital Automation

Since digital consoles already speak the primary language of automation (digital), it's usually much easier and more cost-effective to encode/decode gain, effects, routing and other automation functions in this domain. Translating physical controls (such as faders, rotary pots and switches) into digital values that can be understood by the mixer or console is often done using either VCA technology or digital controls.

- *VCA.* As with analog automation, a physical control can be designed to pass a scaled DC voltage (often 0–5 VDC). The voltage level (which is equal to the parameter's setting) is then converted into a digital value that can be understood by the mixer or console's processor, which varies the parameter value accordingly and stores the value into the device's automation system.

- *Digital controller.* Certain devices and parameters (particularly switches) are able to output their control settings in digital form. This data can then be understood by the system directly and can be fed to the device's automation system.

Grouping

Commonly, console automation systems allow signals to be easily arranged into one or more groups. This is often easily done, as a single control voltage or digital value can be used to control the relative balance of several grouped channels and/or tracks (Figure 11.36). As was said earlier, this feature makes it possible for several instruments to be balanced, while offering control over their relative levels from a single fader (thereby avoiding the need to change each channel volume individually).

Figure 11.36. *A single control voltage or digital value can be used to control the relative balance of several channels that are assigned to the same "group."*

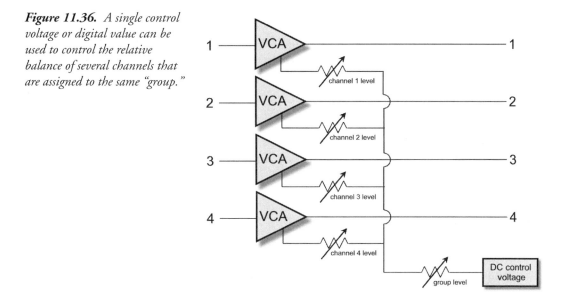

Automation Modes of Operation

Just as different mixer or console designs often vary in design, ability and function, the approach to automating a mix will also differ in form and sophistication from one system to the next. For example, some automation systems offer only a few options for automating a mix, whereas SSL, for example, has at least eight distinct automation modes. Having said this, in addition to functions for merging and joining the various parts of several mixes into a completed take (an option on many larger console automation designs), there are only three basic modes for automating a mix: write mode, update, and read mode.

Write Mode

Once the mixdown process is underway, the process of writing the automation data into the system's memory can begin (of course, a basic automation mix can be made during the recording or overdub phases). During the *write mode*, the system will begin the process of recording all of the mix moves that can be automated into memory in real-time. This mode can be used to globally record all of the settings and moves that are made on all the input strips (in essence, storing all of the mix moves, live and in one pass) . . . or on each input strip (allowing track mixes to be built up individually). The latter lets us focus all of our attention on a difficult or particularly important part and/or passage. Once done, another track or group of tracks can be "written" into the system's memory, and then another, until an initial mix is built up.

Update Mode

As the name implies, the update mode allows us to go back at a later time and alter the mix settings that were originally written into memory. In this mode, the updated settings can be changed by adding or subtracting from the track's control data, rather than by completely rewriting it. An advantage to this is best shown by example. . . . Let's say that we did a really good job at writing several complex and time-consuming moves for a group of tracks. The only problem lay in the fact that a few of the tracks were simply not loud enough. Instead of taking the time to rewrite the tracks from scratch, we can simply go into the update/offset mode and change the "relative" balance levels of one or more tracks. In this way, the complex moves will remain intact, while the relative volume levels are changed (thereby making the overall passage louder or softer).

Level Matching

When writing over previous automation data (or when entering into the update mode), the concept of matching current fader level or controller positions to their previously written position often becomes important. As an example, let's say that we needed to redo several track moves that occurs in the middle of a song. If the current volume fader positions don't match up with their previously written positions, the mix levels could jump during the transition on playback. In order to set the faders to their previous positions, some form of indicator is needed. For those who are working with a console with moving faders, there will be no need to manually match the settings as the faders and/or controls will conveniently and automatically move to their current mix level position. However, those who are working in a purely VCA-based system will often need to match the positions by hand. This is often done with the aid of *nulling indicator lights* (or fader level bars that are shown on a computer or video monitor screen) which read the difference and indicate whether the setting is higher, lower or equal to the current mix position (Figure 11.37). After a level match is achieved, the engineer can switch between the modes without a change in level. After the write mode is entered, the controls can be moved to their new settings.

Figure 11.37. *Readout displays are used to show current fader positions.*
a. Null indicator lights.

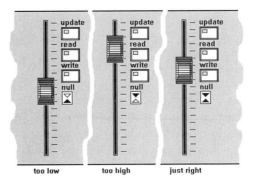

too low   too high   just right

a

b. On-screen bar graph.

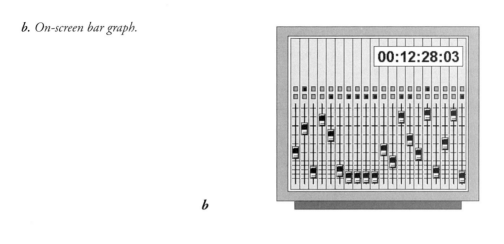

00:12:28:03

b

Read Mode

Consoles that are placed into the *read mode* will playback the mix information directly from the system's automation computer, which will convert the data into a form that can be understood by the system's VCA, DCA, moving fader or digital readout controls. Once the final mix (or mixes) has been achieved, all that's needed is to roll the tape, sit back and listen to the mix.

Moving Fader Automation

Unlike a VCA-based system in which the DC voltage levels control automation levels, a moving fader system actually has motor-driven gain controls that move under the control of the system's automation system (Figure 11.38). Often, this system type relies less on the traditional write, update and read modes, as the faders are touch-sensitive. This means that a fader can easily be programmed to enter into the update mode, simply by touching it and moving it to the desired position. Since it's always physically at the current read position, all that's needed to update the levels to a new position is to take hold of the fader and move it to its new location. The automation computer can then automatically stores this information as updated material. Often, any number of faders can be assigned as a group that can be controlled from a single master or by grabbing any fader in the group (another sight that's often a great source of fun!).

Other Automation Options

Many mixer and console designs that are equipped with automation also allow for snapshots to be taken of the current console settings. Just as a camera "freezes" a Kodak moment in time, a console snapshot can place many or all of the mixing surface's settings into memory for recall at a later time. With a nonautomated mixer or console, one of the most tedious tasks for an assistant engineer or engineer is to "zero" the board. This is the process of setting all of the controls to their initial "zero," null or centered state . . . in other words, one where all the controls are ready for a new session. The process of *zeroing* a console can be a tedious exercise in repetition (particularly if there

Figure 11.38. *Mixers and consoles, such as the Yamaha 01v are designed with moving faders. (Courtesy of Yamaha Corporation of America, www.yamaha.com/proaudio)*

are 96 inputs and tons of controls). With the help of a snapshot, this process could automatically happen at touch of a button. The same kind of recall automation is used to reset an entire console to its former settings when returning to the studio several weeks after a session.

As was mentioned, different automation systems will likely work and perform in very different ways. Most large-console systems will offer extensive editing and merging facilities, letting you mix short sections of a song, project or soundtrack and combine the best of these "takes" into a final mix master. These takes (or different versions of a single or composite mix) could then be saved to disk or disc, for recall at a later time.

MIDI-Based Automation

Many project-oriented mixers and console systems are capable of communicating automation information via MIDI. The immediate advantages to this include the following:

- Automation information can be stored into a conventional MIDI sequencer.
- Most modern effects devices can be MIDI controlled, allowing effects to respond to program-change messages during a mix.
- MIDI data is an established, cost-effective standard in worldwide use for professional and project studio production.
- Automation data can be easily synchronized to a project via MIDI time code.

Most digital mixers, controllers and small console designs can be controlled (at a basic or complex mix level) via MIDI, in either a dynamic or snapshot fashion. These mixer systems include MIDI ports for communicating mix-related data to/from a computer or MIDI sequencing program (which is usually transmitted as a stream of MIDI sys-ex messages). Because of the expense, most analog mixer designs don't include MIDI mixing capabilities (unless they have been specially fitted with a VCA-based automation system that can be communicated via

MIDI). A few small console designs, however, include switching and routing systems that can be dynamically and/or snapshot automated.

Effects Automation via MIDI

One of the most common ways to automate an effects device from a MIDI sequence or during a live performance is through the use of program-change commands. In the same way that electronic instruments can store sound patches into a memory-location register for later recall, most modern effects devices will let you store program patches into memory, where they can be recalled at any time using program change commands.

The use of program change commands (and occasionally continuous-controller messages) make it possible for signal processing patches and parameters to be altered during the playback of a MIDI sequence. By locking a sequence to a mix, it's possible for a series of effect changes to be automatically triggered at the proper time within a sequence and automated mix.

The Art of Mixing . . .

Actually, the heading "The Art of Mixing" could easily fill a book (and I'm sure it has). However, I'd simply like to point out the fact that it is indeed an *art form . . .* and, as such is a very personal process. I remember the first time that I sat down at a console (an older Neve 1604). I was truly petrified and at a loss as to how to approach the making of A MIX. *Am I over-equalizing? Does it sound right? Will I ever get used to this sea of knobs?* Well folks, as with all things . . . the key is simply to sit right down and mix, mix, mix (Figure 11.39)! It's always a good idea to watch others in the process of practicing their art and take time to listen to the work of others (both the known and the not-so-well-known). With practice, it's a foregone conclusion that you'll begin to develop your own sense of the art and style of mixing . . . which, after all, is what it's all about. It's up to you now. Just dive into the deep end—and have fun!

Figure 11.39. Given the fact that an engineer spends a huge amount of time sitting on his/her backside, it's always a good idea to make an investment in your own and your clients' posture and comfort. (Courtesy of McGraw Publishing Peripherals, www.sittingmachine.com)

CHAPTER 12

Signal Processors

Signal processing has become an increasingly strong part of modern audio and music production. It is the function of a signal processor to change, augment or otherwise modify an audio signal in either of the following two ways:

- *In the analog domain.* Audio signal levels can be directly processed without being converted into a digital form.
- *Using digital signal processing (DSP).* Signals that have been converted from analog into a digital, binary form can be mathematically recalculated according to a specific program algorithm, so as to alter the nature of the sound.

Both processing methods are frequently used in all phases of audio production and exert an increasing amount of control over amplitude level processing (dynamic range), the spectral content of a sound (equalization) and effects (the augmentation or recreation of room ambience, delay, time/pitch alteration and tons of other "special effects" that can range from being sublimely subtle to "in yo' face").

As we move into the new millennium, we've witnessed huge advances in the way that integrated personal computer systems and dedicated processors can directly process signals. These same advances also have had an impact on the way complex effects processing can be automated within an audio production environment. In this chapter, we'll take an in-depth look at the many types of traditional and not-so-traditional processing tools that are currently available. We'll begin with a good, hard look at the broad range of analog-based signal processors that have and continue to be the cornerstones of music and sound production. These signal processors include such devices as equalizers, dynamic range controllers, acoustic and electro-acoustic reverb systems . . . and even some tape-based tricks that can come in handy. After this initial discussion, the chapter continues on to study the ever-important field of digital signal processing (DSP).

Equalization

The most common form of signal processing is *equalization* (also known as *EQ*). The audio equalizer (Figure 12.1) is a device or circuit that allows a recording, mix or audio engineer to control the relative amplitude of various frequencies within the audible bandwidth. Put another way, it lets you exercise tonal control over the harmonic or timbral content of a recorded sound. EQ may need to be applied to a single recorded channel, to a group of channels or to an entire program signal (often as a step in the mastering process) for any number of other reasons:

- To correct specific problems in a recording or in a room (possibly to restore a sound to its natural tone)
- To overcome deficiencies in the frequency response of a microphone or in the sound of an instrument
- To allow contrasting sounds from several microphones or tape tracks to blend together better in a mix
- To increase the separation between mics or recorded audio tracks by seeking to reduce those frequencies that excessively "leak" between channels
- To alter a sound purely for musical or creative reasons

Figure 12.1. *Examples of an equalizer.*
a. EQ from an SSL SL 9000J console strip. (Courtesy of Solid State Logic, www.solid-state-logic.com)

a

b. NSEQ-2 stereo parametric equalizer. (Courtesy of Millennia Music & Media Systems, www.mil-media.com)

b

Equalization refers to the alteration in frequency response of an amplifier so that the relative amplitude levels of certain frequencies are more or less pronounced than others. EQ is specified as plus or minus a certain number of decibels at a certain frequency. For example, a signal can be boosted by "+4 dB at 5 kHz."

Although only one frequency was specified in the example, in reality, a range of frequencies above, below and centered around the specified frequency will also often be affected. The amount of boost or cut at frequencies other than the one named is determined by whether the curve is peaking or shelving, by the bandwidth of the curve (a factor that's affected by the Q settings, which determines how many frequencies will be affected around a chosen centerline) and by the amount of boost or cut at the named frequency. For example, a +4 dB boost at 1000 Hz might easily add a degree of boost or cut at 800 Hz and 1200 Hz (Figure 12.2).

Figure 12.2. *Various boost/cut EQ curves centered around 1 kHz. (Courtesy of Mackie Designs, www.mackie.com)*
a. Center frequency: 1 kHz.; bandwidth 1 octave; ±15 dB boost/cut.

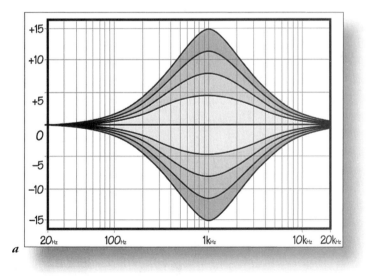

a

Figure 12.2. *(continued)*
b. Center frequency: 1 kHz.;
bandwidth 3 octaves; ±15 dB
boost/cut.

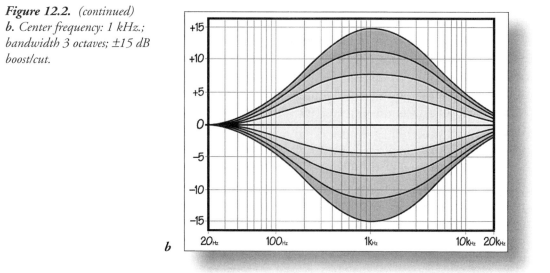

Older equalizers and newer "retro" systems often base their design around filters that employ *passive* components (i.e., inductors, capacitors and resistors) and use amplifiers only to make up for internal losses in level, called *insertion loss*. Figure 12.3a shows typical signal levels in a passive equalizer that's set for a flat response, while Figure 12.3b shows the signal level structure of an equalizer that has a low-end boost.

Figure 12.3. *Typical signal*
levels in a passive equalizer.
a. EQ is set for flat response.
b. EQ filter is set for 6-dB boost
at 100 Hz.

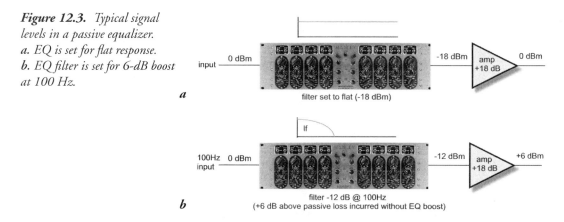

Most equalization circuits today, however, are of the *active filter* type that change their characteristics by altering the feedback loop of an op amp (Figure 12.4). This is by far the most common EQ type and is generally favored over its passive counterpart because of its low cost, small size, lightweight, wide gain range and line-driving capabilities.

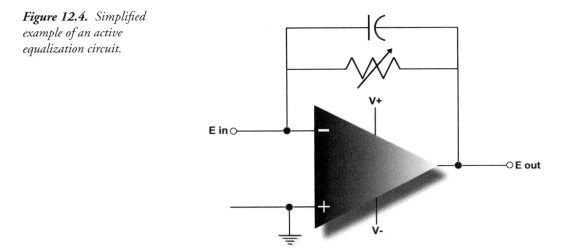

Figure 12.4. *Simplified example of an active equalization circuit.*

Peaking Filters

The most common equalization curve is of the *peaking* type. As its name implies, a peak-shaped bell curve can either be boosted or cut around a selected center frequency. Figure 12.5 shows the curves for a peak equalizer that's set to boost or cut at 1000 Hz. The quality factor (Q) of a peaking equalizer refers to the width of the bell-shaped curve. A curve with a high Q will have a narrow bandwidth with few frequencies outside the selected bandwidth being affected, whereas a low-Q curve is very broadband and can affect many frequencies (or even octaves) around the center frequency.

Figure 12.5. *Peaking equalization curves.*

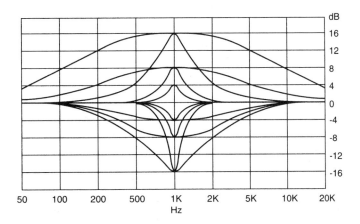

Bandwidth is a measure of the range of frequencies that lie between the upper and lower –3dB (half-power) points on the curve (Figure 12.6). The Q of a filter is an inverse measure of the bandwidth (such that higher Q values mean that fewer frequencies will be affected . . . and vice versa). To calculate Q, simply divide the center frequency by the bandwidth. For example, a filter centered at 1 kHz that's a third of an octave wide will have its –3 dB frequency points located at 891 Hz and 1223 Hz, yielding a bandwidth of 232 Hz (1123 – 891). This EQ curve's Q, therefore, will be 1 kHz divided by 232 Hz, or 4.31.

Figure 12.6. *The number of hertz between the two points that are 3 dB down from the center frequency determines the bandwidth of a peaking filter.*

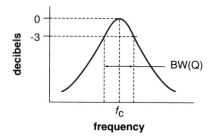

Shelving Filters

Another type of equalizer is the *shelving filter*. Shelving refers to a rise or drop in frequency response at a selected frequency, which tapers off to a preset level and continues at that level to the end of the audio spectrum. Shelving can be inserted at either the high or low end of the audio range and is the curve type that's commonly found on a home stereo's bass and treble controls (Figure 12.7).

Figure 12.7. *High/low, boost/ cut curves for a shelving equalizer.*

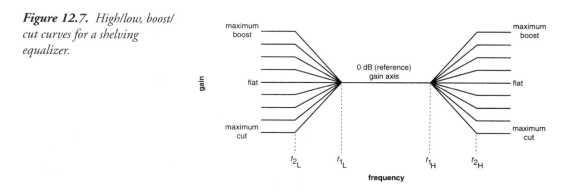

High-Pass and Low-Pass Filters

Equalizer types also include *high-pass* and *low-pass filters*. As their names imply, certain frequencies are passed at full level, while others are attenuated. Frequencies that are attenuated by less than 3 dB are said to be inside the *passband*; those attenuated by more than 3 dB are in the *stop-*

band. The frequency at which the signal is attenuated by exactly 3 dB is called the turnover or *cutoff frequency* and is used to name the filter frequency. Ideally, attenuation would become infinite immediately outside the passband; however, in practice this isn't attainable. Commonly, attenuation increases at rates of 6, 12 and 18 dB per octave. This rate is called the *slope* of the filter. Figure 12.8, for example, shows a 700-Hz high-pass filter response curve with a slope of 6 dB per octave, while Figure 12.9 shows a 700-Hz low-pass filter response curve having a slope of 12 dB per octave.

Figure 12.8. *A 700-Hz high-pass filter with a slope of 6 dB per octave.*

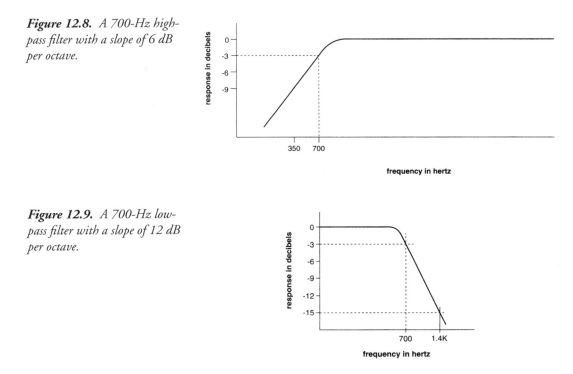

Figure 12.9. *A 700-Hz low-pass filter with a slope of 12 dB per octave.*

High- and low-pass filters differ from shelving EQ in that their attenuation doesn't level off outside the passband. Instead, the attenuation continues to increase. A high-pass filter in combination with a low-pass filter can be used to create a *bandpass filter,* with the bandwidth being controlled by the filter turnover frequencies and the Q by the filter's slope (Figure 12.10).

Figure 12.10. *A bandpass filter is created by combining high- and low-pass filters with different cutoff frequencies.*

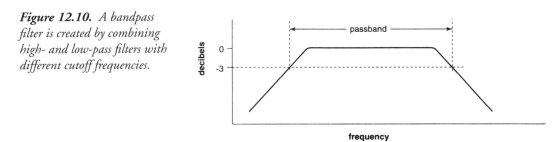

Equalizer Types

Four of the most commonly used equalizer types use one or more of the previously described filter types: the selectable frequency equalizer, the parametric equalizer, the graphic equalizer and the notch filter.

The selectable frequency equalizer (Figure 12.11), as its name implies, has a set number of frequencies from which to choose. These equalizers usually allow a boost or cut at a number of selected frequencies with a predetermined Q. This form of equalization is most often found on older console designs and on certain low-cost production consoles.

Figure 12.11. API 550A *switchable equalizer within the 7600 outboard module. (Courtesy of API Audio, www.apiaudio.com)*

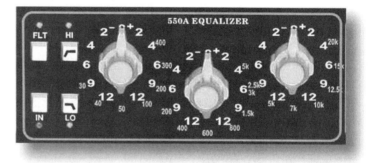

The *parametric equalizer* (Figure 12.12) has continuously variable center frequency adjusts (often on each band), rather than those that are selectable in discrete steps. Control over the center frequency (Q) may be either selectable or continuously variable, although certain manufacturers have no provisions for a variable Q. The amount of boost or cut is also continuously variable. Generally, each set of frequency bands will overlap into the next band section, providing smooth transitions between frequency bands or allowing multiple curves with a high Q to be placed in narrow frequency ranges. Because of its increased flexibility and performance, the parametric equalizer has become the standard design for input strips in most modern consoles.

Figure 12.12. The EQF-100 *Full Range Vacuum Tube Equalizer. (Courtesy of Summit Audio, Inc., www.summitaudio.com)*

A *graphic equalizer* (Figure 12.13) provides boost/cut level control over a series of center frequencies that are equally spaced according to music intervals. An "octave band" graphic equalizer may, for example, have 12 equalization controls spaced at the octave intervals of 20, 40, 80, 160, 320, 640 Hz, and 1.25, 2.5, 5, 10 and 20 kHz, with ⅓-octave equalizers having up to 36 center-frequency controls. The controls for the various equalization bands generally use linear sliders that are vertically arranged side by side. At a glance, the physical positions of these controls provide a "graphic" readout of the overall frequency response curve. This type of equalizer is often used in applications that can help with fine-tuning a system to compensate for the acoustics of various types of rooms, auditoriums and studio control rooms.

Figure 12.13. *Rane GE 130 single-channel, 30-band, [case one-third]-octave graphic equalizer. (Courtesy of Rane Corporation, www.rane.com)*

In addition to modifying sound, an equalizer such as a *notch filter* can be used to remove hum and other undesirable discrete-frequency noises. A notch filter can be tuned to attenuate a particular frequency and has a very narrow bandwidth, so it has little effect on the rest of the audio program (Figure 12.14). Notch filters are used more in film-location sound than in studio recording because the problems encountered in location work aren't usually present in a well-designed studio.

Figure 12.14. *Notch filter response curves. (Courtesy of Orban Associates, Inc.)*

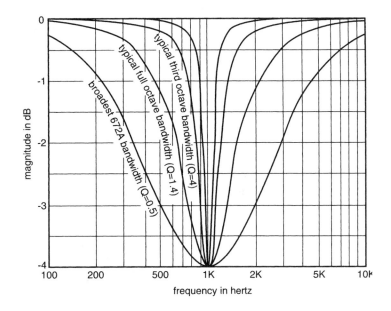

Applying Equalization

Although most equalization is done by ear, it's helpful to have an idea about which frequencies affect an instrument in order to achieve a particular effect. On the whole, the audio spectrum can be divided into four frequency bands: low (20–200Hz), low-middle (200–1000 Hz), high-middle (1000–5000 Hz) and high (5000–20,000 Hz).

When the frequencies in the 20–200 Hz range are modified, the fundamental and the lower harmonic range of most bass information will be affected. These sounds often are felt as well as heard, so boosting in this range can add a greater sense of power or punch to music. Lowering this range will weaken or thin out the lower frequency range.

The fundamental notes of most instruments are within the 200–1000 Hz range. Changes in this range often result in dramatic variations in the overall signal energy, with an increase adding to the overall impact of a program. Because of the ear's sensitivity in this range, a minor level change will often result in a major audible effect. The frequencies around 200 Hz can give the bass a greater feeling of warmth without loss of definition; frequencies in the 500–1000 Hz range might make an instrument sound hornlike. Too much boost in this range can cause listening fatigue.

Higher-pitched instruments are most often affected in the 1000–5000 Hz range. Boosting these frequencies often results in an added sense of clarity, definition, and brightness. Too much boost in the range of 1000–2000 Hz can have a "tinny" effect on the overall sound, while the upper mid-frequency range (2000–4000 Hz) affects the intelligibility of speech. Boosting in this range can make music seem closer to the listener, but too much of a boost often tends to cause listening fatigue.

The high-frequency region (5000–20,000 Hz) is composed almost entirely of instrument harmonics. For example, boosting frequencies in this range can add sparkle and brilliance to a string or woodwind instrument. Boosting too much might produce sibilance on vocals and make the upper range of certain percussion instruments sound harsh and brittle. Boosting at around 5000 Hz has the effect of making music sound louder. A boost of 6 dB at 5000 Hz, for example, can make the overall program level sound as though it's been doubled in level; conversely, attenuation can make music seem more distant.

Do It Yourself Tutorial: Equalization

1. Using a parametric or graphic equalizer, change the relative levels on several music recordings using the previous frequency ranges and experience the effects firsthand.

2. Using a parametric or graphic equalizer, change the relative levels on several isolated instrument tracks using the following frequency ranges and experience the effects firsthand.

Table 12.1 gives a further analysis of the ways that frequencies and equalization can interact with various instruments. (For more information, refer to the section "Microphone Placement Techniques" in Chapter 4.)

Table 12.1 Instrumental frequency ranges of interest.

Instrument	Frequencies of Interest
Kick drum	Bottom depth at 60–80 Hz, slap attack at 2.5 kHz
Snare drum	Fatness at 240 Hz, crispness at 5 kHz
Hi-hat/cymbals	Clank or gong sound at 200 Hz, shimmer at 7.5 kHz to 12 kHz

Instrument	Frequencies of Interest
Rack toms	Fullness at 240 Hz, attack at 5 kHz
Floor toms	Fullness at 80–120 Hz, attack at 5 kHz
Bass guitar	Bottom at 60–80 Hz, attack/pluck at 700–1000 Hz, string noise/pop at 2.5 kHz
Electric guitar	Fullness at 240 Hz, bite at 2.5 kHz
Acoustic guitar	Bottom at 80–120 Hz, body at 240 Hz, clarity at 2.5–5 kHz
Electric organ	Bottom at 80–120 Hz, body at 240 Hz, presence at 2.5 kHz
Acoustic piano	Bottom at 80–120 Hz, presence at 2.5-5 kHz, crisp attack at 10 kHz, "honky tonk" sound (sharp Q) at 2.5 kHz
Horns	Fullness at 120–240 Hz, shrill at 5–7.5 kHz
Strings	Fullness at 240 Hz, scratchiness at 7.5–10 kHz
Conga/bongo	Resonance at 200–240 Hz, presence/slap at 5 kHz
Vocals	Fullness at 120 Hz, boominess at 200–240 Hz, presence at 5 kHz, sibilance at 7.5–10 kHz

One way to zero in on a particular frequency using an equalizer (especially a parametric equalizer) is to increase the amount of boost and change the center frequency until the desired instrument range to be EQed is found. The level can then be decreased until the desired effect is obtained. Attenuation (cut) can be achieved in a similar manner.

If boosting in one instrument range causes you to want to do likewise in other frequency ranges, it's possible that all you're doing is raising the program's overall level. Although it's easy to get caught up in this syndrome, it would be far easier (and better) to simply increase the overall program level. If turning it up doesn't sound satisfactory, it's possible that one of the ranges is too dominant and requires attenuation.

As far as recording with EQ goes, there are a number of different opinions on this subject. Some use EQ liberally to make up for placement and mic deficiencies, whereas others use it sparingly, if at all. One example where EQ might be used sparingly is if one engineer knows that someone else will be mixing a particular song or project. In this situation, the engineer who'll be doing the mix may have a very different idea of how the instruments should sound. If large amounts of EQ were recorded to tape during the session, the mixdown engineer might have to work very hard to counteract the original EQ settings. On the other hand, if everything was recorded flat, the producer and artists might have difficulty passing judgment on a performance or hearing the proper balance during the overdubbing phase. Such a situation might call for equalization in the monitor mix, while leaving the recorded tracks alone. In situations where several mics are to be combined onto a single tape track, the mics can be individually EQed only during the recording phase. In situations where a project is to be engineered, mixed and possibly even mastered, the engineer might know in advance the type and amount of EQ that he/she might want. Obviously, this comes under the "art" category of recording and comes with experience. Having said this, it's wise that any "sound-shaping" should be determined and discussed over with the producer and/or artist before the sounds are committed to tape.

Whether you choose to use EQ sparingly, as a right-hand tool for correcting deficiencies or even at all . . . there's no getting around the fact that the equalizer is a powerful tool. When used properly, it can greatly enhance or restore the musical and sonic balance of a signal. Experimentation and experience are the keys to equalizer use, and no book can replace the trial-and-error process of "just doing it!"

Before closing, it's important to keep one age-old viewpoint in mind—that the equalizer shouldn't be regarded as a cure-all for improper mic technique. Rather, EQ should be used as a tool for correcting the minor problems of room acoustics or to modify a system or pickup's frequency response. Unless it's obvious that corrections will need to be made (because of instrument, room or time limitations) or if you're after a specific effect, you might consider using EQ moderately. Alternatively, you might consider using another mic and/or placement technique in order to get a good instrument sound. If an instrument is poorly recorded in an initial recording session, it's often far more difficult and time-consuming to "fix it in the mix" at a later time. Getting the best possible sound onto tape will definitely improve your chances for attaining a sound and overall mix that you can be proud of.

Dynamic Range

Like most unpredictable things in life that get out of hand from time to time, the level of a signal can vary widely from one moment to the next. For example, if a vocalist lets out an impassioned scream following a soft whispery passage, you can almost guarantee that the mic's signal will jump from its optimum recording level into severe distortion. Conversely, if you set an instrument's mic to accommodate the loudest level, its signal might be buried in the mix during the rest of the song. For these and other reasons, it becomes obvious that it's sometimes necessary to exert some form of control over a signal's dynamic range by using various techniques and dynamic controlling devices.

Metering

Amplifiers, magnetic tape and even digital media are limited in the range of signal levels that they can pass without distortion. As a result, audio engineers need a way to determine whether the signals they're working with will be stored or transmitted without distortion. The most convenient way to do this is to use a visual level display, such as a *meter*. If preventing distortion on the tape was the only concern, peak-indicating meters could be used to display the maximum amplitude fluctuations of a waveform. However, the way humans perceive loudness doesn't bear much relationship to a signal's instantaneous peak level. For example, since the ear's perception of loudness is proportional to the rms (average) value of a signal and not its peak value, a peak meter might read higher at a particular point in the program, while not sounding noticeably louder (Figure 12.15). A far better indicator would be one that would average the level of a signal over time (Figure 12.16). The scale chosen for such a meter is calibrated in volume units—

hence the name *VU meter* (Figure 12.17). Zero VU is considered to be the standard operating level for most consoles, mixers and analog tape machines. (Digital's standard operating levels are far more ambiguous, and you should consult the particular device manual.)

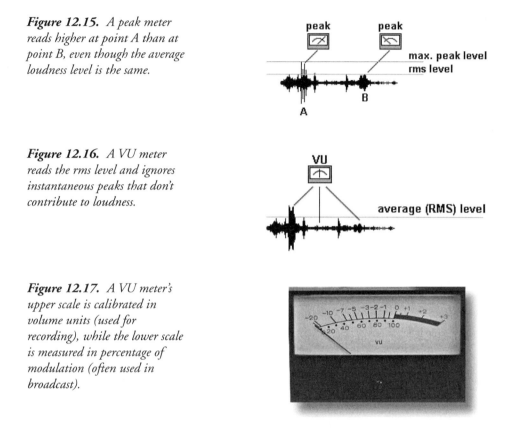

Figure 12.15. *A peak meter reads higher at point A than at point B, even though the average loudness level is the same.*

Figure 12.16. *A VU meter reads the rms level and ignores instantaneous peaks that don't contribute to loudness.*

Figure 12.17. *A VU meter's upper scale is calibrated in volume units (used for recording), while the lower scale is measured in percentage of modulation (often used in broadcast).*

Although VU meters do the job of indicating rms volume levels, they ignore the short-term peaks that can overload tape. These peaks can be from 8–14 dB higher than the indicated rms value. This means that the professional console systems must often be designed so that unacceptable distortion doesn't occur until at least 14 dB above 0 VU. Typical VU meter specifications are provided in Table 12.2.

Table 12.2 VU meter specifications.

Characteristic	Specification
Sensitivity	Reads 0 VU when connected across a +4-dBm signal (1.228 V in a 600-Ω circuit)
Frequency response	@0.2 dB from 35 Hz to 10 kHz; @0.5 dB from 25 Hz to 16 kHz
Overload capability	Can withstand 10 times 0-VU level (+24 dBm) for 0.5 sec and five times 0 VU (approximately + 18 dBm) continuously

In addition to the traditional meter readout, a large number of displays now use light-emitting diodes (LEDs) or liquid crystal displays (LCDs) to indicate levels in ways that are easy to read (and often fun to watch). These readouts often simultaneously display both rms and peak level values (or can be switched between the two), and since no moving parts are used, they can instantaneously display any selected signal level type.

Getting to the heart of the matter, it goes without saying that whenever the signal is too high (hot), it's an indication for you to grab hold of the channel's mic trim, output fader or whatever level control is the culprit . . . and turn it down. In doing so, you actually become the dynamic range changing device. In fact, the main channel fader (which can be controlling a tape track's level during mixdown, an input level during recording, etc.) is by far the most intuitive and most often used dynamic gain changer in the studio.

In practice, the difference between the maximum level that can be handled without incurring distortion and the average operating level of the system is called *headroom*. Some studio-quality preamplifiers are capable of signal outputs as high as 26 dB above 0 VU and thus are said to have 26 dB of headroom. The 3 percent distortion level for analog magnetic tape, by comparison, is typically only 8 dB above 0 VU, whereas console amplifiers have distortion of less than 0.4 percent at this level. For this reason, the best recording level for most program material is around 0 VU, although higher levels are possible (providing that short-term peak levels aren't excessively high). In some circumstances (such as when using higher bias, low-noise/high-output analog tape), it's actually possible to record at higher levels without distortion, as the tape formulation is capable of handling higher magnetic flux levels.

Because I view recording as an art form, I have to rise to the defense of those who prefer to record certain instruments (particularly drums and percussion) at levels that bounce or even "pin" VU needles at higher levels than 0 VU. In recording to a professional analog machine, this can actually give a track a "gutsy" feel that can add impact to a performance. It's probably not a good idea to record instruments that contain high-frequency/high-level signals (such as a snare or cymbals) at these levels (as the peak transients will probably distort in a way that's not always pleasing). Always be aware that you can often add distortion to a track at a later time (using any number of ingenious tricks), but you can't remove it from an existing track. Again, it's always wise to talk such moves over with the producer and/or artist.

The idea of "pinning" the input levels of any digital recording device is a definite no-no. The dreaded "clip" indicator of a digital meter means that you've reached the saturation point, with no headroom to spare above this point. I actually encourage you to slightly pin the meters of digital devices (make sure the monitors aren't turned up too far), just to find out how obnoxious even the smallest amount of clipping can sound. . . . Pretty harsh, huh?

Dynamic Range Processors

The overall dynamic range of music is potentially on the order of 120–130 dB (Figure 12.18), whereas the dynamic range of the digital medium (when working with 16-bit wordlengths) is approximately 90 dB. The range of analog magnetic tape is on the order of 60 dB, excluding the use of noise-reduction systems that can improve this figure by 15–30 dB . . . all which still fall short of music's full 120-dB range. The overall dynamic range of a compact disc is often 80–90 dB; however, when working with 20- and 24-bit digital wordlengths, a system, processor or channel's overall dynamic range will actually begin to approach that of music.

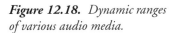

Figure 12.18. *Dynamic ranges of various audio media.*

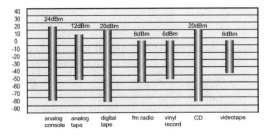

Even with such a wide dynamic range, unless the program is played back in a noise-free environment, either the quiet passages would get lost in the ambient noise of the listening area (35–45 dB SPL for the average home) or the loud passages would be too loud to bear. Similarly, if a program of wide dynamic range were to be reproduced through a medium with a narrow dynamic range (such as the 20- to 30-dB range of an AM radio or the 40- to 50-dB range of FM), a great deal of information would get lost in background noise. To prevent these problems, the musical dynamics can be reduced to a level that's appropriate for the reproduction medium (radio, home system, car, etc.). This gain reduction can be accomplished either by manually riding the fader's gain or by using a *dynamic range processor* that automatically changes the dynamic range of a signal.

The concept of automatically changing the gain of an audio signal (through the use of compression, limiting and/or expansion) is perhaps one of the most misunderstood aspects of audio recording. This can be partially attributed to the fact that a well-done job won't be overly obvious to the listener. Changing the dynamics of a track or overall program can affect the way in which it will be perceived (either unconsciously or consciously) by making it "seem" louder, by reducing its volume range to better suit a particular medium or by making it possible for a particular sound to ride above other tracks within a mix.

Compression

A *compressor* (Figure 12.19), in effect, can be thought of as an automatic fader. It's used to proportionately reduce the dynamics of a signal that rises above a user-definable level (known as the *threshold*) to a lesser volume range that is manageable by the dynamics of a system's electronics and/or amplifiers, that is more appropriate to the overall dynamics of a playback medium, or that matches the dynamics of other recorded tracks within a song or audio program.

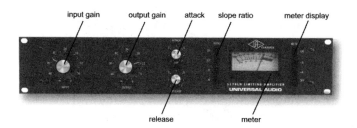

Figure 12.19. Universal Audio 1176LN limiting amplifier. (Courtesy of Universal Audio, www.uaudio.com)

By turning down the louder signal levels, you are, in fact, reducing a track, group or program's overall dynamic range. Because the loudest signals have been automatically turned down (hence the terms *compressed* or "*squashed*"), it's now possible to boost this signal to a level that works better in a mix or as an optimized output level (Figure 12.20). Now, let's think about this process for a moment . . . the louder signals have just been turned down (effectively reducing the dynamic range). If we follow this stage with a boosting amplifier that turns the signal back up in level, we've not only raised the louder signals back to a prominent level—we have also turned up the softer signals. In effect, what this process has *really* done is to turn up the softer signals that would otherwise be buried in the background noise of the other music tracks or in room ambience.

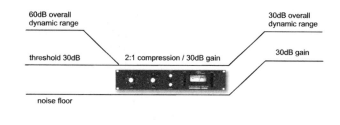

Figure 12.20. A compressor reduces input levels that exceed a selected threshold by a specified amount. This reduced dynamic range signal can then be boosted in level at the output, thereby allowing the softer signals to be raised above other program or background sounds.

The most common controls on a compressor (and most other dynamic range devices include input gain, threshold, output gain, slope ratio, attack, release and meter display.

- *Input gain.* This control is used to determine how much signal will be sent to the compressor's input stage.
- *Threshold.* This setting determines the level at which the compressor will begin to proportionately reduce the input signal level. For example, when the threshold is set to –20 dB, all signals that fall below this level will be unaffected, while signals above this level will be proportionately attenuated. Setting this level to –40 dB will change the signal such that signals that rise above the –40-dB signal will be reduced . . . meaning that the overall dynamics will be reduced. From this, you get the idea that increasing the threshold level will reduce the dynamic range at the device's output. On some systems (as with Figure 12.17), the input gain actually serves to control the threshold level. In this situation, rais-

ing the input level will increase the threshold level . . . and thus lower the overall dynamic range.

- *Output gain.* This control is used to determine how much signal will be sent to the device's output. It's used to boost the reduced dynamic signal into a range where it can best match the level of a medium or be better heard in a mix.

- *Ratio.* Determines the "slope" of the input-to-output gain ratio. In simpler terms, it determines the increase of input signal (in dB) that's needed to cause a 1-dB increase at the compressor's output signal (Figure 12.21). For example, a ratio of 4:1 will produce a 1-dB increase in output for every 4-dB increase at the input, an 8-dB input increase will raise the output by 2-dB, etc. A ratio of 2:1 will produce a 1-dB increase in output for every 2 dB increase at the input. Get the idea?

- *Attack.* This setting (which is calibrated in milliseconds, "ms") determines how fast or how slowly the device will turn down signals that exceed the threshold. It is defined as the time it takes for the gain to decrease to a percentage (usually 63 percent) of its final gain value. In certain situations (as might occur with instruments that have a long sustain, i.e., bass guitar), setting a compressor to instantly turn down a signal might be audible (possibly creating a sound that "pumps" the signal's dynamics). In this situation, it would be best to use a slow attack setting. On the other hand, such a setting might not give the compressor time to react to sharp, transient sound (such as a hi-hat). In this case, a fast attack time would probably work better. As you might expect, you'll need to experiment with this setting to arrive at the fastest attack setting that won't audibly color the signal's sound.

- *Release.* Similar to the attack setting, release (which is calibrated in milliseconds "ms") is used to determine how slowly or how fast the device will restore a signal to its original dynamic level, once it's fallen below the threshold point (defined as the time required for the gain to return to 63 percent of its original value). Too fast a setting will cause the compressor to change dynamics too quickly (creating an audible pumping sound), while too slow a setting might affect the dynamics during the transition from a loud to a softer passage. Again, it's best to experiment with this setting to arrive at the slowest release setting that won't color the signal's sound.

- *Meter display.* Changes the compressor's meter display to read the device's output or gain reduction (whereby the VU or LED meter indicates "GR" by moving either up or down a scale depending on the system's design). In some designs, there's no need for a display switch, as LED readouts are used to simultaneously display output and gain reduction levels.

As was previously stated, the use of compression (and most forms of dynamics processing) is often misunderstood, and compression can be easily be abused. Generally, the idea behind these processing systems is to reduce the overall dynamic range of a track, music or sound program or to raise its overall perceived level . . . without adversely affecting the sound of the track itself. It's a well-known fact that overcompression can actually "squeeze" the life out of a performance by limiting the dynamics and even the transient peaks and valleys that can give life to a

Figure 12.21. *The output of a compressor is linear below the threshold and follows an input/ output gain reduction ratio above this point.*

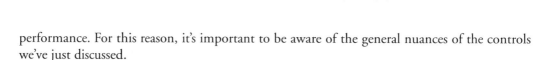

performance. For this reason, it's important to be aware of the general nuances of the controls we've just discussed.

Do It Yourself Tutorial: Compression

1. Record or obtain an *uncompressed* bass guitar track and monitor it through a compressor. Increase the threshold level until the compressor begins to kick in. Can you hear a difference? Can you see a difference on the console or mixer's meters?

2. Set the levels and threshold to a level you like and then set the release time to a fast release setting. Can you hear the signal pumping? Now, set the attack to a fast setting. Does it sound better or worse? Now set both times to a slower setting. Better?

3. Do the same routine and settings using a snare drum track. What did you find?

Limiting

If the compression ratio is made large enough, the compressor will actually become a *limiter* (Figures 12.22 and 12.23). A limiter is used to keep signal peaks from exceeding a certain level in order to prevent the overloading of amplifier signals, recorded signals on tape or disc, broadcast transmission signals, and so on.

Figure 12.22. *The Symetrix 501 peak-rms compressor/ limiter. (Courtesy of Symetrix Inc., www.symetrixaudio.com)*

Figure 12.23. *Behringer Multicom Pro MDX4400 4-Channel Compressor/Limiter. (Courtesy of Behringer International GMBH, www.behringer.de)*

Most limiters have ratios of 10:1 or 20:1 (Figure 12.24), although some have ratios that can range up to 100:1. Because of the fact that a large increase above the threshold at the input will result in a very small increase at its output, the likelihood of overloading any equipment that follows the limiter can be greatly reduced. Commonly, limiters have two basic functions:

- *To prevent short-term peaks from reaching their full amplitude.* When recording to certain media (such as cassette and videotape), these high-energy, transient signals actually don't significantly add to the program's level, relative to the distortion that can result from their presence (i.e., if they saturated the tape) or the noise that would be introduced into the program (if the signal was recorded at a low enough level, so that the peaks wouldn't distort).

- *To prevent signal levels from increasing beyond a specified level.* Certain types of audio equipment (often those used in broadcast transmission) are often operated at or near their peak output levels. Significantly increasing these levels beyond 100 percent would severely distort the signal and possibly damage the equipment. In cases like these, a limiter can be used to prevent signals from significantly increasing beyond a specified output level.

Figure 12.24. *The output of a limiter is linear below the threshold and follows a high input/output gain reduction ratio (10:1, 20:1 or more) above this point.*

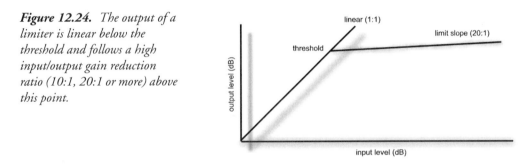

Unlike the compressor, extremely short attack and release times are often used to quickly limit fast transients and to prevent the signal from being audibly "pumped." Limiting a signal during the recording and/or mastering phase should only be used to remove occasional high-level peaks, since excessive use would trigger the process on successive peaks and would be noticeable. If the program contains too many peaks, it's probably a good idea to turn the signal's level down to a point where only occasional extreme peaks would be detected.

Do It Yourself Tutorial: Limiting

1. Feed an isolated track or recorded song through a limiter.
2. With the limiter switched out, turn the signal up until the meter begins to peg (you might want to turn the monitors down a bit).
3. Now reduce the level and turn it up again, with the limiter switched in. Is there a point where the level stops increasing . . . even though you've increased the input signal? Increase the threshold level and experiment with the signal's dynamics. What did you find out?

Expansion

Expansion is the process by which the dynamic range of a signal is proportionately increased. Depending on the system's design, an expander (Figure 12.25) can operate either by decreasing the gain of a signal (as its level falls below the threshold) or by increasing the gain (as the level rises above it). Most expanders are of the first type, in that as the signal level falls below the expansion threshold, the gain is proportionately decreased (according to the slope ratio). In this way, reducing the lower-level signals has the effect of increasing the signal's overall dynamic range (Figure 12.26). These devices can also be used as noise reducers, simply by adjusting them so that the noise is downwardly expanded during quiet passages, while the louder program levels are unaffected and/or are only moderately reduced.

Figure 12.25. *The Aphex Model 622 Logic Assisted Expander/Gate. (Courtesy of Aphex Systems, Inc., www.aphex.com)*

Figure 12.26. *Commonly, the output of an expander is linear above the threshold and follows a low input/output gain expansion ratio below this point.*

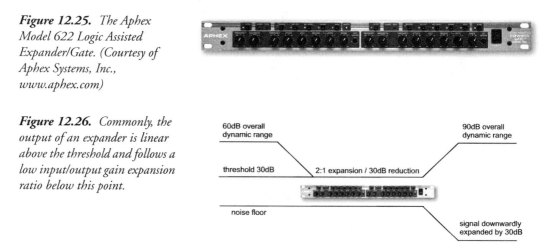

As with any dynamics device, the attack and release settings should be carefully set to best match the program material. For example, choosing a fast release time for an instrument that has a long sustain can lead to an audible pumping effect. Conversely, slow release times on a fast-paced, transient instrument could cause the dynamics to return to its linear state more slowly than would be natural. As always, the best road toward understanding this and other dynamics processes is through experimentation.

The Noise Gate

One other type of expansion device is the *noise gate* (Figure 12.27). This device allows a signal above the selected threshold to pass through to the output at unity gain and without dynamic processing. Once an input signal falls below this threshold level, however, the "gate" acts like an infinite expander and effectively mutes the signal by fully attenuating it (Figure 12.28). In this way, the desired signal is allowed to pass while background sounds, instrument buzzes, leakage or other unwanted noises that occur between pauses in the music aren't.

Figure 12.27. *The Drawmer SDS201 2-channel Gate. (Courtesy of Drawmer, www.transaudiogroup.com)*

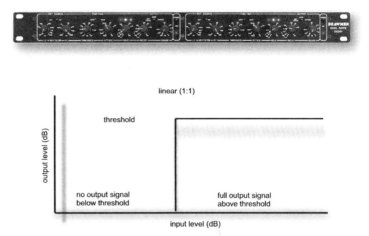

Figure 12.28. *The output of a gate is linear above the threshold and follows an infinite expansion slope (i.e., is turned off) below this point.*

The general rules of attack and release apply to gating as well. Fortunately, these settings are a bit more obvious during the gating process than with any of the other dynamic tools. Setting the attack and release times either too fast or too slow will often be immediately obvious as the sound will be cut off—either when listening to the instrument or vocal track on its own (solo) or within a mix.

Commonly, a special key input (Figure 12.29) is included as a sidechain input to a noise gate. A sidechain input is an external control input that's capable of affecting a device's main audio path, by allowing an external analog signal source (such as a miked instrument, synthesizer or oscillator) to trigger the gate's audio output path. For example, you could use the mic or recorded track of a kick drum to "key" a gate that's set to pass a low-frequency oscillator. Whenever the kick sounds, the oscillator will be passed through the gate. When the two are combined, you might have a deep kick sound that'll make the room shake, rattle and roll.

Figure 12.29. *Diagram of a basic keyed-input noise gate.*
a. *The signal is passed whenever a signal is present at the key input.*
b. *Without a key signal, the signal is "gated" (not allowed to pass).*

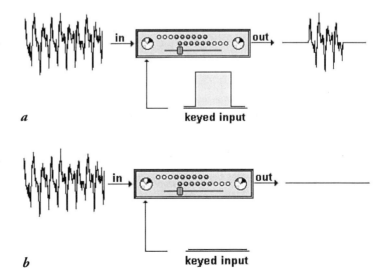

Controlling Dynamics

In multitrack recording, dynamic range changes usually deal only with single instruments or groups of instruments. In radio, television, disc mastering and tape mastering, however, entire songs are commonly compressed and the need to apply the proper parameters is often more critical in order to keep artifacts and "pumping" to a minimum . . . as well as to preserve some form of the performance's natural dynamics.

The process of limiting is usually reserved for the recording of speech or instruments that have loud, sharp transients (momentary high peak levels). During a recording session, it's much more common for compression to be used, in order to balance the dynamics of a track to the overall mix or to keep the signals from overloading preamplifiers, tape, amps and even you ears. Compression should be used *with care* for any of the following reasons:

- To minimize the changes in volume that occurs when an instrumentalist or vocalist has too great a dynamic range for the music or to smooth out momentary source-to-mic distance changes.

- To balance out the different volume ranges of a single instrument. For example, the notes of an electric or upright bass often vary widely in volume from string to string. Compression can be used to "smooth out" the bass line by matching their relative volumes. In addition, some instruments (such as horns) are louder in certain registers, because of the amount of effort that's required to produce the notes. Compression is often a useful tool for equalizing these changes.

- To reduce other frequency bands by inserting a selective filter into the compression chain that causes the circuit to compress frequencies in this band more than those outside the offensive band (frequency-selective compression). A common example of this function is a *de-esser*, which is used to accentuate high frequencies in a compressor's detection circuit, so as to suppress those "SSSS," "CHHH" and "FFFF" sounds that can often distort and/ or stand out in a recording.

- To allow a signal to be made significantly louder in the mix while increasing its overall volume output only slightly (as occurs when a TV commercial "seems" louder than the show). This is accomplished by reducing the ratio of average-to-peak levels.

Although it may not always be the most important, this last application gets a great deal of attention, as many producers strive to cut their recordings as "hot" as possible; that is to say, they want the recorded levels to be as far above the normal operating level as possible without blatantly distorting . . . the logic being that in this competitive business louder recordings, when broadcast, played on a multiple CD player, from an MP3 playlist, etc., will stand out from the softer recordings in a top-40 playlist. In fact, reducing a song or program's dynamic range will make the overall levels appear to be louder, by using a slight (or not-so-slight) amount of compression and limiting to squeeze an extra 1 or 2 dB gain out of a song. This increase in gain will also add to the perceived bass and highs because of our ears' increased sensitivity at louder levels (remember the Fletcher–Munson curve?). To achieve these hot levels without distortion, dual-channel compressors and limiters often are used during the mastering process to remove peaks and to raise the average level of the program.

Compressing a mono mix is done in much the same way as one might compress a single instrument. The adjustment of the threshold, attack, release and ratio controls, however, are more critical in preventing the "pumping" of prominent instruments in the mix. Compressing a stereo mix gives rise to an additional problem: If two independent compressors are used, a peak in one channel will reduce just the gain on that channel and will cause sounds that are centered in a stereo image to shift (or jump) toward the channel that's not being compressed (since it will actually be louder). To avoid this center shifting, most compressors (of the same make and model) can be linked as a stereo pair. This procedure of ganging the two channels together interconnects the signal-level sensing circuits in such a way that a gain reduction in one channel will cause an equal reduction in the other (thereby preventing the center information from shifting in the mix).

Digital Signal Processing

In modern audio production, an ever-increasing amount of signal processing is being carried out in the digital domain through the use of *digital signal processing* (*DSP*). One of the biggest advantages to working with DSP is that software programming can be used to configure a digital processor in order to achieve a wide range of effects, such as reverb, echo, delay, EQ, pitch shifting or gain changing.

The task of processing a signal in the digital domain is accomplished by combining logic circuits in a building-block fashion. These logic circuits follow basic binary computational rules according to a specialized program algorithm. When combined, these logic circuits can be used to alter the numeric values of sampled audio in a predictable way. After a program has been configured (from either internal RAM or the system's software), complete control over a program's setup parameters can be altered and measured as discrete numbers or as percentages of a full value. Because these values are both discrete and digital, the settings can be precisely duplicated or, better yet, saved to disk as a file that can be easily recalled at any time.

The following sections, "Real-Time and Non-Real-Time DSP," "DSP Basics" and "The Real World of DSP Design," are excerpts from the book *Hard Disk Recording for Musicians* by David Miles Huber © 1995, Amsco Publications, NYC and have been reprinted with permission.

Real-Time and Non-Real-Time DSP

The number-crunching process involved in performing DSP calculations can be performed in one of two ways:

- Real time
- Non-real time

Real-time calculations are capable of processing a signal in the here and now. In other words, real-time systems can alter, mix or otherwise process the sampled audio of a live input or as it's being recorded to or reproduced from a digital media. In a hard-disk environment, this process is often non-destructive, that's to say the signal alteration is often done while reproducing audio from disk without affecting the original soundfile data. Non-destructive editing also allows any inadvertent changes to be easily undone at a later time.

Non-real-time signal processing, on the other hand, is often used by hard-disk systems and digital audio workstations that don't have or don't need a dedicated real-time processor block to perform some or all processing tasks. Once a non-real-time processing function is called-up, the system's processor or coprocessor dedicates itself to performing the task in non-real time (which means that you'll have to wait while the system performs the necessary calculations). When these calculations have finished, the final results are usually written to hard disk as a separate file. If this processed section is used to replace an existing non-processed segment (such as a faded or crossfaded area), it's generally attached or "tagged" to marker points in the original, unprocessed soundfile areas. Upon playback, the processed section is butt-joined with the original areas to reproduce the program output in a seamless manner.

DSP Basics

The scope and capabilities of digital signal processing are limited only by speed, number-crunching power and human imagination. Yet the process itself is made up of only three basic building blocks (Figure 12.30):

- Addition
- Multiplication
- Delay

Figure 12.30. The DSP process is made up of only three basic building blocks: addition, multiplication (gain/attenuation), and delay.

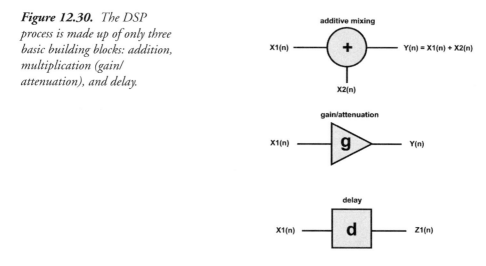

One of the best ways to understand the building blocks of digital logic is through application. Just for fun, let's try building a simple 4-in × 1-out digital mixing "app" that could be used to combine the output channels of a device.

Addition

As you might expect, a digital adder sums together the various bits at the input of the circuit in order to create a single combined result. With this straightforward building block, the word value of each input is mathematically added (at one samplepoint in time) to create a single digital word-value. Now that the inputs can be easily combined, we have the first building block of a simple mixer that combines the four input signals into one output channel (Figure 12.31).

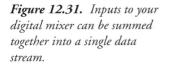

Figure 12.31. Inputs to your digital mixer can be summed together into a single data stream.

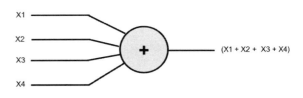

Multiplication

The multiplication of sample values by a numeric coefficient allows the gain (level) of digitized audio to be changed either up or down. Whenever a sample is multiplied by a factor of 1, the result is unity gain or no change in level. Multiplication by a factor of less than 1 yields a reduction in gain (attenuation), while the multiplication by a number greater than 1 will result in an increase in gain.

Now that we have this, we can add gain controls to our mixer. This will finally give us some real control, as digital faders can be used to modify the gain of each of our mixer's channel (Figure 12.32).

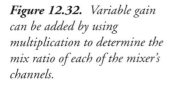

Figure 12.32. Variable gain can be added by using multiplication to determine the mix ratio of each of the mixer's channels.

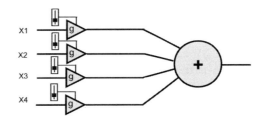

Before moving on, let's take a look at how multiplication can be applied to the everyday production world by calculating how the gain of a recorded soundfile can be changed over time. Examples include: fade-ins, fade-outs, crossfades and (of course) gain changes.

In order to better understand this process, let's look at how a non-real-time fade might be created using a hard-disk-based system. Suppose that we have a song that wasn't faded during a mixdown session but now definitely needs to be faded out over the final chorus section. The first task is to define the part of the song that needs to be faded (Figure 12.33a), call up the fade function and then perform it.

While the fade is being performed, it's the processor's job to continually multiply the affected samples by a diminishing coefficient. The result will be a file that's reduced in gain over the duration of the defined region. After the region has been recalculated, the results are then automatically written to disk as a separate file (Figure 12.33b). Upon playback, the fade is then digitally spliced onto the original soundfile at the appropriate point or points (Figure 12.33c).

Figure 12.33. Example of a 1-second fade.
a. Original soundfile.
b. Defined area to be faded is calculated and written to disk as a separate file.
c. Faded file is tagged to the original file and is reproduced with no audible break or adverse effect.

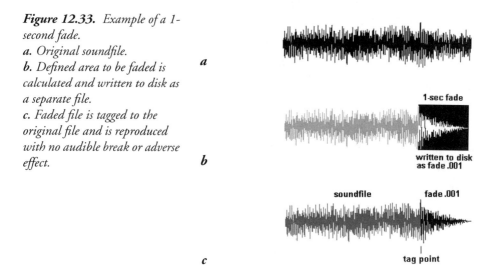

Delay

The final DSP building block deals with time, that is, the use of delay over time in order to perform a specific function or effect. In the world of DSP, delay is used in a wide variety of applications. This discussion, however, focuses on two types:

- Effects-related delay
- Delay at the sample level

Most modern musicians and those associated with audio production are familiar with the way different delay ranges can accomplish a wide range of effects. They're also often familiar with the use of digital delay for creating such sonic effects as doubling and echo. These effects (discussed later in this chapter) are created from discrete delays that are 35 milliseconds or more in length (1 millisecond, or ms, equals 1 thousandth of a second).

In its basic form, digital delay is accomplished by storing sampled audio directly into RAM. After a defined length of time (in milliseconds or seconds), this sampled audio can be read out from memory for output or further processing (Figure 12.34).

Figure 12.34. *A digital delay device stores sampled audio into RAM memory, where it can be read out at a later time.*

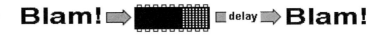

As this delay time is reduced below the 10-ms range, however, a new effect begins to take hold. The effect of mixing a variable short-term delay with the original undelayed signal creates a unique series of peaks and dips in the signal's frequency response. This effect (known as *flanging*) is the result of a selective equalization (Figure 12.35) that occurs when the delay time is varied. If you have a digital delay hanging around, you can check out this homemade flange effect by combining the delayed and undelayed signal and listening for yourself.

Figure 12.35. *The effect of mixing a short-term delay (that varies in delay over time) with the original undelayed signal creates a comb (multiple notch) filter response that goes by the generic effect name of flanging.*

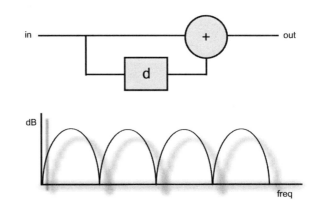

By further reducing the delay times downward into the microsecond range (1 microsecond or ms equals 1 millionth of a second), you can begin to introduce delays that affect the digitized signal at the sample level. In doing so, control over the phase characteristics can be introduced to the point that selective equalization is accomplished. Figure 12.36 shows examples of two basic EQ circuits that provide low- and high-frequency shelving characteristics in the digital domain.

Figure 12.36. *Simple EQ circuits and possible response curves.*

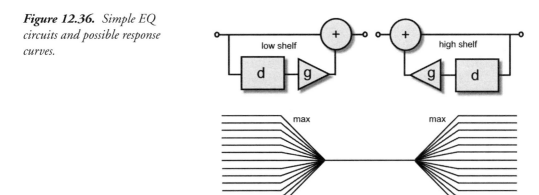

The amount of equalization to be applied (either boost or cut) depends on the multipliers that control the amount of gain that's to be fed from the delay modules. By adding more delay and multiplication stages to this basic concept, it's possible to create digital equalizers that are more complex and parametric in nature.

It should be pointed out that delays of such short durations aren't created using RAM delay-type circuits (which are used in creating longer delays). Rather, sample-level delay circuits (known as shift registers) are used, which are better suited to the task, because they're simpler and more cost-effective in design.

Getting back to our digital mixer example, we can now finish the project by adding some form of equalization to the final stage. For example, Figure 12.37 shows the project as having a very simple high-pass filter. This filter could be used to cut out any low-end rumble that could get into the system courtesy of a loud air conditioner or a local transit authority bus.

Figure 12.37. *Sample-level delays can be added to the mixer to provide equalization.*

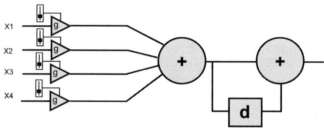

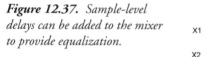

Do It Yourself Tutorial: Digital Delay

1. Get out a digital delay unit and balance its output mix (so that the input signal is set equally with the delayed output signal). Note: If there is no mix control, mix the delayed/undelayed signal at the console.
2. Play back various instrument sources (i.e., solo instruments and vocal tracks) while slowly sweeping the delay settings.
3. First try the 35 ms and greater range. Can you hear the discrete delays?
4. Now slowly vary the settings over the 10–35 ms range. Can you simulate a phasing effect?
5. If the unit has a phase setting, turn it on. . . . Does it sound different?
6. Now change the delay settings a little faster to create a wacky flange effect. If the unit has a flange setting, turn it on. . . . Does that sound different? Try playing with the time-based settings that affect its sweep rate. . . . Fun, huh?

Echo and Reverberation

Now that we've looked at single delay effects, let's add to our bag of effects tricks by successively repeating delays to create echoes. Repeated echoes are created by feeding a portion of a delayed signal's output back into itself (Figure 12.38). By adding a multiplier stage into this loop, it's possible to vary the amount of feedback gain and thus control both the level and the number of repeated echoes.

Figure 12.38. *By adding a simple feedback loop to a delay circuit, it's possible to create an echo, echo, echo effect.*

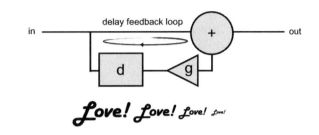

Although reverb could be placed in a separate category, it's actually nothing more than a series of closely spaced echoes. In nature, acoustic reverb can be broken down into three subcomponents:

- Direct signal
- Early reflections
- Reverberation

The *direct signal* is heard when the original sound wave travels directly from the source to the listener. *Early reflections* are the first few reflections that are reflected back to the listener from large, primary boundaries in a given space. Generally, these reflections are the ones that give us subconscious cues as to the perception of size and space.

The last set of reflections makes up the signal's *reverberation* characteristic. These sounds are broken down into zillions of random reflections that travel from boundary to boundary within the confines of a room. They're so closely spaced in time that the brain can't discern them as individual reflections, so they are perceived as a dense, single decaying signal.

By designing a system that uses a number of delay lines which are carefully controlled in both time and amplitude, it's possible to create an almost infinite number of reverb types, parameters and characteristics. To illustrate this point, let's return to our basic building block approach to DSP and build a crude reverb processor (Figure 12.39).

Figure 12.39. *A crude reverb processor design.*

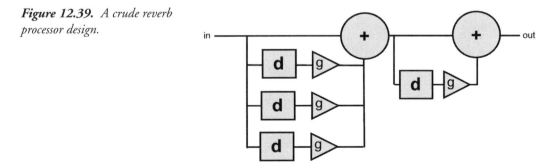

As you now know, the direct signal is the first signal to arrive at the listener. Thus, it can be represented as a single data line that flows from the input to the output. Included with this first DSP stage are a number of individually tunable delay lines. By tuning these various modules to different times between a range of 10 ms to over a second, early reflections can be simulated (being perceived as ranging from bathroom size to the Grand Canyon). Following this section, one or more delay lines with echo feedback loops (which are designed to repeat echoes at a very fast rate and decay over a predetermined time) can be placed into the system.

Such a reverb unit might sound crude since its simplicity would severely limit the sound type and quality. But, by adding more stages and placing the gain controls under microprocessor and user-defined algorithm command, a large library of high-quality room sounds could be created.

The Real World of DSP Design

As we've seen from the preceding discussion, the basics of DSP are fairly straightforward. In application, however, these building blocks can be programmed into a processing block that uses rather complex combinations and elaborate algorithms to arrive at a final result.

In addition to complexities that can exist, there are restrictions and overhead requirements that must be used to safeguard the process from cranking out erroneous, degraded or even disastrous results. For example, whenever a number of digital samples are mixed together using an adding circuit, the results could easily equal a large number that's beyond the system's maximum signal limit. Without proper design safeguards, a condition known as "bit wraparound" could occur, causing the signal to output a loud "pop!" Fortunately, modern design has removed this and other obvious gremlins. However, the deeper you dig into DSP, the more you'll find that the challenge in designing a quality system isn't in getting it to work but in eliminating the pesky glitches that often stem from performing complex sonic functions.

In addition to debugging a system, word accuracy is often of concern. For example, whenever signal processors are called on to add samples together or to multiply numeric values by lengthy coefficients, it's possible for errors to accumulate or for the final results to be greater than 16-bits in wordlength. To reduce these errors to acceptable levels or to prevent the "chopping off" of potentially important least-significant-bit values, most high-quality processors are capable of calculating wordlength values with a resolution of up to 24 or 32 bits.

The remainder of this chapter will examine many of the common signal processing devices and applications that are used in audio production. These include delay, reverberation, pitch- and time-shift related effects and psychoacoustic enhancement.

Delay

One of the most common DSP effect alters the parameter of time by introducing various forms of delay into the signal path. Creating such a delay circuit is a relatively simple task to accomplish in the digital domain. Although dedicated delay devices (often referred to as *digital delay lines* or *DDLs*) are readily available on the market (Figure 12.40), in fact most digital signal processors that can perform multiple functions are capable of creating this straightforward effect.

Figure 12.40. *tc electronic's 2290 dynamic digital delay. (Courtesy of tc electronic, www.tcelectronic.com)*

You might recall that delays in the millisecond (ms) and second range generally rely on the storage of sampled audio in RAM. After a defined length of time, the data is read out and can be mixed with the original, undelayed signal (Figure 12.41). The maximum delay time that can be delivered by such a device is limited only by its sample frequency and memory capacity.

Figure 12.41. *A DDL introduces one or more discrete repeats of the input signal at user-defined intervals.*
a. *Single delay signal.*
b. *Signals fed back into memory can create numerous, repeated delays.*

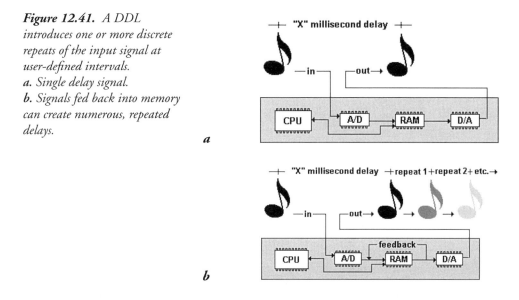

When delays of 35–40 ms and greater are used, the listener perceives the sound as being a discrete echo. When mixed with the original signal, this can add depth and richness to an instrument or range of instruments. Care should be used, however, when adding delay to an entire musical program, since the program could easily begin to sound "muddy" and unintelligible.

Reducing these delay times into the 15–35 ms range will create delays that are spaced too closely together to be perceived by the listener as being discrete delays. Instead, these closely spaced delays create a "doubling" effect (Figure 12.42) when mixed with an instrument or group of instruments. Basically, the brain is fooled into thinking that more instruments are playing than actually are. The effect, at least subjectively, increases the sound's density. This effect (known as *doubling* or automatic double tracking, ADT) can be used on background vocals, horns, string sections and other grouped instruments to make the ensemble sound as though it's double (or even triple) its actual size. This effect also can be used on foreground tracks, such as vocals or instrument solos, to create a larger, richer and fuller sound.

Figure 12.42. In certain instances, doubling can fool the brain into thinking that more instruments are playing than actually are.

Whenever delays that fall below the 15-ms range are slowly varied over time and then are mixed with the original undelayed signal, a combing effect is created. *Combing* is the result of changes that occur when equalized peaks and dips appear in the signal's frequency response (Figure 12.43). Either by manually or automatically varying the time of one or more of these short-term delays, a constantly shifting phase effect known as *phasing* or *flanging* can be created. Depending on the application, this effect can range from being relatively subtle (phasing) to having moderate-to-wild shifts in time and pitch (flanging).

Figure 12.43. Peaks and dips in a signal's frequency response (as shown in gray areas) result from the combination of several short-term delays that shift over time to create the effects of phasing or flanging.

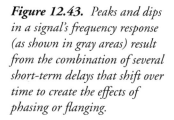
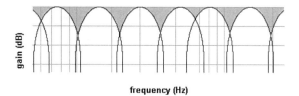

By combining two identical (and often slightly delayed) signals that are slightly detuned in pitch from one another, another effect known as *chorusing* can be created. Chorusing is an effects tool that's often used by guitarists and other musicians to add depth, richness, and harmonic structure to their instruments' sound.

Reverb

In professional audio production, natural acoustic *reverberation* is an extremely important tool for the enhancement of music and sound production. A properly designed acoustical environment can add a degree of quality and natural depth to a recorded sound that often affects the performance as well as its overall sonic character. In those situations where there is little, no or a bad-sounding natural ambience, a high-quality digital reverb device (Figure 12.44) can be used to fill out and add a sense of dimensional space to the production.

Figure 12.44. Sony DRE-S777 sampling reverb processor. (Courtesy of Sony Electronics, Inc., www.sony.com/proaudio)

As we learned in Chapter 3 (Studio Acoustics and Design), reverb is the creation of closely spaced and random multiple echoes that are reflected from one boundary to another within a determined space (Figure 12.45). This effect can help give us perceptible cues as to the size, density and nature of a space (even though it's been artificially generated) and can definitely add to the perceived warmth and depth of a recorded sound.

Figure 12.45. Signal level versus reverb time.

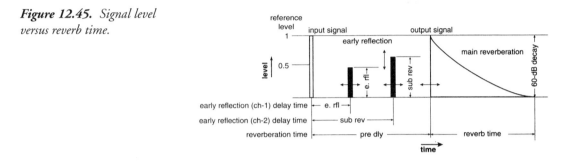

Pitch- and Time-Shift-Related Effects

Certain signal processors are capable of altering the speed and pitch of an audio program. This complicated task often requires that the system carry out a series of complex calculations in either real or non-real time.

It should be noted that not all speed/pitch processors are created equal and (depending upon the device or software) they will best be able to effect sounds within a limited pitch-shift range. That is to say, a hardware system or program might be able to correct or alter program material over a limited shift range without introducing certain amounts of granularity or harmonic distortion into the signal (although these figures have definitely improved in recent years). This range will often vary from system to system and with different types of program material (such as voice, music and complex waveforms). In the final analysis, your ears must judge as to whether the drawbacks of induced distortion (if any) outweigh the benefits of the effect itself.

Currently, the options for pitch and time shifting include changing pitch without changing duration, changing duration without changing pitch and changing both duration and the pitch.

Pitch Shifting

Pitch shifting (Figures 12.46 and 12.47) can be used to vary the pitch of a program (either upward or downward) in order to transpose the relative pitch of an audio program without affecting its duration. This process can take place either in real time or in non-real time. One way this process works is by writing sampled audio data to a temporary storage medium (known as a circular buffer) where it can be resampled at either a higher or a lower sample rate (Figure 12.48). In order to revert the resampled signal back to its original rate, the signal must be altered to match the device's original sample rate. The resulting change in pitch will be a ratio that exists as the difference between the internally resampled rate and the outgoing rate. In short, the process works by interpolating the input signal to a higher sample rate (that's higher in pitch) . . . resampling this internal rate back to the original rate at the device's output will lower the program's pitch (and vice versa).

Figure 12.46. *Digitech's Studio Vocalist. (Courtesy of Digitech, www.digitech.com)*

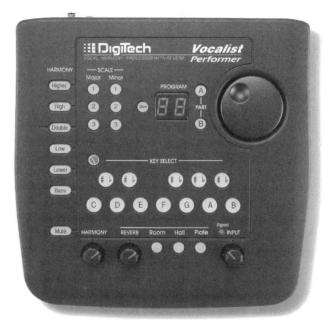

Figure 12.47. *Eventide DSP 7000 Ultra-Harmonizer. (Courtesy of Eventide Inc., www.eventide.com)*

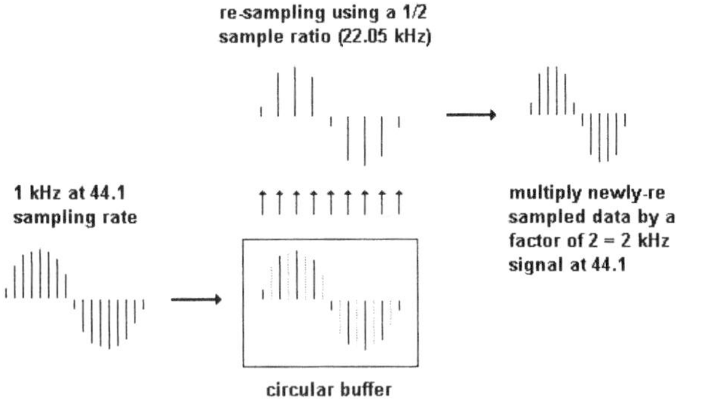

Figure 12.48. *Pitch shifting occurs by resampling data that is temporarily stored in a circular buffer at either a higher or a lower rate.*

re-sampling using a 1/2
sample ratio (22.05 kHz)

1 kHz at 44.1
sampling rate

multiply newly-re
sampled data by a
factor of 2 = 2 kHz
signal at 44.1

circular buffer

Time Compression/Expansion

By combining variable sample rates and pitch shifting techniques, it's possible to alter a program's duration (varying the length of a program by raising or lowering its playback sample rate) or to alter its relative pitch (either up or down). In this way, three possible time/pitch combinations can occur:

- A program's length can be altered without affecting its pitch.
- A program's length can remain the same, while pitch is shifted either up or down.
- Both a program's pitch and length can be altered (using simple resampling techniques).

In addition to their application in music, time compression and expansion (which are available on most digital audio editors, workstations, samplers and signal processors) have also become standard processing options for use in the audio-for-video, film, and broadcast industries. This feature can help give producers control over the running time of film video and audio soundtracks while maintaining the original, natural pitch of voice, music and effects.

Psychoacoustic Enhancement

Certain signal processors rely on psychoacoustic cues in order to function. In other words, these devices operate by affecting the ways in which sound is actually perceived by the brain. The earliest and most common of these devices are those that enhance the overall presence of a signal or entire recording by synthesizing upper-range frequency harmonics and then mixing theses back with the originally recorded signal. Although the addition of synthesized harmonics won't significantly affect the program's overall volume, the effect is a marked increase in its perceived presence.

Other psychoacoustic devices have come onto the market and have become standard production tools that make use of complex harmonic, phase, delay and equalization parameters to help shape the sound into one that's interesting and has a sonic character all its own.

In addition to synthesizing harmonics in order to change or enhance a recording or track, other digital psychoacoustic processors deal exclusively with the subject of *spatialization* (the placement of an audio signal within a three-dimensional acoustic field), even though the recording is being played back over stereo speakers. By varying the parameters of a stereo or multiple input source, this processing function create phase and amplitude paths that can fool the brain into perceiving that the stereo image is actually emanating from a soundfield that's wider than the physical speaker positions. In practice, care should be taken when using these devices, as the effect is often carried off with degrees of success that vary from system to system. In addition, the use of phase relationships to expand the stereo soundfield can actually cause obvious cancellation problems when the program is listened to in mono.

Multiple-Effects Devices

Because most digital signal processors can be easily programmed to perform various functions, an ever-increasing number of signal processors are *multiple-effects devices* (Figures 12.49 and 12.50). Multi-effects, in this case, can have several basic meanings:

- A single-effects device might offer a range of processing functions, but allow only one to be called up at a time.
- A single-effects device might offer a range of processing functions that can be simultaneously called up and performed by several processing blocks.
- An effects device might have multiple ins and outs, each of which can perform several processing functions (effectively giving you multiple processors or one that can be used in a multichannel mixdown environment . . . i.e., surround sound).

Figure 12.49. Controller for tc electronic System 6000 digital effects processor. (Courtesy of tc electronic, www.tcelectronic.com)

Figure 12.50. Lexicon 960L digital effects processor. (Courtesy of Lexicon, Inc., www.lexicon.com)

Both device types are common and are used as invaluable tools in project, recording and audio production facilities where processing versatility is always a big plus. Table 12.3 lists just a few of the possible effects that can be offered by these amazing devices.

Table 12.3.

Reverb	Delay, chorus, phasing and flanging
Equalization	Compression, limiting, expansion and gating
Pitch shift	Time change
Sample rate conversion	Spectral and spatial enhancement
Sampling or one-shot sampling	Overdrive distortion
Wah pedal	Rotary speaker and auto-panning
Tremolo and vibrato	Effects morphing

Plug-Ins

In addition to the DSP effects offered in the hardware domain, a constantly growing list of signal processors are available for personal computer platforms in the form of the ever-popular signal processing *plug-in*.

These software utilities offer virtually every processing function imaginable (often at a fraction of the price of their hardware counterparts) with little or no reduction in quality, capabilities or automation features. These programs (which are programmed and marketed by large and smaller third-party companies, alike) are designed to be integrated into a digital audio editing or workstation production environment, in order to perform a particular real- or non-real-time processing function.

Currently, several plug-in standards exist. If a plug-in conforms to the standard that's supported by a computer's specifications, operating system (OS) and/or digital audio software . . . then it should work, regardless of its form, function and/or manufacturer. As of writing, the most popular standards are *DirectX* (PC), *AudioSuite* (Mac), *VST* (PC/Mac) and *TDM* (Digidesign PC/Mac). Further information and graphic screenshots can be found in Chapter 6.

Dynamic Effects Editing Using MIDI

One of the most common ways of automating effects devices during a mix (either from a MIDI sequence or on-stage during a live performance) is by using *MIDI program-change commands*. In the same way that a favorite sound patch can be stored into an instrument's memory for later recall, most MIDI-equipped effects devices will let you store effects type and parameter "patch" data into memory.

Program-change commands (and, occasionally, continuous controller messages) allow complex signal processing functions to be easily called up during the normal playback of a MIDI sequence. Often, making such a "scene change" is as simple as inserting the desired program change number into the MIDI track that matched the device's MIDI IN assignment number (Figure 12.51).

Figure 12.51. Effects settings can be automated throughout a system with the use of MIDI program-change commands.

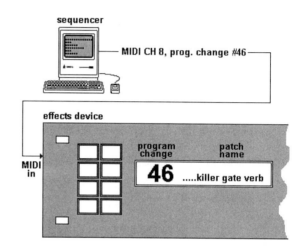

In addition to calling up program changes (either live or from a sequence), effects parameters such as program type, reverb time, depth or EQ can often be changed in real time through the use of system-exclusive messages. Control over these messages is generally carried out through the use of an external MIDI hardware controller or a software controller application (Figure 12.52).

Figure 12.52. Dynamic control over effects parameters by way of an external MIDI command controller.

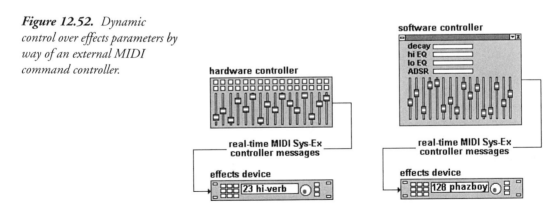

Generic hardware MIDI controllers have gained in popularity, because of the simple fact that you can actually get your hands on a set of knobs or data sliders. By assigning a controller to an individual function, it's possible to tweak the effects (or electronic instrument) settings in real time . . . often a very satisfying and straightforward experience.

A type of computer-based editing software known as a *patch editor* also offers dynamic control over effects parameters by letting you edit and fine-tune effects settings through the use of on-screen, mouse-driven visual graphics and scales or, alternatively, through the display of numeric values that directly represent the device's control parameter settings. Control over these parameters is accomplished in real time through the use of device-specific system exclusive (sys-ex) messages. After the desired effect or multi-effect has been assembled and fine tuned, these settings can be saved into as a preset for recall at a later time.

Once a device's preset bank has been filled, it's often possible to transmit these settings as a sys-ex "bulk dump" or "program dump" back to a MIDI sequencer, where the data can be stored as a named file. Using this method, multiple preset banks can be stored and recalled back to the device, thereby allowing a much larger number of effects "patches" to be stored in a computer-based library.

As with electronic instrument patches, effects patches can be acquired from a number of different sources. Among these are patch data cards (device-specific ROM cards or cartridges that contain patch data from manufacturers or second-party developers), patch data disks (computer files containing patch data from manufacturers or second-party developers) and, of course, the Web (you'd be surprised how many device-specific sys-ex patch dumps are available in cyberspace).

In the final analysis, the underused and often misunderstood use of MIDI effects automation is a supercharged production tool that can help boost your musical and production effectiveness with a minimum of fuss and bucks.

CHAPTER

Noise Reduction

Because of the increased dynamic range and the demand for better quality sound that's been brought about by newer microphone technologies, digital audio, the compact disc and surround sound home movies, it's become more important than ever for those in all forms of audio production to pay close attention to the background noise level that's produced by analog magnetic tape, amplifier self-noise and the like. The overall dynamic range of human hearing roughly encompasses a full 130 dB; however, this range can't be adequately recorded onto analog tape. In order to record an analog dynamic range in excess of 60 dB, some form of dynamic range compensation is required. The limitations imposed on conventional analog ATR and VTR audio tracks are dictated by tape noise (which is perceptible when the overall signal level is too low), or by tape saturation (which is caused by distortion when recording at levels that are too high). Should an optimum recording level produce an unacceptable amount of noise, the engineer is then faced with two options: record at a higher level (with the possibility of increased distortion) or change the signal's overall dynamic range.

Although this chapter focuses largely on reducing noise from one or more analog tracks, not all of the blame can be put on our older technology friends. In fact, noises can crop up from any number of modern-day sources, such as noise from mics, preamps, effects and outboard gear, background noise, analog communication lines . . . not to mention poorly designed digital audio converters, digital jitter (see Chapter 6) and a host of other sources.

In the first part of this chapter, we'll largely focus on reducing tape noise that's a natural by-product of the analog recording/playback process. In the latter half, we'll be dealing with single-ended and digital noise reduction (NR) processors that can help to reduce noises that have been introduced into the recording chain from any number of gremlin sources.

Analog Noise Reduction

Analog tape noise might not be a limiting factor when you're dealing with one or two tracks in an audio production, but the combined noise and other distortions that are brought about by combining 8, 16, 24, 48 or more tracks can range from being bothersome to downright unacceptable. The following types of noises are often the major contributors to the problem:

- Tape and amplifier noise
- Crosstalk between tracks
- Print-through
- Modulation noise

Modulation noise is a high-frequency component that causes frizziness as well as sideband frequencies that can distort the signal (Figure 13.1). This noise is due, in part, to irregularities in the coating of magnetic recording tape and only occurs when a signal is present and increases as signal levels rise. It's interesting to note that this noise is often higher in level than you might expect and actually plays a major role in what could be called the "analog sound."

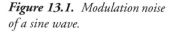

Figure 13.1. *Modulation noise of a sine wave.*

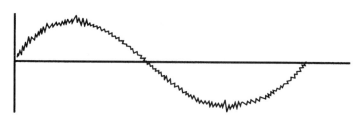

Do It Yourself Tutorial: Modulation

1. Feed a 0-VU, 1-kHz test tone to a track on a professional analog recorder.
2. Listen to the recorder's source (input) signal through the monitors at a moderate level. Sound like 1 kHz?
3. Switch the recorder to monitor the track's playback (tape) signal. Does it sound different?

These analog-based noises can be reduced to acceptable levels by taking any of the following actions:

- Improving the dynamic range of professional recording tape and/or increasing tape speed in order to record at higher flux levels

- Designing the tape heads and electronics with reduced crosstalk between recording channels

- Using thicker tape base to reduce print-through

By using tape formulations that combine low noise and high output (which have generally increased the signal-to-noise ratio by 3 or more dB), noise levels can be reduced even further.

The Compansion Process

When all is said and done, making most or all of the above improvements often proves to be too costly and/or impractical . . . and even if they're implemented, it's not possible to get around the fact that noise is an inherent part of the analog recording process.

In order to reduce the effects noise on an analog recording, a process called *compansion* was created. The compansion encode–decode process gets its name from the fact that the incoming signal is (comp)ressed before it is recorded onto tape; upon reproduction, the signal is reciprocally exp(anded) back to its original dynamic range (with a resultant reduction in background tape noise). This is a reciprocal process, in that the encoder side of the device must be placed between the console (or other desired source) and the tape track, whereas on playback, the decoder must be placed between the tape output and the console (or desired destination).

To better understand this process, let's take a look at Figure 13.2. In this example, the overall dynamic range of an input signal is restricted (compressed) before the signal is sent to the recorder's track input. This is done so that the newly compressed signal can be recorded onto tape at levels that are significantly higher than the tape noise. During playback, the signal is then downwardly expanded back to its original dynamic range, while the lower-level tape noise is likewise expanded down to lower (and ideally inaudible) levels.

Figure 13.2. Example of a full-bandwidth compansion noise reduction process.

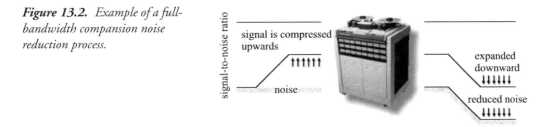

The dbx Noise-Reduction System

The dbx system (Figure 13.3) is a full-bandwidth compansion system that can provide between 20 and 30 dB of noise reduction. The compression chain uses a 2:1 ratio across the entire frequency bandwidth between the levels of –90 dBm and +25 dBm (with its unity gain point occurring at +4 dBm or 0 VU). Because all signals are compressed and expanded at a 2:1 ratio (regardless of signal level), the system isn't level-sensitive and doesn't require precise setup calibrations in order to work properly.

Figure 13.3. *The dbx 140X Type II noise reduction system. (Courtesy of dbx Professional Products, www.dbxpro.com)*

From the example in Figure 13.4, let's assume that we're recording a program that has a 60-dB dynamic range onto a tape machine that also has a 60-dB signal-to-noise (S/N) range. During the softest program passages, the tape noise will be just as loud as the program and therefore will be very audible. With the dbx system, the program passes through the record section and is compressed (by 2:1) into a 30-dB overall dynamic range before being recorded onto tape. (It's important to note that the recorded tape noise is now 30 dB below the softest music passage.) On playback, the expander will proportionally turn down softer signals until the dynamics are back to their original 60-dB range. Since the noise is now 30 dB below the softest passage, it will be reduced further to a level 30 dB below this point . . . at a near inaudible level of –90 dB.

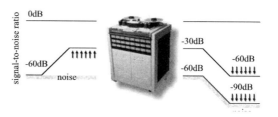

Figure 13.4. *Expansion of dynamic range from 60 dB to 90 dB when using dbx noise reduction.*

A 2:1 ratio was chosen over a higher value because of the reduced effect that tape dropouts have on the expansion process. Since the expander reduces its output signal twice as much as the signal being played back from the tape, a tape dropout of 2 dB would cause a 4-dB dropout to occur. A 3:1 ratio would cause a 6-dB dropout. A ratio of 1.5:1 would make the dropout problem less noticeable, but it would do so at the cost to the noise reduction process (only 20 dB would be possible). Therefore, a 2:1 ratio was considered as a compromise between the best noise reduction levels and oversensitivity to dropouts.

Although dbx noise reduction processors are used less common today than in past eras, they're still available in either of two flavors: Type I and Type II. Type II is an improvement over its predecessor, in that its filters prevent mistracking caused by end-of-band frequency response errors (which can be caused by head irregularities, azimuth error and overaggressive filtering). The Type II detector preemphasis circuit (which boosts the high end on recording and reciprocally cuts it on playback to further reduce noise) is greater than in Type I. This helps to overcome the reduced headroom limitations that are often inherent in broadcast and cassette production equipment.

The Dolby Noise-Reduction System

Although similar in theory to the preceding example, another system type operates by breaking the audio signal into several, separate frequency bandwidths. This makes it possible for bandwidths that contain louder passages to be unaffected, while softer passages (especially those in the upper noise-perception range) are appropriately processed. An example of such a system is the Dolby family of analog companders.

Dolby companders are available for audio production in four popular flavors: the professional Dolby SR, Dolby A, and the consumer Dolby B and C.

Dolby spectral recording (SR) is an encode–decode process that readily lends itself to any analog audio recording or transmission application. This system (Figure 13.5), which is popular amongst professional studios that use analog multitrack recorders, practically eliminates the influence of noise and nonlinearity on reproduced sound by improving tape noise figures by as much as 24 dB.

Figure 13.5. *The Dolby XPSR SR 24tk noise reduction rack. (Courtesy of Dolby Laboratories, Inc, www.Dolby.com)*

The Dolby SR signal shaping processor is a sidechain that runs parallel to the device's main audio path. The output of this sidechain is either added or subtracted from the main signal, depending on whether the circuit is enabled as an encoder or decoder. At the lowest signal levels (or in the absence of a signal), Dolby SR applies a fixed-gain/frequency characteristic that reduces noise and other low-level disturbances. Although most of its electronics are used for spectral analysis, Dolby SR's principal operating system consists of five groups of fixed- and sliding-band filters with gentle slopes that are arranged by level and frequency. Those with fixed bandwidths are electronically controlled to vary their gain; those with fixed gain can be adjusted to cover different frequency ranges. These filters are crosslinked by a technique known as action substitution, which allows both type of filters to be selected in the proper proportion.

By selecting and combining filters, the SR control circuit can create an infinite number of filter shapes through which the signal can pass in the encoding process. During decoding, filter shapes are automatically created that are the exact opposite of those used during encoding. This results in an accurate, linear response in level, phase and frequency.

In practice, below a specified threshold, an SR channel will optimize these gain and EQ settings to dynamically compress the signal during recording and conversely expand the signal on playback. Louder signals within any of the bands will significantly alter the processing curves and will be companded less.

Each tape track has its own, dedicated SR channel that is used for both the record and playback processing modes. Switching between these modes is usually carried out automatically by the recorder's transport switching logic.

Dolby's original professional compansion system, Type A, reduces tape noise approximately 10–15 dB by dividing the audio spectrum into four separate bands (Figure 13.6). Each frequency band has its own dynamic-range processor, such that the presence of a high-level signal in one band won't interfere with detection circuits in another band.

Figure 13.6. The four filter bands of the Dolby A system.

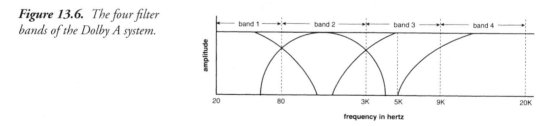

The outputs of the four filters and signal processors are combined in such a manner that low-level signals (below −40 dBm) are boosted by 10 dB from 20 Hz to 5 kHz, with the boost gradually rising between 5 kHz and 14 kHz to yield a maximum noise reduction of 15 dB. These bands aren't sharply defined, so when the noise reduction in Band 2 is disabled by the presence of loud signals between 80 Hz and 3 kHz, some noise reduction (in addition to masking) will be provided by Band 1 and above 1.8 kHz, by Band 3. If Band 3 also has its noise reduction turned off by a loud signal between 3 kHz and 9 kHz, Band 4 will continue to add noise reduction above 5 kHz. Bands 1 and 4 rarely have their noise reduction shut down completely (except by very loud organ tones or by cymbal crashes). Although the actual amount of noise reduction throughout the audio spectrum changes from one moment to the next, the noise level perceived by the ear will remain constant.

To ensure that the tape is played back at the same level at which it was recorded, a 400- or 700-Hz, 0-VU level tone is recorded at the beginning of each tape, allowing levels to be adjusted properly during playback. This ensures that signals below the threshold during recording will be the same amount below the threshold on playback. A ±3 dB tolerance can exist before a difference in the playback signal will become noticeable. If the Dolby signal is played back too loudly, too little expansion will take place because the limiter is blocked by the high-level signal and the signal will sound compressed and overly bright. If the signal is played back at too soft a level, the expander will expand too much and the signal will sound dull and have too great a dynamic range. The recorded signal level doesn't matter as long as the playback level is the same.

Dolby B (which is commonly found in consumer cassette decks) is designed solely to reduce tape hiss. The Dolby B system acts only to compand the upper-frequency component and has no operating effect on lower-frequency noises, such as hum or rumble. The effect of type B noise reduction is 3 dB at 600 Hz, rising to 10 dB at 5 kHz, at which point it levels off in a shelving fashion. Dolby C is a more recent consumer version of Dolby SR that offers up to 20 dB of overall noise reduction when used on cassette tapes.

Single-Ended Noise-Reduction Process

Since a compansion systems must process a signal before recording and after playback to achieve noise reduction, noise can't be removed from a source signal; it can only be prevented from entering the recording medium. The noncomplementary or *single-ended* noise reduction process is different, in that it extracts noise from an audio source by combining a downward dynamic-range expander in conjunction with a program-controlled dynamic low-pass filter. The expansion and dynamic processing can often be used either together or separately to provide the greatest possible amount of noise reduction.

Both analog and digital single-ended noise reduction systems are currently available. Digital systems are most commonly found as a program algorithm that can be called up from a multi-effects processor (Figure 13.7) or as a plug-in option for a digital audio editor (Figure 13.8); analog systems are often devices that are designed around specialized voltage-controlled amplifier (VCA) and level/frequency detection circuitry.

Figure 13.7. *Behringer's Denoiser system Model SNR 2000. (Courtesy of Behringer International GMBH, www.behringer.de)*

Figure 13.8. *Ray Gun noise-reduction plug-in. (Courtesy of Arboretum Systems, Inc., www.arboretum.com)*

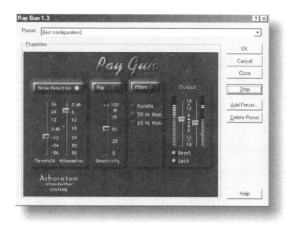

Single-ended noise reduction systems work by breaking the audio spectrum up into a number of frequency bands, such that whenever the program's signal level within each band falls below a user-defined threshold, the signal is attenuated. This downward expansion/filtering process accomplishes noise reduction by taking advantage of several basic psychoacoustical principles:

- Music is capable of masking noise that exists at lower levels within the same bandwidth.
- Reducing the bandwidth of an audio signal reduces the perceived noise.

It's a psychoacoustical fact that our ears are more sensitive to noises that contain a greater number of frequencies than to those containing fewer frequencies. In light of this, the dynamic filter examines the incoming signal for high-frequency content and, in its absence, the filter's bandwidth is decreased. When high-frequency energy returns, the filter opens back up as far as necessary to pass the entire or required signal.

Noise Gates

A *noise gate* can be a very effective noise-reduction device when used to reduce background noise on certain program material. A noise gate effectively passes signals that fall above a user-defined threshold at unity gain, while turning off signals that fall below this threshold. This makes it a useful tool for eliminating noise, leakage and other gremlins from a track within a mix. Obviously, this process wouldn't work well on a final output channel, as the entire program would simply cut out below any reasonable threshold.

When "gating" a specific track, it's often necessary to take time out to fine-tune the device's attack and release controls, in order to eliminate any unwanted "pumping" or "breathing" of the noise floor, as the signal falls and rises around the threshold point.

Digital Noise Reduction

Advanced forms of digital signal processing (DSP) have recently been designed that can reduce the noise content of previously recorded material. These noises include artifacts such as tape hiss, hum, obtrusive background ambience, needle ticks, pops and even certain types of distortion that are present in the original recording. A number of algorithms are even capable of smoothing out the grainy side effects of digital recordings that have been recorded at a lower bit-rate resolution (dither).

Although stand-alone *digital noise processors* do exist, by far the most popular systems exist as plug-in software applications for computer-based digital audio editors and workstations. These algorithms can be used to reduce the noise on one or more recorded tracks within a multitrack session, or they can be used during a mastering setting to reduce the noise of a final master tape . . . also, let's not forget a major application of digital noise processing for professional and novice users, alike: cleaning up and restoring older 45s, LPs and analog tapes for transfer to CD.

Most noise reduction plug-ins and programs (Figures 13.9 and 13.10) can help to remove hum, tape hiss and other extraneous noises from your recordings by analyzing a segment of the offending noise (a brief, isolated noise passage without music will yield the best result). This analysis is then used to create a "sonic footprint" or noise template, which can then be digitally subtracted (by a user-definable amount) from the original, offending soundfile or region. Once the footprint has been analyzed, the noise reduction process can be called into action (in real or non-real time, depending upon the program), after which the file can be saved to disk for further editing or mixing . . . or burned to CD during the mastering process.

Figure 13.9. *Cool Edit Pro's Noise Reduction application. (Courtesy of Syntrillium Software Corp., www.syntrillium.com)*

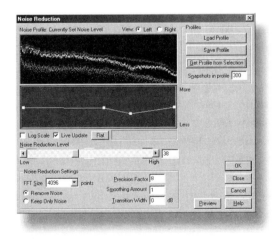

Figure 13.10. *Digidesign's*
Intelligent Noise Reduction
(DINR) system. (Courtesy of
Digidesign,
www.digidesign.com)

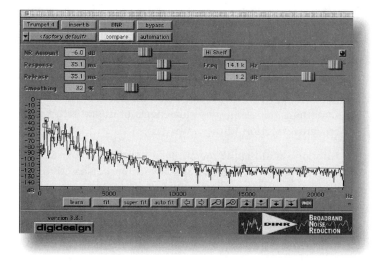

Before closing this section, it's important that we briefly discuss a few of the unfortunate artifacts that can occur as a result of digital NR. The most notable of these is "chirping." This audible artifact most often occurs whenever too much NR processing is applied to a soundfile. The
effect literally sounds like a huge flock of small chirping birds that can either be heard in the
background or experienced a bit too loud (as in the Alfred Hitchcock movie). If you find yourself running for cover, it's best to undo the process (it's always wise to keep the original file
intact as a future backup reference) and begin again with a lower NR setting or a higher quality
level of processing (if this option is available).

De-clicking and De-popping

The process of removing *clicks* is actually slightly different from that of noise reduction. This
two-step process first goes about the task of detecting high-level clicks (or those that exceed a
defined threshold), either one by one or by detecting all of the clicks within a file in a single
detection pass. Once the offending clicks are detected, the program can then reconstruct and
repair them by performing a frequency analysis, both before and after the click. In the latter
process, most programs will sample enough of the surrounding material to make plausible
guesses as to what the original waveform should sound like. Finally, it pastes this resynthesized
"guess" over the nasty offender, ideally rendering it less noticeable or gone.

The process of reducing *pops* is often similar in form and function, with the exception that the
natures of clicks and pops are quite different—both in their duration and in their frequency
makeup. As a result, noise reduction plug-ins will usually offer processing sections that are specifically suited to performing either task.

CHAPTER 14

Monitoring

In the recording process, our ability to judge and adjust sound is primarily based on what's heard through the monitor speaker system (Figure 14.1). In fact, within the professional audio and video industries, the word *monitor* refers to a device that acts as a subjective standard or reference by which program material can be judged.

Figure 14.1. *Example of a professional monitoring system. (Mega Studios—Paris, Courtesy of Solid State Logic LTD., www.solid-state-logic.com)*

Despite steady advances in design, speakers are still one of the weakest links in the audio chain. This weakness is generally due to potential nonlinearities that can exist in a speaker system's frequency response. In addition to this, interactions with a room's frequency response often lead to peaks and dips that can affect a speaker's sonic character in ways that are difficult to predict . . . Add to this, the factors of personal "tastes" in speaker sound, size and design types, and you'll find that speakers are one of the most subjective tools in a production studio.

Speaker and Room Considerations

Unless you have several rooms that have identical dimensions, materials and furnishings (an unlikely scenario), you can bet your bottom buck that the same speaker system will sound different in different room environments. That is to say, it'll interact with each environment to exhibit a different frequency response curve.

Although variations from one control or production room to the next often play a big part in giving a facility its overall sound, extreme variations in a room's frequency response can lead to production difficulties that can be heard in the final product. For this reason, certain basic principles (which are covered in the control room design section of Chapter 3) have become common knowledge to many who attempt the art of control room design. A few examples of these include:

- Reducing standing waves to help reduce erratic frequency response characteristics in a room
- Reducing excessive bass buildup in room corners through the use of bass traps
- Keeping the room/equipment layout symmetrical throughout a room so the left/right, front/back image is consistent
- Using absorptive and reflective surfaces to help "shape" a room's sonic character

With an increased awareness of careful room design and the availability of acoustical products that can help shape a room's sound, production and mixdown room designs have greatly improved over the past several decades. Yet even when high acoustical construction standards are followed, no two rooms will sound exactly alike. Because of the untold number of acoustic variables, a project that's been recorded in one facility will often sound quite different when played and/or mixed in another.

To help reduce or eliminate these variations, many professional studios will go one step further and *tune* (equalize) their speakers to the room's acoustics so that the adjusted frequency response curve will be reasonably flat and, therefore, reasonably compatible with most other control rooms.

Tuning a room can be roughly carried out by placing a ⅓-octave bandwidth graphic equalizer between each of the console's control-room monitor outputs and the power amplifier, and by feeding pink noise (which contains a flat energy spectrum curve throughout the audio range) into the speaker system. Pink noise, rather than individual sine waves, is used for testing because it has a random nature and doesn't stimulate standing waves within a room. The presence of such standing waves would introduce inaccurate analyzer readings that would vary with the microphone's position in the room.

Each speaker is then acoustically measured and adjusted (in ⅓-octave increments) using an instrument known as a spectrum analyzer, which is used to visually display the speaker's frequency response as measured through a specially calibrated omnidirectional condenser microphone (Figure 14.2). Initial readings are taken by placing the mic at the console's central listening position and also throughout the listening area (as the response curve of room will almost always vary from one spot in a room to another).

Figure 14.2. *A real-time spectrum analyzer can be used to help adjust a graphic equalizer's response curve in order to determine a speaker's optimum frequency response at the listening position.*

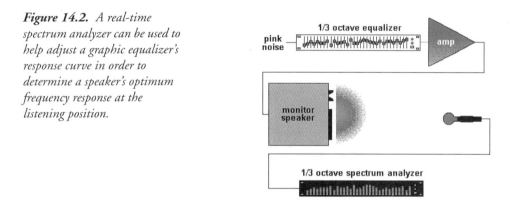

More accurate readings of both a speaker/room's frequency response and delay/reflection response can be measured by using a TEF Time Delay Spectrometer (TDS). This computer-based system is capable of accurately reading frequency- and time-related response curves in control rooms, studios, performance halls and auditoriums, and then can assist with or automatically tune the room using the proper response variables (Figure 14.3).

Figure 14.3. *Frequency- and time-related measurements can be made with great precision using modern DSP Hardware and a personal computer. (Courtesy of Gold Line, www.gold-line.com)*

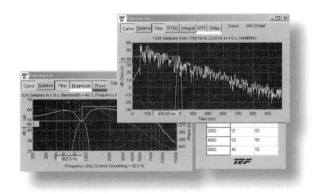

Speaker Design

Just as the sound of a speaker system will vary when heard in different acoustic environments, speakers of different designs will likewise often sound very different from one another. Enclosure size, number of components and driver size, crossover frequencies and design philosophy contribute greatly to these differences in sound quality.

Professional speaker enclosures usually are of one of two design types: air suspension and bass reflex. An air-suspension speaker enclosure is an airtight system that seals the air in its interior from the outside environment. This system type (Figure 14.4a) generally provides a strong, "tight" bass response, while often being rolled off at the extreme low end. In the bass-reflex or vented-box design (Figure 14.4b), a tuned bass porthole is designed into the front or rear of the speaker enclosure. This allows the air mass inside the enclosure to mix freely with the outside air in such a way as to act as a tuned resonator, which serves to acoustically boost the speaker's output at the extreme lower octaves.

Figure 14.4. *Speaker enclosure designs.*
a. Air suspension.
b. Bass reflex.

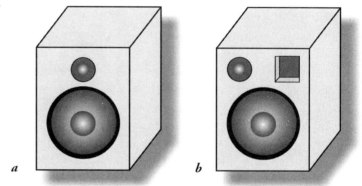

a   *b*

With so many variables to consider in a speaker and room combination, it quickly becomes clear that there's no such thing as the "ideal" monitor system. The choice is often more of a matter of personal taste and current marketing trends than one of subjective measurements. Monitors that are widely favored over a long period of time tend to become regarded as the industry standard; but this can easily change as preferences vary. Again, the best judge of what works best for you should be your own ears and personal sense of style.

Speaker Polarity

One of the most common oversights that can have a drastic effect on the sound of a multi-speaker system is to wire them out-of-phase with respect to each other. *Speaker polarity* is said to be electrically in-phase (Figure 14.5a) whenever one signal that's equally applied to both speakers will cause their cones to move in the same direction (either positively or negatively). When the speakers are wired out-of-phase (Figure 14.5b), one speaker cone will move in one direction while the other moves in the opposite direction.

Speaker polarities can be easily tested by applying a mono signal to both or all of the speakers at the same level. If the signal's image appears to originate from directly between the speakers, they have been properly wired in-phase. If the image is hard to locate and appears to originate

Figure 14.5. *Relative in-phase and out-of-phase cone motions.*
a. In-phase.
b. Out-of-phase.

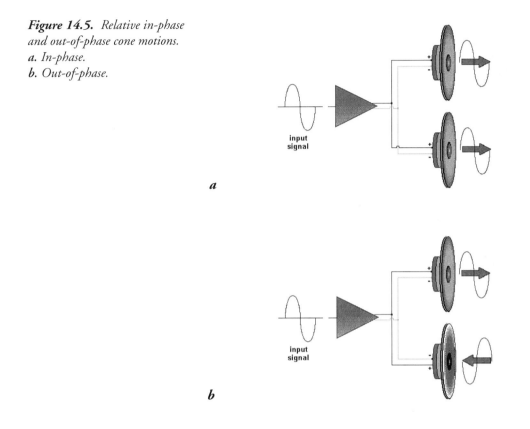

a

b

beyond the outer boundaries of a stereo speaker pair, or shifts as the listener moves his or her head, it's a good bet that the speakers have been improperly wired out-of-phase.

An out-of-phase speaker condition can be easily corrected by checking the speaker wire polarities. The "hot" lead (+ or red post) leading from each amp channel should be secured to the same lead on its respective speaker (Figure 14.6). Likewise, the negative lead (– or black post) should be connected to its respective speaker lead.

Figure 14.6. *Banana plug binding posts are often used to properly connect wires to the speaker leads.*

Speaker cable is generally color coded, with white or red being positive (+) and black being negative (–). If no color-coding is present, heavy-duty power cable and other cabling types that are suitable for speakers can be used. These cables will often have a notched ridge (or set of ridges) or have a printed white band that is generally connected to the negative lead post.

Speaker wire gauges should always be as heavy duty as is possible or practical. #18 wire is considered to be the minimum for lengths of less than 25' to 50', while #14 is considered the minimal length that should be used for 50' and 100' runs. (Note: The smaller the gauge number, the thicker the wire; therefore #14 is thicker than #18.) Two reasons for increasing the thickness of the conductor as cable length increases are as follows:

- All cable has resistance, which will increase with length. Thinner cables generally have greater resistance values, meaning that more power will be dissipated in the cable and therefore will be unavailable to drive the speaker.

- The higher the cable resistance, the lower the amplifier's effective damping factor. Damping factor is related to how well the amplifier is able to control the motion of the speaker cone. The lower the damping factor, the less control the amp will have over the speaker (often resulting in a loss of tightness, definition and clarity in the low end). Again, thicker conductors will have less resistance and thus help to minimize damping problems.

Crossover Networks

Because individual speaker elements (drivers) are more efficient in some frequency ranges than in others, different drivers are often used in combination to give the desired frequency response and level output. Large-diameter drivers (such as 15" and 30" units) produce low-frequency information more efficiently than high frequencies; medium-sized speakers (such as 4" and 5" units) produce midrange frequencies better than their high- or low-frequency counterparts; and small speakers (such as ½" to 1½" diaphragm sizes) reproduce highs better than any other range.

These speakers are often connected by *passive crossover networks*, which prevent any signals outside a certain frequency range from being applied to a specific speaker. Passive networks make use of frequency-selective inductors and capacitors to split the frequency range into several bands, which are then sent to their respective driver speakers. This design provides a smooth transition from speaker to speaker by routing input signals above the crossover frequency to the mid- and/or high-frequency driver, while routing signals below the crossover frequency to the bass driver or drivers (Figure 14.7).

Figure 14.7. Example of a passive two-way crossover system. a. Crossover/amp layout.

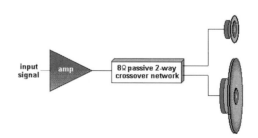

a

b. Frequency response curves showing crossover frequencies of 1500Hz.

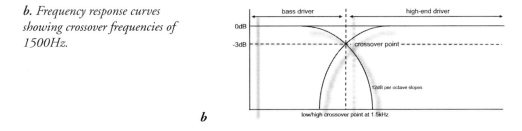

b

If a speaker system has only one crossover frequency, it's called a *two-way system* because the signals are divided into two bands. If the signal has two crossover frequencies, it's called a *three-way system*. The Westlake Audio BBSM-8 monitor speaker (Figure 14.8), for example, is a ported three-way system that uses two 8" woofers for the bass, a 3.5" cone midrange driver for the mid-frequencies, and a 1" soft dome tweeter for the highs. The crossover frequencies are 600 Hz and 5 kHz, respectively.

Figure 14.8. *Westlake Audio BBSM-8 monitor speaker. (Courtesy of Westlake Audio, www.westlakeaudio.com)*

Certain designs incorporate crossover level controls that determine how much energy is to be sent to the middle- and high-frequency drivers. This lets you compensate for various room environments and/or deficiencies (for example, an absorptive room might require more high-frequency energy than would a live room).

Electronic crossover networks (Figure 14.9), called *active crossovers*, differ from conventional passive crossover systems in that each line level audio signal is split into various frequency bands (depending on whether the speaker is a two- or three-way system). Each equalized signal is then fed to its own power amp, which in turn is used to drive the respective bass, mid and/or high driver elements. These systems are generally referred to as being bi-amplified or tri-amplified, depending on the number of power amps that are used per channel. Such systems have several advantages:

- The crossover signals are low in level, meaning that inductors (which can introduce audible ringing and intermodulation distortion) can be eliminated from the design.

- Power losses (due to the inductive resistance in the passive crossover network) can be eliminated.

- Each frequency range has its own power amp, so the full power of each amplifier in the respective speaker efficiency range will be available (meaning that excessive current draws in one range won't affect the sound of other drivers in the speaker system).

Figure 14.9. *Example of an active two-way crossover system.*

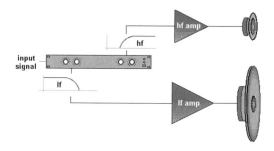

For example, let's assume that we're feeding a 100-watt power amp through a passive crossover network to a low- and high-frequency speaker system. If the low frequencies are pulling 100 watts (W) of power from the amp and a high-frequency signal comes along that requires an additional 25 W of power, the amplifier won't be able to supply it. Both the lows and the highs will become distorted. These monitoring requirements, however, could be met without incurring distortion by using an active crossover network to feed a 100-W amp for the low-end speaker and a separate 25-W amp for the highs.

The crossover points for either a passive or an active crossover network are generally 3 dB down from the flat portion of the response curve. Frequency ranges outside the filter's passband may have various rolloff slopes (usually 6, 12, 18, or 24 dB per octave with a 12 dB/octave slope being the most common). Depending on the speaker's design, almost any frequency can be selected to be a crossover point; however, a few of the most commonly selected frequencies are 500, 800, 1200, 5000, and 7000 Hz.

Monitoring

When mixing, it's important that the engineer be seated as closely as possible to the center of the soundfield (making allowances for the producer, musicians and others who are doing their best to be in the "sweet spot") . . . and that all the speaker volumes are adjusted equally. For example, if the engineer is closer to one speaker than another, that speaker will be louder and the engineer may be tempted either to pan the instruments toward the far speaker or boost that entire side of the mix to equalize the volumes. The resulting mix will sound centered when played in that room, but when listened to in another environment, the mix might sound off-center. As a quick check against this, the engineer should always make sure that an audible volume difference between speakers is accompanied by a corresponding visual difference on the VU meters, which are monitoring the signal sent to tape. Another guard against off-center levels is to monitor pink noise (or a test tone signal) from each speaker in the soundfield to check that they're equally loud (either by doing a quick audible check or by placing a microphone in the center listening position). In the latter case, the resulting levels from each speaker can be read and matched using an SPL meter or VU meter from a spare console input.

Here are a few additional pointers that can help you get the best sound from your control room monitors:

- Make sure that the room's reverb time is both low and smooth over the audible range.
- Keep large reflections to a minimum within the room (at least 20 dB down from the direct signal).
- Keep all room boundaries and reflections as symmetrical as possible along the side and front/back mixing axis.
- If the speakers are mounted in soffits, make sure the front wall has a hard, smooth surface.
- If diffusers are used, place them at the rear part of the room.
- Angle the monitors toward the listening position in both the horizontal and vertical planes.
- When checking monitor levels, ensure that your stereo/surround output levels are equally balanced by calibrating the signal using the console's main output meters.

Monitor Volume

Before continuing, I'd like to bring up another important factor in monitoring . . . *volume*. It's important to keep in mind that the Fletcher–Munson curves will always have an effect on the frequency balance of a mix, as our ears will perceive recorded sound differently at various monitoring levels. If you have set a balance while listening at loud levels, your ears will easily perceive the extreme high and low frequencies in the mix. When the mix is played back at lower levels (such as over the radio, TV or computer), your ears will be much less sensitive to these frequencies and the bass and extreme highs will probably be deficient (leaving the mix sounding distant and lifeless). Conversely, if you set a balance while listening at levels that are too low, the extreme frequencies will be unduly exaggerated when played back at moderate to high listening levels.

Unlike the 1970s, when excruciatingly high SPLs tended to rule in most studios, more recent decades have seen the reduction of monitor levels to a more moderate 75–90 dB SPL. These levels offer a good compromise level for mixing because they more accurately represent listening levels that are encountered in the average home (meaning that the Fletcher–Munson curves will be more closely matched). Ear fatigue and potential ear damage due to prolonged exposure to high SPLs by industry professionals can also be avoided at these levels.

Monitoring Configurations

In addition to getting the best overall sound, another monitoring concern that has gained importance is the need to tailor the mix to the intended room/speaker configuration (i.e., mono-stereo, mono-surround and stereo-surround).

It's important to remember that a large percentage of your potential customers may first hear your mix over a computer or FM and AM radio in mono. Therefore, if a recording sounds good in stereo but poor in mono, it might not sell well because it failed to take these media into account. The same might go for a surround sound mix of a music video or feature release film in which proper attention wasn't paid to phase cancellation problems in mono and/or stereo (or vice versa). The moral of this story is simply this: To prevent potential problems, a mix should be carefully checked in all its release formats in order to ensure that it sounds good and that no out-of-phase components are included that would cancel out instruments and potentially degrade the balance.

The most commonly accepted speaker configurations are mono, stereo and surround sound.

Mono

Even in this day and age, much of the buying public will first experience a mix in monaural (*mono*) sound (Figure 14.10). That is to say, they'll hear your song over the radio, on TV, in an elevator, on the computer, etc. . . . For this reason record companies, producers and everyone else involved in the process will often place a great deal of importance on mono compatibility and the overall sound of a mono mix. In fact, it's not uncommon for a separate mono mix to be made to ensure that it'll sound as good as it can for the intended medium.

Figure 14.10. Example of a mono monitoring configuration.

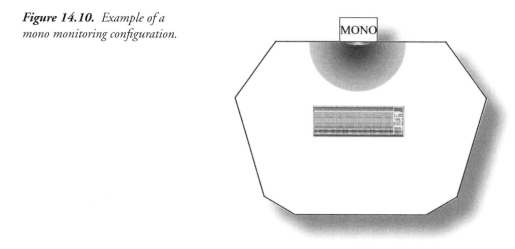

Stereo

Ever since the practical development of the 45°/45° record cutting process, stereophonic or *stereo* sound (Figure 14.11) has ruled the turntable, radio, CD player and TV. The creation of a quality stereo mix is extremely important, with relation to L/R balance, overall frequency balance, dynamics, depth and effect.

Figure 14.11. Example of a
stereo monitoring configuration.

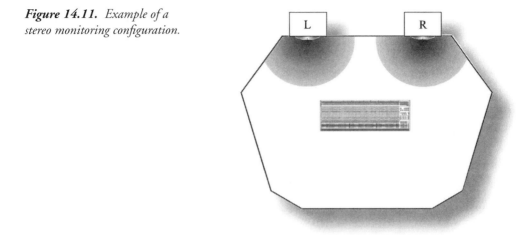

When mixing in stereo, it's always a good idea to check for mono compatibility. Phase cancellations can cause instruments or frequencies in the spectrum to simply disappear. The best tools for reducing phase errors are: good mic techniques, a phase meter or XY oscilloscope display (where the left channel is sent to the vertical trace and right to the horizontal trace inputs) . . . and, of course, your ears.

Surround Sound

With the advent of 5.1 surround playback in home and audio "theaters," *surround sound* (Figure 14.12) has grown into a major professional and consumer entertainment market. The 5.1 name refers to the five, full-range channels (left, center, right, surround left and surround right), plus a sixth sub-bass channel (containing a narrow frequency response of 5–125 Hz).

Figure 14.12. Example of a
surround sound monitoring
configuration.

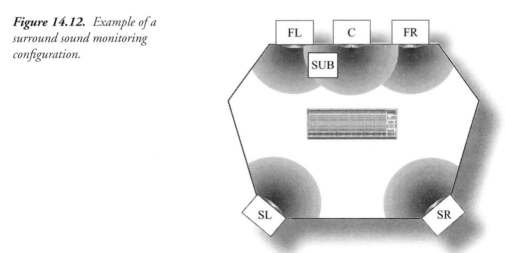

DVD videos and audio discs commonly use 5.1 encoding in the form of Dolby Digital (a scheme that encodes the discrete 5.1 information into a single bitstream, known as AC-3). Most players are able to decode or to route this serial bitstream to an external decoder in various digital surround formats.

The debate over how to mix for surround has raised the temperature of many a panel forum and Web chat discussion. These tend to fall into three camps:

- Those who advocate that a traditional L/C/R mix be created in the front field, while placing the ambient and special effects into the rear surround field
- Those who prefer to leave such traditional concepts behind and look at the soundfield as a full 360° environment, where instruments and mix elements can fall anywhere in the surround soundfield
- Those who tend to leave the decision up to the artist, producer, engineer and record company.

When dealing with discrete 5.1, compatibility issues aren't usually a major problem, as it's understood that media will be played back on a surround sound system. In such a situation, it's common for a separate stereo/mono-compatible mix to be built up from the soundtrack or production mix.

Should the mix be encoded into Dolby ProLogic (a scheme that encodes the surround information into a L/R stereo track using complex phase relationships), care should be taken to ensure surround, stereo, and mono playback compatibility.

Mix References

Even if your monitor speakers have perfect time/frequency response readings, few people who buy recordings will have flat speaker/room curves and, as a result, they won't hear the exact same mix that was heard in the control room. The buying public will often hear different frequency balances, due to response variances between the almost limitless types of speaker/listening room combinations. Faced with this fact, the best we can do as professionals is to rely on our judgment, our experience and our ears to create a mix that'll do the best possible justice to a project under a wide range of listening conditions.

To obtain the best possible compromise in the overall sound balance, several alternative monitoring choices are available during both the recording or mixdown process. In most facilities, several monitoring combinations are available as a reference during a session and/or mix (Figure 14.13): far-field, near-field, small-speaker and headphone monitoring.

Quite often, a console will let you select between speaker/monitor options, with each set commonly having its own associated amplifier for power and level matching flexibility.

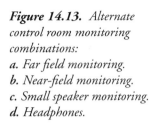

Figure 14.13. *Alternate control room monitoring combinations:*
a. Far field monitoring.
b. Near-field monitoring.
c. Small speaker monitoring.
d. Headphones.

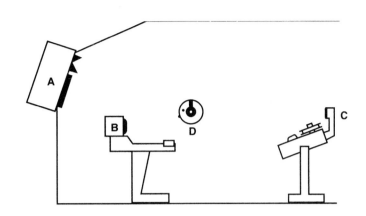

Far-Field Monitoring

Far-field monitors often involve large loudspeaker systems that are capable of delivering relatively accurate sound at moderate to high volume levels. Because of their large size and basic design, the enclosures are generally soffit mounted (built into the control room wall to reduce reflections around and behind the enclosure and to increase overall speaker efficiency . . . further reading can be found in Chapter 3).

These large-driver systems (Figures 14.14 through 14.16) are often used during the recording phase because of their ability to safely handle high sound output levels (which can come in handy should a microphone drop or a vocalist decide to be cute and scream into a mic). They're also great for listening to a mix at loud levels in order to hear the impact that it'll have on the dance floor or in a souped-up car system. In fact, certain types of music rely on bass levels that only these systems can deliver at moderate-to-high SPLs (although it's important to be aware of the danger of long-term exposure to sound levels like this).

Figure 14.14. *Westlake TM-3 Reference Series monitors. (Courtesy of Westlake Audio, www.westlakeaudio.com)*

Figure 14.15. Tannoy System
215 DMTII main reference
monitor. (Courtesy of Tannoy/
TGI North America, Inc.,
www.Tannoy.com)

Figure 14.16. Genelec 1039A
main monitor system. (Courtesy
of Genelec OY,
www.genelec.com)

Near-Field Monitoring

Although far-field monitors are generally the best reference at high listening levels, few systems are equipped with speakers that can deliver such high SPLs (nor would you always want them to). For this reason, many professionals use of monitor speakers that more realistically represent the type of listening environment that John and Jill Q. Public most likely have.

The term *near-field* refers to the placement of small to medium-sized "bookshelf" speakers to each side of the working environment or on (or slightly behind) the metering bridge of a production console. These speakers (Figures 14.17 through 14.20) are generally placed at closer working distances, allowing us to hear more of the direct sound and less of the room's overall acoustics.

Figure 14.17. Yamaha NS-10M studio monitor speakers. (Courtesy of Yamaha Corporation of America, www.yamaha.com)

Figure 14.18. KRK K-RoK passive studio monitor. (Courtesy of KRK Systems, Inc., www.krksys.com)

Figure 14.19. Event 20/20bas Active biamplified system. (Courtesy of Event Electronics, www.event1.com)

Figure 14.20. *Mackie HR824 active monitor speaker. (Courtesy of Mackie Designs, Inc., www.mackie.com)*

In recent times, near-fields have become an accepted standard for monitoring in almost all areas that relate to audio production for the following reasons:

- Quality near-field monitors more accurately represent the sound that would be reproduced by the average home speaker system.

- The placement of these speakers at a position closer to the listening position reduces unwanted room reflections and resonances. In the case of untuned rooms, this creates a more accurate monitoring environment.

- These moderate-sized speaker systems cost significantly less than their larger studio reference counterparts (not to mention the reduced amplifier cost since less wattage is needed).

As with any type of speaker system, near-fields differ widely in both construction and fundamental design philosophy. It almost goes without saying that care should be taken when choosing the speaker system that best fits your production needs and personal tastes.

Small Speakers

As radio, television and Web airplay are increasingly becoming major forces in audio production and in the boosting of recording sales, it's often a good idea to monitor your final mix through a small, inexpensive speaker set that mimics the nonlinearities, distortion and poor bass response of those media. Such speakers can either be bought or easily made. Occasionally, these small speakers are incorporated into console and two-track ATR designs for easy monitoring.

Before listening to a mix over such small speakers (Figures 14.21), it's often a good idea to take a break in order to allow both your ears and your brain to recover from the prolonged exposure of listening to higher sound levels over larger speakers.

Figure 14.21. *Altec Lansing ASC-22C multimedia speakers. (Courtesy of Altec Lansing Technologies, Inc., www.alteclansing.com)*

Headphones

Headphones (Figure 14.22) are also an important monitoring tool because they remove you from the room's acoustic environment and ideally let you hear only what has or is being recorded. Headphones also offer excellent spatial positioning in that they let the artist, engineer or producer place a sound source at critically specific positions within the stereo field without reflections or other environmental interference from the room. Since they're portable, you can take your favorite headphones with you to quickly and easily check out a mix in an unfamiliar environment.

Figure 14.22. *Sony MDR-7506 professional dynamic stereo headphones. (Courtesy of Sony Electronics, Inc., www.sony.com/proaudio)*

Open air (the most common type) and sealed headphone types each have their advantages. When recording, it's generally best to use the sealed type, so as to prevent or minimize the monitor feed from leaking into the recorded track.

Your Car

Last, but not least, your (or any other) car can be a big help in determining how a mix will sound in one of modern society's most popular listening environments. You might try it out on a basic car system or a souped-up bass bomb . . . hey, you might even want to take your mix out for a spin.

CHAPTER

Product Manufacture

One of greatest misconceptions surrounding the music, visual and other media-related industries is the idea that once you walk out the door of a studio with your final master in hand (or nowadays, in your hip pocket), the creative process of producing a project is finally over. All that you have left to do is hand the DAT, CD or other medium over to a duplication facility and ta-dah . . . the buying public will be clamoring for your product. Obviously, this scenario is often far from the truth. Now that you have the program content in hand, you have to think through and implement the following additional stages if your product is to make it into the hands of the consumer.

- Mastering
- Product manufacture
- Marketing and sales

After the recording and mixing phases of a project have been completed, the next step toward getting the product out to the people is to transform the final master into a form that can be mass-produced, distributed, marketed and sold. Given the various technologies that are available today, this could take the form of a compact disc, DVD, CD-ROM, cassette tape, vinyl record or encoded Internet file. Each of these medium types has its own set of manufacturing and distribution needs that require a great deal of careful attention throughout each step of the manufacturing and/or creation process.

Choosing the Right Facility and Manufacturer

Just as recording studios have their own unique personalities and particular "sound," the right mastering and duplication facilities may also have a profound effect on a project.

If a project is being underwritten and distributed by an independent or major record label, they will generally be fully aware of their production needs and will certainly have an established production and manufacturing network in place. If, however, you are distributing the project yourself, the duty of choosing the best facility or manufacturing organization that'll fit your budget and quality needs is all yours.

A number of resources exist for finding such manufacturers. For starters, Billboard Online (www.billboard.com) provides numerous services for searching out media mastering and manufacturing facilities. They offer resource magazines (such as the Billboard Tape/Disc Directory) as well as a free online search database for *Billboard* magazine subscribers. The Mix Master Directory (Intertec Publishing, 6400 Hollis St., Suite 12, Emeryville, CA 94608; 1-510-653-3307, www.mixonline.com, e-mail: mmd@intertec.com) publishes an annual directory of industry-related products and services. This directory provides a comprehensive listing of manufacturers, recording studios, producers, engineers, music business services, etc., and is cross-referenced by product and service categories. The Recording Industry SourceBook (artistpro.com, 236 Georgia Street, Suite 100, Vallejo, CA 94590; 1-707-554-1935, www.artistpro.com) also includes a full listing for these companies. Another simple but effective resource is to look at the back-page ads in most music- and audio-related magazines.

Manufacturing facilities come in two types: those that perform and offer all of their services *in-house* (on the premises), and those that hire other business or individuals to perform various services.

Neither of these types is good or bad. They may, however, affect how and when the various stages of production will be performed. On one hand, in-house facilities are able to handle a project from beginning to end; however, these facilities are often large and expensive (meaning that there may not be one near you). On the other hand, manufacturers and duplicators that farm out projects may not have total control over their production timeline, but often offer local one-on-one service.

It's always wise to do a full background check on any type of facility. You should always ask for a promotional pack (which includes product and art samples, facility and service listing, as well as a price sheet), and you might want to ask about former customers and their contact information (so you call and check up on their experiences). Most importantly, once you've chosen a mastering and/or manufacturing facility, it's very important that you be given art proofs and test printings BEFORE the final products are mass duplicated.

Making a *test pressing* is well worth the time and money; the alternative is to receive a few thousand copies at your doorstep, only to find that they're not what you wanted—a far more expensive, frustrating, and time-consuming option. It's never a good idea to assume that a manufacturing or duplication process is perfect and doesn't make mistakes . . . remember, Murphy's Law can pop up at any time!

Mastering

The *mastering process* is an art form that uses special, high-quality audio gear in conjunction with one or more sets of critical ears to help you attain the particular "sound" and feel that you want. Working with such tools (Figure 15.1), a mastering engineer or experienced user can go about the task of ordering and shaping the various cuts on a project into a form that can be replicated into a final saleable product.

Figure 15.1. Example of a professional mastering facility. (Courtesy of Colossal Mastering, Chicago, IL— www.colossalmastering.com)

In past decades, when vinyl records ruled the airwaves and spun on everyone's sound system, the art of transferring high-quality sound from a master tape to a record was as much art as it was technology—and it still is. Since this field of expertise is well beyond the abilities and equipment of almost every recording engineer and producer, the field of mastering is left to a very select few (who are known as vinyl mastering engineers).

The art of transferring sound to CD is also still very much an art form that's often best left to those who are familiar with the tools and trade of getting the best sound out of a project. However, recent advances in computer and effects processing technology have made it much easier for producers, engineers and musicians to own high-quality hard- and software tools that are capable of creating a professional-sounding final product in the studio or on a desktop computer.

The Mastering Process

In addition to the concept of capturing the pure artistry of a production onto tape, hard disk or other medium, one of the many goals during the course of recording a project is the idea that final product will have a certain "sound." This sound might be "clean," "punchy," "gutsy" or any other sonic adjective that you're striving for. You hope that once all of the cuts have been mixed, you'll be able to sit back and say, "Yeah, that's it!" If this isn't the case, having an experienced mastering engineer "shape" the sonic character of your project through the careful use of level balancing, dynamics and EQ can save the day.

Another factor that can affect a project's sound is the reality that recordings might need to be made and/or mixed in several studios, living rooms, bedrooms and/or basements over the course of several months or years. This could mean that the various cuts would actually sound different from each other. In situations like this, it'll be even more important that you seek out someone who's experienced at the art of mastering.

The process of successful mastering can help a project:

- *Sound "right."* This is often accomplished through the use of careful EQ and overall EQ matching. As was mentioned, this process takes not only the right set of processing gear, but also experienced ears that intuitively know how the project will sound under a wide range of playing conditions.

- *Be in the right order.* Choosing a project's song order is best done by the artist and/or producer to convey the overall "feel." In addition to order, the process of setting the gap times between songs can make the difference between awkward pauses and a project that "flows" from one cut to the next.

- *Playback at the best level.* Traditionally, levels are set to the highest possible level that the medium will allow before distortion. Given the fact that record companies will always want their music to "stand out" above the rest when played on TV, the radio or the Web, compression is often used to maximize this level. Again, this is an artistic choice that often requires experience. Overcompression often leads to audible artifacts that can affect the "sound" that you worked so hard for. In some cases, light compression or even no compression is a definite alternative. Classical music lovers, for example, often spend big bucks to hear a project's full dynamic range. To tamper with this might affect sales. . . . Although the choice is up to you, the producer and/or the record company, care must be taken when dealing with the subject of dynamics.

- *Match in level.* In addition to getting the best overall level; it's often important that levels match from song-to-song. Keeping songs from sticking out like a sore thumb help to improve the flow and professionalism of a project.

Equipment that deals with the art and technology of balancing these aspects of creating a finished "master" is available in many different guises. Often, top-level mastering engineers will use specially designed EQ, dynamics and level matching gear that often isn't found in recording studio environments. Having said this, with the advent of CD burning software, dedicated processing hardware and software systems (Figures 15.2 and 15.3) are currently on the market that give musicians, producers and engineers a greater degree of control over the final mix and/or finished master than ever before.

Figure 15.2. *Focusrite MixMaster stereo dynamics, EQ and image processor. (Courtesy of Focusrite, www.focusriteusa.com)*

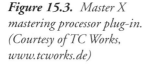

Figure 15.3. *Master X mastering processor plug-in. (Courtesy of TC Works, www.tcworks.de)*

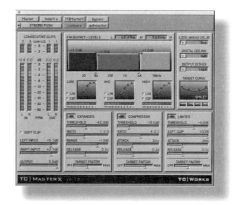

When approaching the question of "Who'll do the mastering?" it's really important that you take a long hard look at the experience level of the one who will be doing the job. Have they mastered projects before? Are you familiar with a few of these projects and did you like their sound? Can they help you to get the sound you want? Bottom line: beware of the inexperienced mastering engineer . . . even if that person is you. Make sure that there are several ears around to listen to the project and listen over several types of systems . . . and above all, be patient, be critical of the project's sound and listen to the opinions of others. Sometimes you get lucky and the mastering process can be quick and painless; other times it takes the right gear, a keen ear and lots of attention to detail.

CD Mastering

Although a project can wind up in any number of final medium forms, as of this writing, the CD (Compact Disc) is still the easiest and most widely recognized medium for distributing music. These 4¾" silvery discs (Figure 15.4) contain digitally encoded information (in the form of microscopic pits) that's capable of yielding playing times of up to 74 minutes, at a standard sampling rate of 44.1 kHz.

Figure 15.4. *The compact disc. (Courtesy of 51bpm.com, www.51bpm.com)*

A pit is approximately ½ micrometer wide, and a standard disc can hold about 2 billion pits. These pits are encoded onto the disc's surface in a spiraling fashion, similar to that of a record, except that 60 CD spirals can fit in the groove of a single long-playing record. The CD spirals also differ from a record in they travel outward from the center of the disc and are impressed into the plastic substrate, which is covered with a thin coating of aluminum (or sometimes gold) so that the laser light can be reflected. When the disc is placed in a compact disc player, a low-level infrared laser is alternately reflected and not reflected back to a photosensitive pickup. In this way, the data is modulated on the disc so that each pit edge represents a binary 1, and the absence of a pit edge represents a binary 0 (Figure 15.5). On reproduction, the data is demodulated and converted back into analog form.

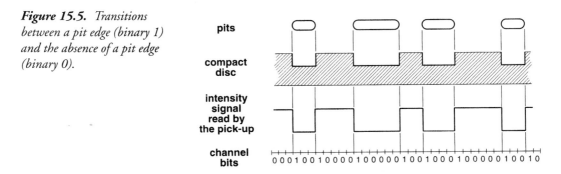

Figure 15.5. *Transitions between a pit edge (binary 1) and the absence of a pit edge (binary 0).*

Songs or other types of audio material can be grouped on a CD as index "tracks." This is done by including a subcode channel lookup table, which makes it possible for the player to identify and quickly locate tracks with frame accuracy.

Subcodes are event pointers that tell the player's microprocessor how many selections are on the disc, as well as their exact locations. At present, eight subcode channels are available on the CD format, although only two (the P and Q subcodes) are used.

Functionally, the CD encoding system splits the 16 bits of information into two 8-bit words and applies error correction in order to correct for lost or erroneous signals. The system can then translate these 8-bit words into a 15-bit word format for ease of recording onto disc (a process known as eight-to-fourteen modulation or EFM). The system then begins by constructing a methodical system for compact-disc operation known as a data frame. Each data frame contains a frame-synchronization pattern (27 bits) that tells the pickup beam where it is on the disc. This is followed by a 17-bit subcode word, 12 words of audio data (17 bits each), 8 parity words (17 bits each), 12 more words of audio, and finally 8 more parity words.

The Process

In order to translate the raw PCM of a music or audio project into a format that can be understood by a CD player, a compact-disc mastering system must be used. Modern mastering systems often come in two flavors: those that are used by professional mastering facilities, and

CD-ROM/CD hardware/software systems (Figure 15.6) that can be installed into a desktop computer to easily and cost-effectively burn CDs (see Chapter 6 for more information on CD burning technology).

Figure 15.6. *Tascam CD-R624 CD Burning System. (Courtesy of Tascam Corporation, www.tascam.com)*

Both system types allow the user to enter audio into the system, assemble tracks into the proper order and enter proper gap times between tracks (in the form of index timings). Depending on the system, cuts might also be processed using crossfades, volume, EQ and other parameters. Once assembled, the project can be "finalized" into a media form that can be directly accepted by a CD manufacturing facility. In the case of a professional system, this media could take the form of either an Exabyte type data tape or ¾" U-matic videotape (using a Sony PCM-1630 digital processor). Although the use of digitally encoded tapes has proven to be highly reliable over the years, more and more, CD pressing plants are receiving final CD master discs that have been recorded directly onto a user-created CD-Recordable (CD-R) disc.

Once the manufacturing plant has received the medium, the next stage in the process is to cut the original CD master disc. The heart of such a CD cutting system is an optical transport assembly that contains all the optics necessary to write the digital data onto a reusable glass master disc that has been prepared with a photosensitive material.

After the glass master has been exposed to a special record laser, it's placed in a developing machine that etches away the exposed areas to create a finished master. An alternative process, known as non-photoresist, etches directly into the photosensitive substrate of the glass master without the need for a development process.

After the glass or CD master disc has been cut, the compact disc manufacturing process will begin under extreme clean-room conditions. First, the glass disc is electroplated with a thin layer of electroconductive metal. From this, a negative metal master is used to create a metal mother, which in turn can be used to replicate a number of metal stampers (metal plates that contain a negative image of the CD's data surface). The resulting stampers make it possible for machines to replicate clear plastic discs (containing the positive encoded pits), which are then coated with a thin layer of foil (for increased reflectivity) and encased in clear resin for stability and protection (Figure 15.7). Once this is done, all that remains is the screen-printing process and final packaging. The rest is in the hands of the record company, the distributors, marketing and you.

Figure 15.7. Tapematic Classic 8010 Optical Disc Replicator. (Courtesy of Tapematic USA, Inc., www.tapematic.com)

CD Burning

Before the availability of CD-recording hardware and software (Figures 15.8 through 15.10), the only way to hear how your final CD would sound was to press a "one-off" disc. This meant that the CD manufacturer had to go through the entire process of creating a glass master and "cut" a single or set of CD for use by the producer, artist and record company as a reference disc. As you could guess, this was a time-consuming and expensive process that was available only to those who had big bucks resting on a project.

Figure 15.8. CDR-850 Compact Disc Recorder. (Courtesy of HHB Communications Ltd., www.hhb.co.uk)

Figure 15.9. *MasterList CD burning software. (Courtesy of Digidesign, www.digidesign.com)*

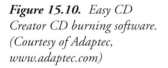

Figure 15.10. *Easy CD Creator CD burning software. (Courtesy of Adaptec, www.adaptec.com)*

In addition to the use of CD burning soft/hardware systems, dedicated CD duplicating systems (Figure 15.11) make it possible for discs to be easily and cost-effectively created for use by clients or for direct artist CD-R sales.

Figure 15.11. *StartRec 400 CD Duplicator. (Courtesy of MicroBoards Technology, Inc., www.microboards.com)*

Nowadays, the process of burning a CD on personal computers has become so widespread and straightforward that not only do artists, producers and engineers make limited disc runs for use as demos and reference discs, but this process has become popular for producing master CDs that are used to create the final glass master.

It's interesting to note that although most pressing plants receive master CDs that have been burned onto a CD-R, many of these discs don't pass the basic requirements that have been set forth for creating an acceptable *Red Book-Audio CD* (the standard industry specification).

Some of the problems associated with CD-Rs that have been burned on a desktop system include the following:

- *Excessive data errors.* This can lead to mass-produced CDs that have problems when being played on older or less reliable CD players. These errors could crop up because of such factors as hardware/software reliability problems, or media integrity.

- *Discs that haven't been "closed."* It's very important that the master disc be closed (a process that ensures that no other sessions or data can be added to the disc). Most CD mastering software packages will give you the option of closing the disc on calling up the burn to disc window.

- *Multisession discs.* Final master discs should never contain multiple sessions (in which music cuts or program material is added at a later time to an existing CD-R). The disc should be recorded in the "disk-at-once" mode (meaning the disc was burned from beginning to end, without any interruptions in the laser burning process).

- *Inaccurate index marker points.* Index markers tell the CD player where the tracks begin and end on a disc. If the markers are wrong, the program could begin early or cut off parts of a song. Once a disc has been cut, always listen to a disc to check for accurate index markers.

In fact—once you've checked the beginning and end marker points, it's always wise to critically listen to the disc from beginning to end. Never forget that Murphy's Law lurks around every corner! Once you've agreed that the CD sounds great, it's always a good idea to burn an extra master that can be set aside for safekeeping, in case something happens to the original production master.

On a final note, many CD burning programs allow you to enter disc title, artist name/copyright and track name field code information directly into the CD's subcode area (Figure 15.12). This means that important identifiers can be directly embedded within the CD itself. As a result, illegal copies will still contain the proper copyright and artist info . . . and discs that are loaded into a computer will often display these fields in a way that identifies the disc title, artist and/or copyright info, as well as the individual song titles.

Figure 15.12. *Windows CD Player showing user subcode information.*

CD Labeling

Once you've burned your own CDs, there are a number of options for printing labels onto your newly burned discs:

- *Using a felt tip pen.* This is the easiest and fastest way to label a CD-R. However, you should use water-based ink pens, since permanent markers use a solvent that can permeate the disc surface and cause damage to either the reflective layer or dye layer.

- *Label printing kits.* "Stick-on" labels that have been printed using specially designed software and an inkjet or laser combination (Figure 15.13) are one of the least expensive and most professional-looking options. You should be aware that some adhesives could leak over time or contain solvents that can adversely affect the disc.

- *CD printers.* Specially designed inkjet or laser printers (Figure 15.14) are able to print four colors onto the face of a printable (while, blank-faced) disc. This is a great option for those who burn lots of discs and want a professional look and feel.

- *Professional printing jobs.* Finally, companies exist that can custom silkscreen your discs. This involves the same screening process as for mass-replicated discs and makes sense when dealing with quantities of 100 or more. Be aware that it takes time to have them printed and is the most expensive "personalized" option, but the results are crisp, clean and professional!

Figure 15.13. *Neato 2000 CD Labeler kit. (Courtesy of Neato LLC, www.neato.com)*

In addition to label printing, programs (as well as many word processing templates) exist for creating and printing the books and trays that are part of the process of making a professional-looking CD (or DAT, cassette or almost any other medium you can think of). Most of the CD labeling kits will include software that lets you import graphics, position text, etc., to create and print out personalized, professional-looking labels (Figure 15.15).

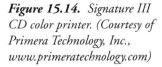

Figure 15.14. *Signature III CD color printer. (Courtesy of Primera Technology, Inc., www.primeratechnology.com)*

Figure 15.15. *Mediaface label printing program. (Courtesy of Neato LLC, www.neato.com)*

DVD

When discussing the compact disc, it would be an oversight to ignore another technology that deeply affects the video, audio and multimedia industries: the DVD.

These CD-drive compatible discs differ from the standard CD format in several ways. The most basic of these are DVDs' increased data density due to a reduction in pit size (Figure 15.16); their double-layer capability (due to the laser's ability to focus on two layers of a single side); and their double-side capabilities (which again doubles the available data size).

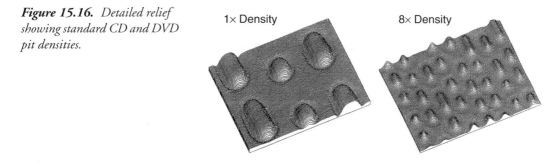

Figure 15.16. *Detailed relief showing standard CD and DVD pit densities.*

1× Density   8× Density

In addition to the obvious benefits that are gained from increasing the data density of a standard CD from 650 Mbyte to up to 17 Gbyte, DVD discs allow for much higher data transfer rates, making DVD the ideal medium for the following applications:

- The simultaneous encoding of digital video and surround-sound audio
- Multichannel surround sound
- Data- and access-intensive video games
- High-density data storage

Using standard compression techniques, this CD-based technology has breathed new life into the home entertainment industries, allowing computer fans to have access to increased game and multimedia storage capacity and home viewers to enjoy master-quality audio and video programming in a digital surround-sound environment.

Although desktop mastering software and hardware systems are available for DVD (and will almost certainly become more affordable and easy to use in the future), much of the art of mastering for DVD is often best left to professionals who are familiar with the finer points of this complex technology.

Cassette Duplication

On a worldwide basis, the prerecorded music cassette is still a strong, cost-effective medium for getting commercial music out to the masses. It's evident from the previous section that a great deal of artistry and quality control goes into the manufacture of CDs. Contrary to public misconception, the same amount of care and quality control is often upheld in cassette duplication. Currently, there are three basic methods of cassette duplication: real-time duplication, bin-loop high-speed duplication and high-speed, in cassette duplication.

Real-Time Duplication

In real-time duplication, cassette slave machines are used in a system to record a program at their normal speed of 1⅞ ips. Thus, the single side of a program lasting 30 minutes will take

30 minutes to duplicate, with the number of copies being dependent on the number of slave decks.

Most industry insiders agree that this format yields the highest cassette reproduction quality, due to the fact that the slave machines are operating at the optimum speed for the medium. The audio signal is kept within the audio bandwidth and isn't shifted into a higher one (as it is with the high-speed processes).

Using this method, the recorded tape (called the duplication master or *dupe master*) is played back from a copy of the original master onto any number of duplication tape drives (Figure 15.17). The final dupe master can exist in any format (including hard disk, DAT, reel-to-reel tape or cassette). Dual-cassette tape recorder decks or multitransport setups are often the easiest to use, as you simply place a cassette into the master drive and place the slave(s) into record.

Figure 15.17. *Example of a hard disk-based real-time cassette duplication system.*

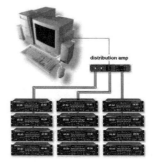

Often, professional facilities will format the duplication (dupe) master onto hard disk or onto an open-reel, 4-track tape format to ensure high quality. Recording in a 4-track stereo format means that side A's stereo program material can be played on the forward direction, while side B will play backwards. By fitting the slave recorders with 4-track record heads, both directions of the tape can be recorded at once—thereby cutting the duplication time in half (Figure 15.18).

Figure 15.18. *By recording side A of a dupe master in the forward direction and side B backwards, it's possible to duplicate both sides of a cassette tape in a single pass.*

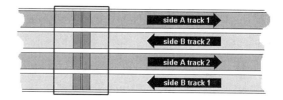

In-Cassette High-Speed Duplication

In-cassette high-speed duplication makes use of high-speed ratios (2×, 8× and 16×) by copying a dupe master (often a cassette) to a set of cassette slave recorders (Figure 15.19).

Figure 15.19. *Telex XGEN in-cassette duplicator. (Courtesy of Telex Communications, Inc., www.telex.com)*

These duplication units often are self-contained, with the master and several slaves being located in the same unit. Extra slaves can often be added on, letting you cost-effectively expand the system. The one drawback to this method is the fact that it might be limited and have trade-offs in frequency, distortion, and wow/flutter.

High-Speed Duplication

High-speed duplication takes place before the tape is loaded into the cassette shells. Using this method, the duplicated tape is recorded on reel-to-reel machines (Figure 15.20) that handle tape better and offer a higher quality than many in-cassette units.

Figure 15.20. *1000 Series tape duplicating system. (Courtesy of Versadyne International, www.versadyne.com)*

High-speed dupe masters are recorded to hard disk or a 4-track master tape that includes a recorded tone signal that corresponds to a 5–15 Hz tone at 1⅞ ips. The program is then repeatedly recorded onto any number of open-deck reel-to-reel recorders (which are designed to accept ⅛" cassette-grade tape on bulk 10½" pancake reels) at extremely high-speed ratios of up to 160×.

Since the duplication process occurs at ratios that are many times the normal speed, the frequency spectrum is also shifted upward into a high-frequency range that's well beyond the audio spectrum. The record heads, frequency response and bias currents must be specially tailored for this demanding application.

The next stage in the duplication process is to load the prerecorded programs (which are now repeatedly recorded onto bulk tape) into their cassette housings. This is accomplished using a machine known as a self-feeding cassette loader (Figure 15.21). The duplicated bulk tape is loaded into the device and a cassette-feed magazine is filled with C-0 cassettes (a cassette that's only loaded with leader tape). Next, the C-0s are dropped into the loading section and the recorded tape is automatically spliced onto the leader at the point at which the beginning sensing tone of the program appears. The loader then fast-forwards the tape, loading it into the cassette until the next tone is sensed. At that point, the loader splices the program's end onto the cassette's tail leader and ejects the cassette. The process then repeats.

Figure 15.21. *Tapematic 2002 self-feeding cassette tape loader. (Courtesy of Tapematic USA, Inc., www.tapematic.com)*

After the tapes have been loaded into their shells, the final stage in the process is to label and package the cassette for sales and distribution. With a large-scale bin-loop production system, it's possible to produce tens of thousands of cassettes each day.

Product quality control is of great importance during this process. The major emphasis often rests on the quality of the duplication master and the master–slave alignment. Distortion and saturation can often be dealt with by using peak limiting and compression sparingly when creating the master. If dynamic or EQ changes are required, the producer should be consulted.

The problem of noise might be reduced by using a digital master or by encoding a tape with Dolby noise reduction. Whenever possible (as with all manufactured products), you should always insist on listening to a duplicated "proof" copy before large quantities are made that might not be what you expected or hoped for.

Vinyl Disc Manufacture

Although the popularity of this medium has waned in recent years (as a result, of course, of the increased marketing, distribution and public acceptance of the CD), the vinyl record isn't dead. In fact, for consumers from Dance DJ hip-hipsters to die-hard classical buffs, the record is still a viable reproduction and production medium. But the truth remains that many record pressing facilities have gone out of business over the years, and there are far fewer mastering labs that currently cut master lacquers. It may take a bit longer to find a facility that fits your needs, budget and quality standards, but it's definitely not a futile venture.

Disc Cutting

The first stage of production is the disc-cutting process. As the master tape is played on a specially designed tape playback machine, its signal output is fed through a disc-mastering console to a disc-cutting lathe. Here the electrical signals are converted into the mechanical motions of a stylus and are cut into the surface of a lacquer-coated recording disc.

Unlike the compact disc, a record rotates at a constant angular velocity, such as 33⅓ or 45 revolutions per minute (rpm), and has a continuous spiral that gradually moves from the disc's outer edge to its center. The time relationship of the recorded material can be reconstructed by playing the disc on any turntable that has the same constant angular velocity as the original disc cutter.

The system of recording used for stereo discs is the 45/45 system. The recording stylus cuts a 90° angle groove into the disc surface, so that each wall of the groove forms a 45° angle with the vertical axis. Left-channel signals are cut into the inner wall of the groove and right-channel signals are cut into the outer wall, as shown in Figure 15.22. The stylus motion is phased so that a signal that's in-phase in both channels (a mono signal or a signal centered between the two channels) will produce a lateral groove motion (Figure 15.23a), while out-of-phase signals (channel difference information) will produce a vertical motion that changes groove depth (Figure 15.23b). Because the system is compatible with mono disc systems (which use only lateral groove modulation), a mono disc can be accurately reproduced with a stereo playback cartridge.

Figure 15.22. The 45/45 cutting system encodes stereo waveform signals into the grooves of a vinyl record.

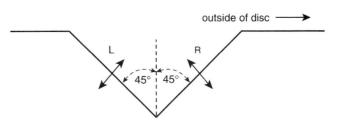

Figure 15.23. *Groove motion in stereo recording. (The solid line is the groove with no modulation.)*
a. *In-phase.*
b. *Out-of-phase.*

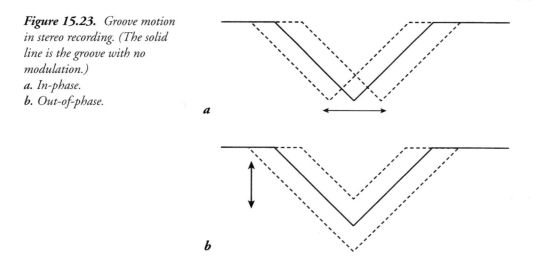

Disc-Cutting Lathe

The main components of the modern disc-cutting lathe (Figure 15.24) are the turntable, lathe bed and sled, pitch/depth control computer and cutting head. Basically, the lathe consists of a heavy, shock-mounted steel base (A). A 65-pound turntable (B) is isolated from the base by an oil-filled coupling (C), which reduces wow and flutter to extremely low levels. The lathe bed (D) moves perpendicular to the turntable in a sled fashion and is used to support and house the cutter suspension (E) and the cutter head (F).

Figure 15.24. *A disc-cutting lathe with automatic pitch and depth control.*

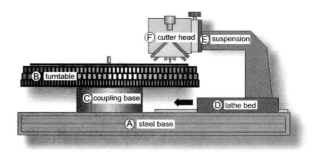

Cutting Head

The cutting head translates the electrical signals that are applied to it into mechanical motion at the recording stylus. The stylus gradually moves in a straight line toward the disc's center hole as the turntable rotates, creating a spiral groove on the record's surface. This spiral motion is achieved by attaching the cutting head to a sled that runs on a spiral gear (known as the lead screw), which drives the sled in a straight track.

The stereo cutting head (Figure 15.25) consists of a stylus that's mechanically connected to two drive coils and two feedback coils (which are mounted in a permanent magnetic field) and a stylus heating coil (that's wrapped around the tip of the stylus). When a signal is applied to the drive coils, the alternating current flowing through them creates a changing magnetic field that alternately attracts and repels the permanent magnet. Because the position of the permanent magnet is fixed, the coils move in proportion to the created field strength and move the stylus. The drive coils are wound and mounted so that energizing either one causes the stylus to move in a plane that's 45° to the left or right of vertical (depending on which coil is being driven).

Figure 15.25. *Simplified drawing of a stereo cutting head.*

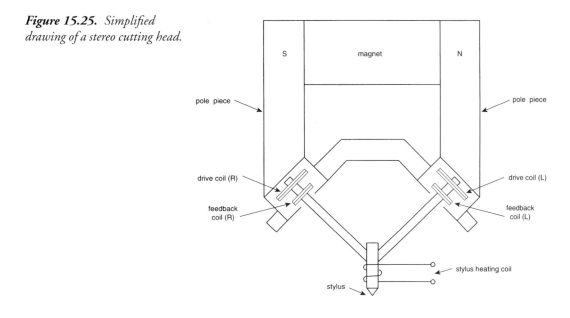

Pitch Control

The head speed determines the "pitch" of the recording and is measured by the number of grooves, or lines per inch (lpi) that's cut into the disc. As the head speed increases, the number of lpi will decrease, with a corresponding decrease in playing time. There are several methods for changing pitch:

- The lead screw can be replaced by one with a finer or coarser spiral.
- The gears that turn the lead screw can be changed to alter the lead screw's rotation speed.
- The lead screw rotation can be varied directly by changing the motor's speed (a common way to continually vary the program's pitch).

The space between grooves is called the land. Modulated grooves produce a lateral motion that's proportional to the in-phase signals between the stereo channels. If the cutting pitch is too high (too many lines per inch, making the grooves very closely spaced) and high-level signals are cut, it's possible for the groove to break through (cut over) the wall into an adjacent groove or for the grooves to overlap (twinning). The former is likely to cause the record to skip when played; the latter causes either distortion or a signal echo from the adjacent groove (due to wall deformations). Groove echo can occur even if the walls don't touch and is directly related to groove width, pitch and level.

These cutting problems can be eliminated either by reducing the cutting level or by cutting fewer lines per inch. A conflict can arise here because a louder record will sound brighter, punchier, fuller and more present (because of the Fletcher–Munson curve effect). As a result, record companies and producers are concerned about the competitive levels of their discs relative to those that are cut by others, so they often don't want to reduce the cutting level.

The solution to these level problems is to vary the pitch to cut more lines per inch during soft passages and fewer lines per inch during loud passages. This is done by splitting the program material into two paths: undelayed and delayed. The undelayed signal is routed to the lathe's pitch/depth-control computer (which determines the pitch needed for each program portion and varies the lathe's screw motor speed). The delayed signal (which usually is achieved by using a high-quality digital delay line) is fed to the cutter head, thereby giving the pitch/depth control computer enough time to change the pitch.

Pitch is divided into two categories: coarse, which refers to spacing between 96 and 150 lpi, and microgroove, which is between 200 and 300 lpi (or more). Microgroove records have less surface noise, wider frequency range, less distortion, and greater dynamic range than coarse-pitch recordings. They can also be tracked with lower stylus pressure, which results in longer life, but makes the stylus more likely to skate across the record if the turntable isn't level. The playback stylus for a stereo microgroove record must have a tip radius of 0.7 mil or less, compared to 2.5 mils ± 0.1 for coarse-groove records. The old 78-rpm and early 33⅓ rpm records were recorded with a coarse pitch; however, virtually all current records are microgroove (having an average pitch of 265 lpi). At maximum pitch, the playing time of one side of a 12-inch disc, with no modulation in the grooves, is about 23 to 26 minutes. The duration of modulated 12-inch discs cut at average levels is about 45 minutes per side, when cut with a variable-pitch lathe.

Recording Discs

The recording medium used on the lathe is a flat aluminum disc that's coated with a film of lacquer, dried under controlled temperatures, coated with a second film and dried again. The quality of these discs (called *lacquers*) is determined by the flatness and smoothness of the aluminum base. Any irregularities in the surface, such as holes or bumps, will cause similar defects in the lacquer coating. Lacquers are always larger in diameter than the final record, which makes it easy to handle them without damaging the grooves. A 12-inch album is cut on a 15-inch lacquer and a 7-inch single is cut on a 10- or 12-inch lacquer. As always, it's wise to cut a reference test lacquer to hear how the recording will sound after being transferred to disc.

The Mastering Process

Once the mastering engineer sets a basic pitch on the lathe, a lacquer is placed on the turntable and compressed air is used to blow any accumulated dust off the lacquer surface. Chip suction is started and a test cut is made on the outside of the disc to check for groove depth and stylus heat. Once the start button is pressed, the lathe moves into the starting diameter, lowers the cutting head onto the disc, starts the spiral and lead-in cuts and begins playing the master production tape. As the side is cut, the engineer changes the previously determined console settings. A photocell mounted on the tape deck senses the white leader tape between the selections on the master tape and signals the lathe to automatically expand the grooves to produce bands. After the last selection on the side, the lathe cuts the lead-out groove and lifts the cutter head off the lacquer.

This master lacquer is never played because the pressure of the playback stylus would damage the recorded soundtrack (in the form of high-frequency losses and increased noise). Reference lacquers, also called reference acetates or simply *acetates*, are cut to hear how the master lacquer will sound.

After the reference is approved, the record company assigns each side of the disc a master (or matrix) number that the cutting room engineer scribes between the grooves of the lacquer's ending spiral. This number identifies the lacquer and any metal parts made from it and eliminate the need to play the record in order to identify it. If a disc is remastered for any reason, some record companies retain the same master numbers; others add a suffix to the new master to differentiate it from the previous one.

When the final master arrives at the plating plant, it's washed to remove any dust particles and is electroplated with nickel. When the electroplating is complete, the nickel plate is pulled away from the lacquer. If something goes wrong at this point, the plating plant must order a new master from the cutting room, as the plating process damages the master.

Vinyl Disc Plating and Pressing

The nickel plate that's pulled off the master is called the *matrix* and is a negative image of the master lacquer (Figure 15.26). This negative image is then electroplated to produce a nickel positive image called a *mother*. Because the nickel is stronger than the lacquer disc, several mothers can be made from a single matrix. Because the mother is a positive image, it can be played to test it for noise, skips and other defects. If it's accepted, the mother is electroplated several times, producing stampers that are the negative images of the disc that are used to press the record.

The stampers for the two sides of the record are mounted on the top and bottom plates of a hydraulic press. A lump of vinylite record compound (called a *biscuit*) is placed in the press between the labels for the two sides. The press is then closed and heated by steam to make the

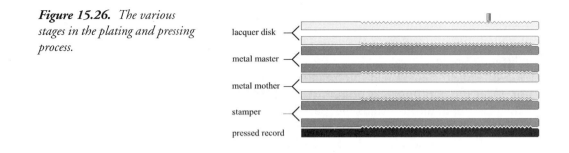

Figure 15.26. *The various stages in the plating and pressing process.*

lacquer disk

metal master

metal mother

stamper

pressed record

vinylite flow around the raised grooves of the stampers. The pressed record is too soft to handle when hot, so cold water is circulated through the press to cool it before the pressure is released. When the press opens, the operator pulls the record off the mold. The excess (called flash) is trimmed off after the disc is removed from the press . . . Once done, the disc's edge is buffed smooth and the product is finally ready for packaging, distribution and sales.

Producing for the Web

In this day of surfing and streaming media, it almost goes without saying that the Web is a huge tool that can potentially get a product or promotional word out to mass audiences. As with other media, mastering for the Internet can either be complicated, requiring professional knowledge and experience . . . or it can be a straightforward process that can be carried out from a desktop computer.

In this world of MP3/MP4, Windows Media, RealAudio, Liquid Audio and who knows what other types of streaming media, the rule that all cyber-producers live by is *bandwidth*. Basically, the bandwidth of a media and delivery/receiving system refers to the ability to squeeze as much data (often compressed data) through a wire, wireless or optical pipeline in as short a time as is possible. Transmitting the highest audio and/or video feed over a limited bandwidth will often take specialized (and often accessible) production tools. Beyond this, an even more important tool is mastery of the medium, mass marketing and good eyes and ears for design layout and media management.

Marketing and Sales

Although this section is mentioned last, it's by far one of the most important areas to be dealt with when contemplating the time, talent and financial effort that are involved in creating a recorded product. Who is my audience? Will this be distributed by a record company or will I try and sell it myself? What should the final product look and sound like? How much is this going to cost me? . . . All of these questions (and more) should be answered long before the record button is pressed and the first downbeat is played.

In this short section, I won't even attempt to cover this extremely important and complex topic. It has been fully discussed in a number of well-crafted and highly recommended books, including the following:

How to Make and Sell Your Own Recording, Diane Sward Rapaport, Jerome Headlands Press, 1999

Releasing an Independent Record, Gary Hustwit, Rock Press, 1995

The Musician's Guide to the Internet, Gary Hustwit, Rock Press, 2000

"The Sound of Money," in *Audio Recording for Profit*, Chris Stone, Focal Press, 2000

These and many other books on this subject discuss in detail the three primary methods by which a finished recording can be distributed and sold: through a major-label record company, through an independent record label, or by selling the product yourself. Each of these marketing and sales tactics represents varying degrees of financial and time outlay, as well as artistic and distribution control. No matter which avenue you choose, it's important that you fully and carefully investigate every aspect of a deal before any final commitments are made.

The Business of Music!

Although most in the industry realize that music in the modern world is a business, it's a done deal that once you get to the phase of getting your or your client's band out to the buying public, you'll quickly realize just how true this is. Building and maintaining an audience with an appetite for your product can easily be a full-time business—one where you'll encounter well-intentioned people, and also others who would think nothing of taking advantage of you or your client.

Whether you're selling your products on the street, at gigs or over the Internet, or whether you're shopping for (or have) a record label—it's often a wise decision to retain the counsel of a trusted music lawyer. The music industry is fraught with its own special legal and financial language, and having someone on your side who has insight into the language, quirks and inner workings of this unique business can be an extremely valuable asset.

CHAPTER 16

Studio Session Procedures

One of the most important concepts to be gained from this book (beyond an understanding of the technology and tools of the trade) is the fact that *there are no rules for the process of recording*. This rule holds true insofar as inventiveness and freshness tend to play a major role in keeping the creative process of making music (and music productions) alive and exciting. In the recording process, however, there are guidelines and procedures that, when followed, can help you have a smooth, professional recording session. At the very least, these procedures can help you solve potential problems when used in conjunction with three of the best tools for guiding you towards a successful project: preparation, creative insight, and common sense.

Preparation

By far, one of the most important steps to be taken when approaching a project that involves a number of creative and business stages, decisions and risks is *preparation*. Without a doubt, the best way to avoid pitfalls and to help get you, your client or your band's project off the ground is to discuss and outline the many factors and decisions that will affect the creation and outcome of that all-important "final product." Just for starters, a number of basic questions need to be asked, long before anyone presses the "REC" button:

- What is the budget for this project?
- How are we going to recoup the production costs?
- How is it to be distributed to the public . . . self-distribution? Indy? Record company?

- Will other musicians be involved?
- Do we need a producer or will we self-produce?
- How much practice will we need? . . . Where? When?
- Should we record it in the drummer's project studio or at a commercial studio?
- If we use the project studio and it works out, should we mix it at the commercial studio?
- Who's going to keep track of the time and budget . . . is that the producer's job? Or will he/she be strictly in charge of creative and contact decisions?
- Are we going to need a music lawyer with contacts and contracts? Do we know someone who can handle the job?

These are but a few of questions that should be asked before tackling a project. Of course, they'll change from project to project and will depend on the final project's scope and purpose. However, in the final analysis, asking the right questions (or finding someone who can help you ask the right questions) can help keep you from having to store 10,000 unsold CDs in your basement.

Once you've tackled as many of the important questions as you can, the next step is to sit down with the producer and discuss the many artistic, financial and business facets that are to go into the making of the project. Before beginning a recording session (possibly a week or more before), it's always a good idea to mentally prepare yourself for what lies ahead. The best way to do this is for you, the producer and the group to sit down with the engineer and discuss instrumentation, studio layout, musical styles and production techniques. This meeting lets everyone know what to expect during the session and lets everyone become familiar with the engineer, studio and staff.

Recording

From an engineering standpoint, it's a foregone conclusion that no two sessions will ever be exactly alike. In fact, in keeping with the "no rules" rule, they're often radically different from each other.

Before beginning a session, it's always a good idea to have a basic checklist that can help answer what type of equipment will be needed, the number and type of musicians/instruments, their particular miking technique (if any) and where they'll be placed (by drawing out a studio floor plan, having a prepared sheet can help the engineer/assistant engineer set up the studio in less time and help get the session off to a good start, as shown in Figure 16.1).

If analog tape machines are to be used, they should be cleaned, demagnetized, and (if necessary) aligned for the specific tape formulation that's to be used for the session (it's preferable to use the actual session tape itself). Generally it's a good idea to record a 100-Hz, 1-kHz, and 10-kHz tone at 0 VU (–10 VU if 7½ ips) on all tracks at the beginning of the tape to indicate the proper operating level. If it's necessary to overdub or mix at another studio, the tones will make it easier for the engineer to calibrate the unknown tape machine to your reference tones.

Figure 16.1. *Floor plan layout for Capitol Studios A and B (which can be opened up to create a shared space). (Courtesy of Capitol Studios, Hollywood, CA, www.capitolstudios.com)*

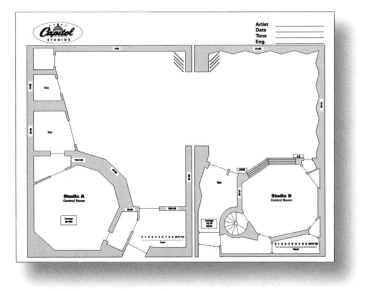

If tape-based digital recorders are to be used, make sure that the tapes have been properly formatted. Although it's true that certain recorders can record onto unformatted tapes in a single pass or in a special update mode, it's never as easy as having tapes that are preformatted. Remember, the idea is to simplify technology as much as possible (Murphy's Law is and always will be alive and well in any production facility).

Instrument placement will often vary from one studio and/or session to the next because of the acoustics of the room, the number of instruments, isolation (or lack thereof) among instruments and the degree of visual contact that's needed. If additional isolation (beyond careful microphone placement) is needed, flats and baffles can be placed between instruments in order to prevent loud sound sources from spilling over into other open mikes. Alternatively, the instrument or instruments could be placed into separate isolation (iso) rooms . . . or could be overdubbed at a later time.

During a session that involves several musicians, the setup should allow them to see and interact with each other as much as possible, so they can give and receive visual cues . . . and otherwise "feel the vibe." The instrument and mic placement, baffle arrangement and possibly room acoustics (which can often be modified by placing absorbers in the room) will depend on your personal preferences, as well as on the type of sound the producer wants. If the mics are close to the instrument and the baffles are packed in close, a tight sound with good separation will often be achieved; a looser, more "live" sound (along with an increase in leakage) will occur when the mics and baffles are placed farther away. An especially loud instrument can be isolated by putting it in an unused iso-room or vocal or instrument booth. Electronic amps that are played at high volumes can also be recorded in such a room. Alternatively, the amps and mics can be surrounded on several sides by sound baffles and (if needed) a top can be put on the "box." Another approach is to cover both the amplifier and the mic with a blanket or other flexible sound-absorbing material (Figure 16.2), so that there's a clear path between the amplifier and the mic. Separation can also be achieved by placing the softer instruments in an iso-room or by plugging otherwise loud electronic instruments directly into the console via a D.I. (direct injection) box, thereby bypassing the miked amp. Piano leakage can be similarly reduced by placing one or more mics inside it, putting the lid on its short support stick and covering it with blankets (Figure 16.3).

Figure 16.2. Isolating an instrument amplifier by covering it with a sound-absorbing blanket.

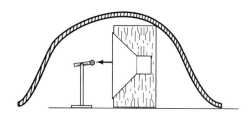

Figure 16.3. Preventing leakage from entering into a piano mic.

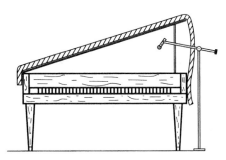

Obviously, these examples can only suggest the number of possibilities that occur during a session. For example, you might collectively choose not to isolate the instruments, and instead, place the instruments in an acoustically "live" room. This approach will require that you carefully place the mics in order to control leakage; however, the result will often yield a live, present sound. As artists, the choices belong to you, the producer, the group and (possibly) the production/distribution company.

The microphones for each instrument are selected either by experience or by experimentation and are then connected to the desired console inputs. The input used for each mic should be noted on a track sheet or piece of paper, so you can easily match each mic and instrument with its console input number.

Some engineers find it convenient to standardize on a system that uses the same mic input and tape track for an instrument type at every session. For example, an engineer might consistently plug the kick drum mic into input #1 and record it onto track #1, the snare mic onto #2, and so on. That way, the engineer instinctively knows which track belongs to a particular instrument without having to think much about it.

Electric and Electronic Instruments

Electric instruments (such as guitars) generally have mid-level, unbalanced, high-impedance outputs that can be directly recorded in the studio without using their amplifiers, via a direct box. As we've already learned, these devices convert line-level, high-impedance output signals into a low-impedance, balanced signal that can be fed directly into a console's mic preamp. Instruments often are recorded "direct" in order to avoid instrument leakage problems in the studio, to reduce noise and distortion that can occur when miking an amp or simply to get a "clean and tight" direct sound.

Electronic instruments (such as synths, samplers and effects boxes) more closely match the impedance and levels of studio equipment and can be plugged directly into a console's line-level input within the control room. In the studio, direct boxes are generally still considered to be the best way to insert a signal into the console.

Any of these instrument types can be played in the control room while listening over the main studio monitor speakers without fear of leakage. If you prefer, many direct boxes are able to split a signal into two paths: one that can be directly inserted into the console and another that can be fed to an instrument speaker in the studio (Figure 16.4). This technique makes it possible for the both direct and miked pickup to be combined and blended at the console, giving you the benefits of a clean, direct sound . . . with the added rougher, gutsier sound of a miked amp.

Figure 16.4. *Schematic for a direct (DI) box.*

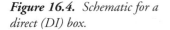
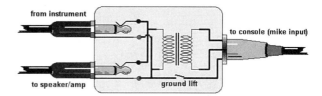

On most guitars, the best tone and lowest hum pickup for a direct connection occurs when the instrument volume control is fully turned up. Because guitar tone controls often use a variable treble rolloff, maximum control over the sound is often achieved by leaving the tone controls at the treble setting and using a combination of console EQ and different guitar pickups to vary the tone. If the treble is rolled off at the guitar, boosting the highs with EQ will often increase the pickup noise.

Drums

During the past few decades, drums have undergone a substantial change with regard to playing technique, miking technique and the choice of acoustic recording environment. The 1960s and 1970s placed the drum set in a small isolation room called a drum booth. This booth acoustically isolated the instrument from the rest of the studio and had the effect of tightening the drum sound because of the limited space (and often the dead acoustics). The drum booth also isolated the musician from the studio, and this physical separation often caused the musician to feel removed and less involved in the action. Today, many engineers and producers have moved the drum set out of smaller iso-rooms and back into larger open studio areas where the sound can fully develop and combine with the studio's own acoustics. In many cases, this effect can be exaggerated by placing a distant mic pair in the room (a technique that often produces a fuller, larger-than-life sound).

Before a session begins, the drummer should tune each drum while the mics and baffles for the other instruments are being set up. Each drumhead should be adjusted for the desired pitch and for constant tension around the rim by hitting the head at various points around its edge and adjusting the lugs for the same pitch all around the head. Once the drums are tuned, the engineer should listen to each drum individually to make sure that there are no buzzes, rattles or resonant after-rings. Drums that sound great in live performance may not sound nearly as good when being close miked. In a live performance, the rattles and rings are covered up by the other instruments and are lost before the sound reaches the listener. Close miking, on the other hand, picks up the noises, as well as the desired sound. Because of these differences, some studios will have their own drum set that's optimized for studio recording.

If tuning the drums doesn't bring the extraneous noises or rings under control, duct or masking tape can be used to dampen them. Pieces of cloth, dampening rings, paper towels or a wallet can also be taped to the head in various locations (which is determined by experimentation) to eliminate rings and buzzes. Although head damping has been used extensively in the past, present methods use this damping technique more discreetly and will often combine dampening with proper design and tuning styles (all of which are the artist's personal call).

For studio recording, it's best to remove the damping mechanisms that are built into most drum sets because they apply tension to only one spot on the head and unbalance its tension. These built-in dampers often vibrate when the head is hit and are a chief source of rattles. Removing the front head and placing a blanket or other damping material inside the drum so that it's pressing against the head can damp the kick drum. This dampening effect can be varied by adjusting how much the material is pressed against the head (which can be varied from being a resonant boom to a dull thud). Kick drums are usually recorded with their front heads removed, while other drums are recorded with their bottom heads either on or off. Tuning the drums is more difficult if two heads are used because the head tensions often interact; however, two heads will often produce a more resonant tone. After the drums are tuned, the mikes can be put into position. Remember to keep the mics out of the drummer's way, or they might be hit by a stick or moved out of position during the performance.

Setup

After the instrument, baffle and mikes have been roughly placed, headphones that are equipped with enough extra cord to allow free movement should be distributed to each player. Before assigning tape tracks, it's always a good idea to confer with the producer and/or musicians to find out how many instruments are to be used on the song, including overdubs. This helps to determine how many tracks will need to be left open. The number of tracks often influences the number and way that mics are to be assigned to the available tracks . . . this holds especially true for drums. If a large number of instruments are to be recorded and overdubbed to a recorder with a limited number of tracks, you might want to consider options such as the following:

- Grouping several instruments and record the composite mix to a limited number of tracks.

- Deciding to record the instruments onto separate tracks and then mix the composite mix (or mixes) to a new master tape (this will at least give you a more options should the group mixes need to be changed at a later date).

- Deciding to use (or rent) a recorder that has more tracks or integrate one or two additional modular digital multitracks into the recording setup. In this day and age, there are often several options for adding on tracks.

When all the mics have been set up, the engineer can use his/her setup sheet to label each input strip with the name of the corresponding instrument. Label strips (which are often provided just below each channel input fader) can be marked with an erasable felt marker, or you could use the age-old tactic of rolling out and marking on a strip of paper masking tape (ideally, a kind that doesn't leave a tacky residue on the console surface).

Once this is done, you can assign the mic/line channels to their respective tracks, making sure to fully document the assignments and other session info on the song or project's *track sheet* (Figure 16.5), which from this point on must be stored in its proper tape box.

Figure 16.5. *Example of a studio track log used for instrument/track assignments. (Courtesy of Ocean Way Recording, www.oceanwayrecording.com)*

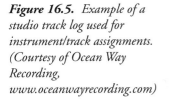

After all the labeling has been completed, the engineer can begin the process of setting levels for each instrument/mic input by asking each musician to play solo or by asking for a complete run-through of the song. By placing each of the channel and master output faders to their unity (0 dB) setting and starting with the EQ settings at the flat position, the engineer can then check each of the track meter readings, adjust the mic preamp gain and listen for preamp overload and (if necessary) insert a gain pad to eliminate distortion.

After these levels have been set, a rough headphone mix can be made so that the musicians can hear themselves. Mic choice and/or placements can be changed or EQ settings can be adjusted (if necessary) to obtain the sound the producer wants on each instrument, and dynamic limiting or compression can be carefully inserted and adjusted for those channels that require dynamic attention. (It's important to keep in mind that it's easier to change the dynamics of a track later during mixdown—particularly if the session is being recorded digitally—than to undo any changes that have been made during the recording phase.)

Once this is done, the engineer and producer can listen for and eliminate any extraneous sounds (such as buzzes or hum from guitar amplifiers or squeaks from drum pedals). This process of selective listening can be eased by soloing the individual tracks as needed. If several mics are to be grouped to one or more tracks, the balance between them should be carefully set at this time.

After this procedure has been followed for all the instruments, the musicians should do a couple of practice ("rundown") songs so that the engineer and producer can listen to how the instruments sound together before being recorded to tape (if tape's not a major concern, you might consider recording these tracks, as they might turn out to be your best takes . . . you just never know). During the rundown, you might consider soloing the various instruments and instrument combinations and finally, monitor to all the instruments together. Careful changes in EQ can be made (or noted in the track sheet for future reference) to compensate for one instrument's covering up another . . . thereby helping them to blend better.

While the song is being run down, the engineer can make final adjustments to the recording levels and the headphone monitor mix. He/she can then check the headphone mix either by putting on a pair of headphones connected to the cue system or by routing the mix to the monitor loudspeakers. If the musicians can't hear themselves properly, the mix should be changed to satisfy their monitoring needs, regardless of their recorded levels. If several cue systems are available, multiple headphone mixes can be built up to satisfy those who need different balances. During loud sessions, the musicians might ask you to turn their levels or the overall headphone level up, so they can hear the mix above the ambient room leakage. It's important to note that high sound-pressure levels can cause the pitch of instruments to sound flat, so musicians might have trouble tuning or even singing with their headphones on. To avoid these problems, tuning shouldn't be done while listening through phones. The musicians should play their instruments at levels that they're accustomed to and adjust their headphone levels accordingly.

The importance of proper headphone levels and a good cue balance can't be stressed enough, as they can either help or hinder a musician's overall performance. The same situation exists in the control room with respect to high monitor-speaker levels: some instruments might sound out of tune, even when they aren't . . . and ear fatigue can easily impair your ability to properly judge sounds and relative balance.

During the practice rundown, it's also a good idea to ask the musician(s) to play through the entire song so you'll know where the breaks, bridges and any point that's of particular importance might be. Making notes and even writing down or entering the timing numbers (possibly into the transport autolocator) can help speed up the process of finding a section during a take or overdub. You can also pinpoint the loud sections and possibly avoid any overloads. If compression or limiting is used, you might keep an ear open to ensure that the instruments don't trigger an undue amount of gain reduction. Even though an engineer might ask each musician to play as loudly as possible, they'll often play even louder when performing together. This fact may require further changes in the mic preamp gain, record level and compression/limiting threshold. Separation between the instruments can be checked by soloing each mic and listening for leakage. If necessary, the relative positions of mics, instruments and baffles can be changed at this time.

Once recording is underway, at the beginning of each performance, the name of the song and a take number are recorded to tape for easy identification (a process that's often referred to as *slating* the tape). A *take sheet* should be carefully kept to note the position of the take on a tape. Comments are also written onto this sheet to describe the producer's opinion of the performance, as well as other information of importance.

During the recording, the engineer watches the level indicators and (only if necessary) controls the faders to keep from overloading the tape. It's also a part of the job to act as another set of production ears by listening both for performance and quality factors. If the producer doesn't notice a mistake in the performance, the engineer just might catch it and point it out. The engineer should try to be helpful and remember that the producer (and/or possibly the band) will have the final say, and that their final judgment of the quality of a performance or recording must be accepted.

When a take is played back, the tape or other media is recued and the monitor system is switched from the program into to the playback/monitor mode. The musicians can then listen to the performance over their headphones or through the studio speakers, or they may come into the studio for a well-deserved listening break.

Overdubbing

Overdubbing (Figure 16.6) is used to add more instruments to a performance after the basic tracks have been recorded. These additional tracks are added by monitoring the previously recorded tape tracks (usually over headphones) while simultaneously recording new, doubled or augmented instruments and/or vocals onto one or more available tracks.

In an *overdub* (*OD*) session, the same procedure is followed for mic selection, placement, EQ and levels as occurs during the recording session. If only one instrument is to be overdubbed, the problem of having other instrument tracks leak into the OD track won't exist. However,

Figure 16.6. *Overdubbing*
allows instruments to be added
to existing tracks on a multitrack
recording medium.

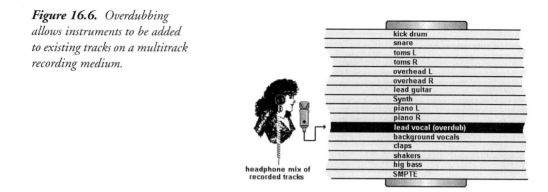

headphone mix of
recorded tracks

leakage can occur if the musician's headphones are too loud or aren't seated properly on his or her head.

If the recorder to be used is analog, it should be placed in the master sync mode (thereby reproducing the previously recorded tracks from the record head in sync). The master sync mode is set either at the recorder or using its autolocator/remote control. Monitor switching between source (monitoring signals being fed to the recorder or console) and tape/sync (monitor signals coming from the playback or record/sync heads) will often be done automatically by the tape machine. The control room monitor mix should prominently feature the instrument(s) being recorded, so mistakes can be easily heard. As during the initial session, the multitrack headphone mix can be adjusted to fit the musician's personal taste.

Should a mistake or bad take be recorded onto an overdub track, it's a simple matter to rewind the tape and re-record over the unwanted track. If only a small part of the take was bad, it's easy to *punch-in* (silently enter the record mode on that track while the tape is rolling in the record-ready mode) and record over the unwanted portion of the take (Figure 16.7). After the section has been corrected, the track can be *punched-out* of record, thereby silently exiting record mode and allowing the originally recorded track signal to play out.

Figure 16.7. *Punch-ins let you*
selectively replace material and
correct mistakes.

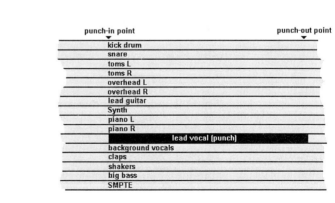

An additional technique that can be used in conjunction with or as an alternative to punching-in over a recorded track is to overdub the instrument onto another available track or set of tracks. The advantage of recording onto another track (if they're available) is that a good or marginal take can be saved and the musician can try to improve the performance, rather than having to erase the previous take in order to improve it. When several tracks of the overdub have been saved, different sections of each take might be better than others. These can be combined into a single, composite take by playing the tracks back in the sync mode, mixing and muting them together at the console (often with the help of automation) and recording them on another tape track (Figure 16.8). It should also be noted that the job of recording multiple takes and then combining then into a single composite is also well suited to a digital audio editor, which can often record unlimited takes (without using up physical tape tracks) and then combine them into a composite soundfile with relative ease.

Figure 16.8. A single composite track can be created from several partially acceptable takes.

Another procedure that's an extension of the composite process (when using a tape-based system) is the concept of *track bouncing* or *ping-ponging*. The process of bouncing tracks is often used to mix a group or an entire set of basic tracks down to one track, a stereo track pair or several grouped tracks. Bouncing can be performed either to make the final mixdown easier (by grouping instruments together onto one or more tracks, as seen in Figure 16.9a) or to open up needed tape tracks by bouncing similar instrument groups down to one or more tracks. This frees up the originally recorded tape tracks for overdubs.

Suppose that we have two ADAT MDMs (giving us a total of 16 tracks) with which to do a demo recording. During the basic tracks, let's say that we decided to record the drums onto all 8 tracks of the ADAT #1 and then record a stereo piano, bass, and lead guitar (4 tracks) onto separate tracks of ADAT #2. After these 12 tracks have been recorded, we could then go back and mix the drums down to a stereo pair of the 4 available tracks on ADAT #2. Once this is done, the original drum tape can be set aside (whenever possible, it's always a good idea to keep the original track mixes, in case you need them in the future or in case of an accidental punch error . . . Murphy's Law, you know!). Once we've put a new tape into ADAT #1, we can go about the task of recording onto the 10 newly opened tracks (Figure 16.9b).

Figure 16.9. *Track bouncing is a common production technique that's used in multitrack recording.*
a. *Instruments can be grouped together onto one or more tracks.*
b. *Bouncing can be used to expand the number of available tracks and thus allow for additional overdubs.*

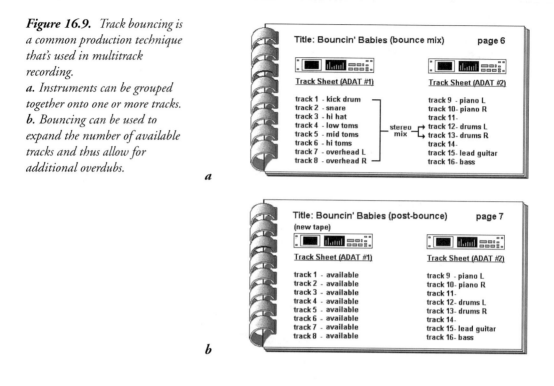

a

b

Mixdown

After all the tracks for a song have been recorded, the multitrack tape (or other medium) can be *mixed down* to mono, stereo or surround for subsequent duplication and distribution to the consumer. When beginning a mixdown using analog multitrack and/or mastering machines, it's customary to demagnetize, clean and align them. Once this is done, the engineer should record 0-VU level tones at the head of the mixdown reel at 1 kHz, 10 kHz, and 100 Hz. This makes it possible for the mastering engineer to align the tape playback machine at these reference frequencies, resulting in a proper playback EQ and levels. If Dolby noise reduction is used, the Dolby level tone should be added to these alignment tones. When mastering to a digital medium (such as DAT), it's only necessary to record a 1-kHz reference tone at the manufacturer's recommended reference level (usually 0 dB or –12 dB).

Once you're ready, the console can be placed in the mixdown mode (or each input module can be switched to the line or tape position) and the fader label strips can be labeled with their respective instrument names. Channel and group faders should be set to unity gain (0 dB) and the master output faders should likewise be set, with the monitor section being switched to feed the mixdown signal to the appropriate speakers.

The engineer can then set up a rough mix of the song by adjusting the levels and the spatial pan positions. The producer then listens to this mix and might ask the engineer to make specific changes. The instruments are often soloed one by one or in groups, and necessary EQ changes can be made. The engineer and producer then begin the cooperative process of "building" the mix into its final form. Compression and limiting can be used on individual instruments as required (either to make them sound fuller and more consistent in level or to prevent them from overloading the mixdown tape when raised to the desired level in the mix). At this point, the console's automation features can be used (if available). Once the mix begins to take shape, reverb and other effects types can be added to shape and add ambience in order to give close-miked sounds a more "live," spacious feeling and to help blend the instruments.

If the mix isn't automation-assisted and the fader settings have to be changed during the mix, the engineer will have to memorize the various fader moves (often noting the tape counter to help keep track of transition times). If more changes are needed than the engineer can handle alone, the assistant, producer or artist (who probably knows the transition times better than anyone) can help by controlling certain faders or letting you know when a transition is coming up. It's best, however, if the producer is given as few tasks as possible, so that he/she can concentrate fully on the music rather than the mechanics of the mix. The engineer listens to the mix from a technical standpoint to detect any sounds or noises that shouldn't be present in the mix. If noises are recorded on tracks that aren't used during a section of a song, these tracks can be muted until needed. After the engineer practices the song enough to determine and learn all the changes, the mix can be recorded and faded at the end. The engineer might not want to fade the song during mixdown, as the mix can be transferred to a digital audio editor (which can perform a fade much more smoothly than even the smoothest hand can).

The different takes of a mix should be identified by slating the tape (which audibly identifies the recorded take) and a complete, detailed master take sheet should be maintained (which is used to note any useful comments). When an acceptable take is recorded on an analog machine, the engineer might want to splice white paper leader tape at both the take's head and tail, so it can be found easily. The leader can be inserted with either rough or tight accuracy depending on whether the producer is in a hurry to go on to the next mix. It should be noted that the practice of leadering tape has declined with the advent of DAT recorders (which have built-in index search and absolute time features), as well as digital audio editors.

Levels should be as consistent as possible between different takes and songs, and it's often wise to monitor at consistent, moderate listening levels . . . this is because variations in the ear's frequency response change at different sound-pressure levels might result in inconsistencies between song balances. Often it's a simple matter to smooth out differences in level when using a digital audio editor for the final order sequencing. However, it's always wise to pay attention to the details of relative volume. Ideally, the level used should be the same as might be heard at home, over the radio or in the car. Because most people listen to music at a moderate volume, monitor levels between 70 and 90 dB SPL should be used (although certain music styles might "want" to be listened to at higher levels). Once the final mix or completed project master is made, you might want to listen to it over different speaker systems (ranging from the smallest to the biggest you can find). If time and budget permits, you could even make copies for the producer and band members to listen to at home and in their cars. In addition, the mix should be tested for mono–stereo compatibility (when using any mix format) to see what changes in instrumental balances might have occurred. If there are any changes in frequency balances or if phase becomes a problem when the mix is played in mono, the original mix might have to be modified.

Editing

After all the final mixes for a recording have been completed, a final edited master can be assembled. The producer and artists begin the process by listening to the songs and deciding on a final sequence, based on their tempos, musical keys, how they flow into one another, and which songs will best attract the listener's attention. Once this is done, the process of assembling the final master can begin.

Analog Sequence Editing

Although the process of assembling a final master entirely in the analog domain is done less frequently than that of loading the final mixes into a digital audio editor, it *is* still done. During this process, the engineer edits the original mixes out from their reels and begins the process of splicing them together into a final sequence on a master reel set. At this time, the level test tones (which were laid down at the beginning of the mixdown session) should be placed at the beginning of side one. Once this is done, the mix master in/out edits should be tightened (to eliminate any noise and silence gaps). This is done by listening to the intro and outro at high volume levels, while the heads are in contact with the tape (this might require that you place the transport into the edit mode). The tape can then be moved back and forth (a process known as "jogging" or "rocking" the tape) to the exact point where the music begins (for the intro) and after it ends (for the outro). Once the in (or out) point is positioned over the playback head, the exact position is marked with a grease pencil. If there's no noise directly in front of this spot, it's a good practice to cut the tape half an inch before the grease pencil mark as a safety precaution against editing out part of the first sound. If there is noise ahead of the first sound, the tape should be cut at the mark and the leader should be inserted at that point. Paper (rather than plastic) leader tape is used because plastic often causes static electricity pops. Blank tape can be also used, but it doesn't let you see the divisions between the songs on a reel and produces tape hiss during the breaks, rather than silence.

The tail of the song might need to be monitored at even higher volume levels because it's usually a fade-out or an overhang from the last note, and is therefore much softer than the beginning of the song. The tape is marked and cut just after the last sound dies out to eliminate any low-level noises and tape hiss.

The length of time between the end of the song and the beginning of the next can be constant, or the timings can vary according to the musical relationship between the songs. Decreasing the time between them can make a song seem to blend into the next (if they're similar in mood), or could create a sharp contrast with the preceding song (if the moods are dissimilar). Longer times between songs help the listeners get out of the mood of the previous song and prepare them for hearing something that might be quite different.

When the sequencing is complete, one or two analog or DAT backup copies should be made of the final, sequenced master, before it leaves the studio. These copies serve as a backup in case the original mixes are lost or damaged. Several CD-R, DAT and/or cassette copies should also be made for the producer, artist and record company executives for final approval.

Digital Sequence Editing

With the advent of digital audio editing systems, the relatively cumbersome process of sequencing music tracks in the analog domain using magnetic tape has given way to the faster, easier and more flexible process of editing the final masters from hard disk. Using this system, all the songs and initial test tones in a given project can be recorded to disk (either as a single, continuous file or as a number of individually named soundfiles). When a computer-based editor is used (Figure 16.10), the start and end points can be located for each song and can define a region (or set of edited regions) that can then be assembled into a final song version. Once this is done, each song can be individually processed using EQ, overall level, dynamics, etc. . . . in order to arrive at a finely tuned final master take. Finally, the proper fades and silence gaps can be entered into the edit session and the completed master can be transferred to a master DAT and/or CD-R. It's always a good idea to make at least one master copy of the final product from the original files as a backup (as well as any copies for those involved in the project).

Figure 16.10. *A digital audio editor can be used to edit, sequence and process final takes into a finished project. (Courtesy of Syntrillium Software Corp., www.syntrillium.com)*

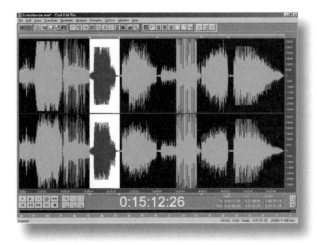

If your system is capable of backing up the soundfile and session data, it's always a good idea to do so, just in case the record company, producer or artist wants to make changes at a later date. This simple precaution could save you a lot of time and frustration as the program data, automation and playlist instructions would be fully restored to disk.

Once all this is done and you've left the studio, then the real work begins . . . getting the final product into the hands of the buying public—a process that's truly not for the faint of heart!

A Final Word on Professionalism

Before we close this chapter, there's one more subject that I'd like to touch on—perhaps the most important one of all . . . professional demeanor. Without a doubt, the life and job of a typical engineer isn't always an easy one. It often involves long hours and extended concentration with people who more often than not are new acquaintances . . . in short, it can be a high-pressure job. On the flip side, it's one that's often full of new experiences (one of the aspects that people like most about the business of sound production is the fact that the demands often change every day) and people who feel passionately about their art and chosen profession.

It's been my observation (and that of many I've known) that the best qualities that can be exhibited by an engineer, assistant engineer and others in the biz are a willingness to experiment and be open to new ideas (flexibility); a sense of humor; an even temperament (this often translates as patience); and communication and psychology (being able to convey and understand the basic nuances of people from all walks of life and with many different temperaments).

The best advice I can give is to be open, be patient . . . and above all, *be yourself.* Be extra patient with yourself. If you don't know something, ask. If you made a mistake (trust me, you will . . . we all do), admit it and don't be hard on yourself. It's all part of the process of learning and gaining experience.

This last piece of advice might not be as popular as the others: It's important to be open to the fact that there are many, many aspects to the music and sound production . . . and you may find that your calling might be better served in another branch of the biz. That's totally OK. Change is an important part of any creative process . . . even life!

17

CHAPTER

Yesterday, Today and Tomorrow

I'm sure you've heard the phrase, "Those were the good old days." I've usually found it to be a catch-all term that refers to a time in one's life that had a sense of great meaning, relevance and all-around fun. Personally, I've never met a group of people who seem to bring that sense of relevance and fun with them into the present more than music and audio professionals, enthusiasts and students . . . and the fact that many of us refer to the tools of our profession as "toys" says a lot about the way we view our work. Fortunately, I was born into that clan and have reaped the benefits all my life.

Music and audio industry professionals, by necessity, keep their noses to the grindstone. But market forces and personal visions often cause them to keep one eye focused on future technologies—whether they are new developments (such as advances in digital audio technologies), rediscovered ones that are decades old (such as the reemergence of tube technology and the reconditioning of older console designs that sound far too good to put out to pasture) or future vaporware technologies that excite the imagination. Such is the time paradox of a music and audio professional, which leads me to the final task . . . addressing the people and technologies in the business of sound recording: yesterday, today and tomorrow.

Yesterday

I've always looked at the history of music and sound technology and applied techniques with a sense of awe and wonder . . . although I can't really explain why. Like so many in this industry, I tend to feel shivers run up my back when I see a wonderful old mic or an original tube compressor. When I read about the Redd 37 (the console at Abbey Road that recorded many of the early Beatles albums, including *Sergeant Pepper*) or see an original Ampex 200 (the first commercially available professional tape machine), I get all giggly and woozy. I experience the same sense of awe when I read about my personal historical heroes such as Alan Dower Blumlein (Figure 17.1), who was instrumental in developing stereo, the television camera, radar . . . his list of accomplishments is second only to those of Edison; or Mary C. Bell (Figure 17.2), who was probably the first woman sound engineer; or the late, unsung hero John (Jack) T. Mullin (Figure 17.3), who stumbled onto a couple of German Magnetophones at the end of WWII and brought them back to the United States. With the help of Alexander Poniatoff and Bing Crosby, John and his machines played a crucial role in bringing the tape recorder into commercial existence (Figure 17.4).

Figure 17.1.   *The life of Alan Dower Blumlein (truly, one of my long-time heroes) has been published by Focal Press. (Courtesy of Focal Press, www.focalpress.com)*

Figure 17.2.   *Mary C. Bell in NBC's dubbing room #1 (April 1948)—inspecting broadcast lacquer discs for on-air programs. (Courtesy of Mary C. Bell)*

Figure 17.3. *John T. Mullin (on the left) proudly displaying his two WWII vintage German Magnetophones, which were the first two tape-based recorders in the United States. (Courtesy of John T. Mullin)*

Figure 17.4. *Early Ampex tape machines. (Courtesy of Mary C. Bell)*

Fortunately, this newfound interest in technological history has manifested itself in the form of an almost frenzied lust for old gear or new toys based on decades-old technologies. At the forefront of this latest craze is the return to tube technology, particularly in the form of new and used tube condenser microphones, mic preamps and signal processing gear (such as EQ, compressors and limiters).

Beyond the simple fact that many of the older mic designs were carefully crafted and have a particular sound that's very different from that of many of their modern counterparts, tube electronics have a sound that's inherently different from those that are based on IC or transistor technology. For starters, when a tube is overdriven to the point of clipping . . . instead of having sharply distorted edges, the tube's distortion will generally have a smoother, more "rounded" edge. This results in a signal that's musical sounding, has less odd-harmonic (square wave) distortion and generally yields a "fatter," less grating sound that's sought after by most musicians, producers and engineers.

In addition to the resurgence of tubes, older recording and studio design techniques have slowly come back into vogue. For example, major music studio designs have swung back towards the style of the 1930s through 1950s of having larger rooms that sound more "live." This swing back in time is seen as a backlash against dry, lifeless acoustic recordings that often resulted from recording in a small, acoustically dead environment (circa 1960s through 1970s, as shown in Figure 17.5). As a result, these larger and acoustically "live" designs allow distant miking techniques to be used in the miking of drums, strings . . . virtually any acoustic instrument that can benefit from the live, ambient quality of a distant-miked recording.

Figure 17.5. *Gilfoy Sound Studios, Inc., circa 1972. (Courtesy of Jack W. Gilfoy)*

In short, there are a lot of benefits to be gained from looking to the past as well as to the future when choosing the tools and toys of our trade. A wealth of experience in design and application has been laid out for us. It's simply there for the taking . . . all we have to do is search it out and put it to good use.

Today

Every so often, major milestones in technological development come along that affect almost every facet of technology. Such milestones have ushered us from the Edison and Berliner era of acoustic recordings, into the era of broadcasting, electrical recording and tape, then into the environment of the multitrack recording studio (Figure 17.6) and finally into the age of the computer, digital media and the Web.

When you get right down to it, the foundation of the information and digital age was laid with the invention of the integrated circuit. The IC has drastically changed the technology and techniques of present-day recording by allowing circuitry to be easily designed and mass produced at a fraction of the size and cost of equipment made with tubes or discrete transistors.

Figure 17.6. *Little Richard at the legendary LA Record Plant's Studio A (circa 1985) recording "It's a Matter of Time" for the Disney film* Down and Out in Beverly Hills. *(Courtesy of the Record Plant Recording Studios, photo by Neil Ricklen)*

When this is combined with advances in digital technology, we've seen the development of new equipment and media that have affected the ways in which music is produced. Integrating cost-effective yet powerful production computers with digital mixing systems, modular digital multi-tracks, MIDI synths/samplers, music-related software, digital signal processors, etc., gives us the recipe for having a powerful production studio in your homes, apartments or personal places of business. Such project and desktop music studios have made it possible for more and more people to create and distribute their own music with an unprecedented degree of ease, quality and cost-effectiveness.

Peter Gotcher (Digidesign co-founder) was one of the first to envision the creation of a cost-effective "studio-in-a-box" (Figure 17.7). This conceptual spark, which started a present-day Fortune 500 company, helped to create a system that would offer the power of professional hard-disk-based audio at a price that most music, audio and media producers could afford (you have to realize that previous systems started at over $100,000!). His goal (and that of countless others since) has been to create an integrated system that would link together the many facets that go into audio and audio-for-visual production, via a personal computer. Years later, this concept has transformed the very way in which music, studio and desktop audio production is being produced.

Figure 17.7. *Digidesign's Pro Control console/Pro Tools-based recording system. (Courtesy of Digidesign, www.digidesign.com)*

It goes without saying that another present-day mover and shaker is the Internet. It made the creation of this book *much* easier for me as a writer (researching equipment is a relative breeze, and companies are able to e-mail me photos in a day, so I no longer need to gather them via snail-mail over a period of months). Also, personal and corporate Web sites can be instantly browsed (well, almost instantly) . . . music can be up- or downloaded. . . . In short, the Internet is an amazing tool that's changing the face of market forces and the way we communicate.

As I write this, one of the major issues facing the music industry is related to the distribution of copyrighted commercial music over the Web. Even the cloudiest crystal ball can foresee that cyberspace will be the next frontier for distributing media (of both the "for fee" and "for free" types). The question is simply a matter of how, using what medium and how fees and royalties would be collected. The early part of this millennium should prove to be a time of shakeup for the many ways in which independent artists can create and distribute their music, as well as a time of restructuring the ways in which established record companies will market, protect and distribute their wares. Even though I have strong feelings about the need to protect the artist's intellectual property, I can't help but feel that cyberspace will be (and already is) a source of increased visibility, viability and hope for the budding individual as well as for the established corporate world of the music biz.

Before I make the jump into the realm of tomorrow, I'd like to take a moment to honor one of the greatest forces driving humanity today (besides sex) . . . information and the dissemination thereof. Through the existence of quality books, trade magazines, university programs, work-shops and the Web, a huge base of information on almost any imaginable subject is now being distributed to and understood by a greater number of aspiring artists and technicians than ever before. These resources often provide a strong foundation for those who are attending accredited schools, as well as those attending the school of hard knocks. No matter what your goals are in life (or in the business of music), I urge you to jump in and read, surf, skip through pages . . . just keep your eyes and ears open for new sounds, technologies and experiences. I promise, the increased knowledge will be well worth the time and effort. So dive in and have fun!

Tomorrow

Usually, I tend to have a decent handle on the forces that might or might not help shape the sounds and toys of tomorrow—but, wow . . . it's next to impossible for anyone to make specific predictions in this fast-paced world. Today, there are simply more choices in information and entertainment than reading a book or watching Lucy and Desi on the tube. Now, we can inter-act with others in a multimedia computer environment (Figure 17.8) that allows us to be more than just spectators—it lets us participate and share our thoughts with others . . . and that leads to a faster pace of communication and growth.

This idea of intercommunication through multiple media has already begun the drive of almost every developing technology toward an e-based commerce that's based on the distribution and sharing of media, marketing and information. In addition to the quest for broadband informa-

Figure 17.8. *Reaching out into cyberspace.*

tion (allowing for the distribution of large amounts of data in short periods of time), the obvious idea of copy protection, privacy and accountability will be paramount in the shaping of e-commerce in the not-so-distant future.

Will we still have Mom and Pop stores? . . . You bet! Will fledgling artists still need and want to gig on the streets and in the bars? . . . Yup! Will there still be record stores with tons of CDs in racks? . . . I sure hope so. It's important that we humans interact—whether in a shop, at the mall, at a convention . . . now we can simply have one more place where people of all ages can go to see and be seen by the masses.

On the technological front, digital has personally given me several fully professional recording/playback systems . . . A few of these are small enough to be placed in a backpack. My music computer has 36 analog inputs, 12 analog outputs, 4 digital I/Os, tons of MIDI ports, as well as a partridge in a pear tree. My favorite thing about my system is that it has a digital, fully configurable, automated computer-based mixer that can run more plug-ins than you can shake several sticks at. Personally, I love the computer's ability to be a chameleon . . . or a horse of a different color . . . one moment it's a sequencer, the next it's a hard disk recorder, next it's a word processor, then a CD player . . . and on and on. It frees me to be creative in an amazing number of ways. But that's just me . . . other ways of working and expressing one's art are equally valid.

Happy Trails

On a final note, I'd like to paraphrase Max Ehrmann's "Desiderata," when he urges us to keep interested in our own career, as it's an important possession in the changing fortunes of time. Through my work as a musician and a writer, I've been fortunate enough to know many fascinating, talented and fun people. For some strange reason, I was literally born with a fascination and love for music and music technology and have developed a strong interest in the music biz. By "keeping interested in my own career" and working my butt off—while having brushes with several cases of extreme luck—I've been able to turn this love for music and sound technology into a successful career.

To me, all of this comes from following your bliss (as some might call it), listening to reason (both your own and that of others you trust) and doing the best work that you can (whatever it might be). As you know, hundreds are waiting in line to make it as an engineer, a successful musician, a producer, etc.

So how does that *one* person make it? By following the same directions as it takes to get to Carnegie Hall—practice! Or as the tee-shirt says . . . Just do it! Through perseverance, a good attitude and sheer luck, you'll follow paths and gather fortunes that you never thought were possible.

Index